Springer Complexity

Springer Complexity is a publication program, cutting across all traditional disciplines of sciences as well as engineering, economics, medicine, psychology and computer sciences, which is aimed at researchers, students and practitioners working in the field of complex systems. Complex Systems are systems that comprise many interacting parts with the ability to generate a new quality of macroscopic collective behavior through self-organization, e.g., the spontaneous formation of temporal, spatial or functional structures. This recognition, that the collective behavior of the whole system cannot be simply inferred from the understanding of the behavior of the individual components, has led to various new concepts and sophisticated tools of complexity. The main concepts and tools – with sometimes overlapping contents and methodologies – are the theories of self-organization, complex systems, synergetics, dynamical systems, turbulence, catastrophes, instabilities, nonlinearity, stochastic processes, chaos, neural networks, cellular automata, adaptive systems, and genetic algorithms.

The topics treated within Springer Complexity are as diverse as lasers or fluids in physics, machine cutting phenomena of workpieces or electric circuits with feedback in engineering, growth of crystals or pattern formation in chemistry, morphogenesis in biology, brain function in neurology, behavior of stock exchange rates in economics, or the formation of public opinion in sociology. All these seemingly quite different kinds of structure formation have a number of important features and underlying structures in common. These deep structural similarities can be exploited to transfer analytical methods and understanding from one field to another. The Springer Complexity program therefore seeks to foster cross-fertilization between the disciplines and a dialogue between theoreticians and experimentalists for a deeper understanding of the general structure and behavior of complex systems.

The program consists of individual books, books series such as "Springer Series in Synergetics", "Institute of Nonlinear Science", "Physics of Neural Networks", and "Understanding Complex Systems", as well as various journals.

Springer Series in Synergetics

SSSyn – An Interdisciplinary Series on Complex Systems

The success of the Springer Series in Synergetics has been made possible by the contributions of outstanding authors who presented their quite often pioneering results to the science community well beyond the borders of a special discipline. Indeed, interdisciplinarity is one of the main features of this series. But interdisciplinarity is not enough: The main goal is the search for common features of self-organizing systems in a great variety of seemingly quite different systems, or, still more precisely speaking, the search for general principles underlying the spontaneous formation of spatial, temporal or functional structures. The topics treated may be as diverse as lasers and fluids in physics, pattern formation in chemistry, morphogenesis in biology, brain functions in neurology or self-organization in a city. As is witnessed by several volumes, great attention is being paid to the pivotal interplay between deterministic and stochastic processes, as well as to the dialogue between theoreticians and experimentalists. All this has contributed to a remarkable cross-fertilization between disciplines and to a deeper understanding of complex systems. The timeliness and potential of such an approach are also mirrored – among other indicators – by numerous interdisciplinary workshops and conferences all over the world.

W. Horsthemke R. Lefever

Noise-Induced Transitions

Theory and Applications in Physics, Chemistry, and Biology

With 56 Figures

Springer

Dr. Werner Horsthemke
Southern Methodist University
Dallas, USA

Professor René Lefever
Université Libre de Bruxelles
Brussels, Belgium

Library of Congress Cataloging in Publication Data. Horsthemke W. (Werner), 1950- Noise-induced transitions. (Springer series in synergetics ; v. 15) Bibliography: p. Includes index. 1. Phase transformations (Statistical physics) 2. Noise. 3. Stochastic processes. I. Lefever, R., 1943- . II. Title. III. Series. QC175.16.P5H67 1983 530.1'36 83-10307

2nd printing 2006

ISSN 0172-7389

ISBN 978-3-642-05719-9 e-ISBN 978-3-540-36852-6

Springer is a part of Springer Science+Business Media

springer.com

Softcover reprint of the hardcover 1st edition 1984

Cover design: Erich Kirchner, Heidelberg

To Brenda
and Anne Marie

Preface

The study of phase transitions is among the most fascinating fields in physics. Originally limited to transition phenomena in equilibrium systems, this field has outgrown its classical confines during the last two decades. The behavior of far from equilibrium systems has received more and more attention and has been an extremely active and productive subject of research for physicists, chemists and biologists. Their studies have brought about a more unified vision of the laws which govern self-organization processes of physico-chemical and biological systems. A major achievement has been the extension of the notion of phase transition to instabilities which occur only in open nonlinear systems. The notion of phase transition has been proven fruitful in application to nonequilibrium instabilities known for about eight decades, like certain hydrodynamic instabilities, as well as in the case of the more recently discovered instabilities in quantum optical systems such as the laser, in chemical systems such as the Belousov-Zhabotinskii reaction and in biological systems. Even outside the realm of natural sciences, this notion is now used in economics and sociology.

In this monograph we show that the notion of phase transition can be extended even further. It applies also to a new class of transition phenomena which occur only in nonequilibrium systems subjected to a randomly fluctuating environment. In other words, for these systems the environment is not constant in time as is usually assumed in the study of nonequilibrium phenomena but displays random temporal variations, also called external noise. These new transition phenomena present a fascinating subject of investigation since, contrary to all intuition, the environmental randomness induces a more structured behavior of the system. We have called this new type of nonequilibrium transition phenomena *Noise-Induced Transitions* in order to stress the essential role of the noise. Noise-induced transitions can occur only if a certain amount of randomness is present in the environment. Remarkably they amount to a symbiotic relationship of order and randomness in stark contrast to the commonly held view that order and randomness form an antagonistic pair. The existence of noise-induced transitions clearly forces us to reappraise the role of randomness in natural phenomena.

In this monograph we present the formalism for the description of nonlinear systems coupled to a random environment and give a detailed study of noise-induced transitions. In Chaps. 1, 3 and 6 we expound the theoretical formalism for the case of extremely rapid external noise. Such noise corresponds to an environment with an extremely short memory. In this case it is legitimate and useful to consider the limit of zero memory. This is the so-called white noise idealization. We use this idealization in Chap. 6 for the discussion of noise-induced transitions and noise-induced critical points. This chapter deals with the steady state properties as well as the time dependent features of noise-induced phenomena. Chapters 2, 4 and 5 contain the mathematical underpinnings of our formalism. We have included these chapters to achieve a self-contained text and to provide, for the nonspecialist in probability theory, an easier access to the modern mathematical literature on random processes, essential for further theoretical progress in this field. Furthermore, we concur wholeheartedly with *Doob* [Ref. 4.2, p. 352]: "It will be seen that the use of rigorous methods actually simplifies some of the formal work, besides clarifying the hypotheses". Indeed, the theory of nonlinear systems coupled to the environment in a parametric way has been plagued by ambiguities and confusions in the past, precisely because of a lack of rigorous methods. Chapters 2, 4 and 5 provide the basic amount of mathematical tools needed for a proper and transparent discussion of systems with parametric noise and noise-induced transitions. The reader who is not interested in the more mathematical aspects of the formalism may skip these chapters or consult them as the need arises. In Chaps. 7, 8 and 9 the formalism is applied to concrete systems as representative examples in physics (electrical circuits, optical bistability, nematic liquid crystals, turbulent superfluid ^{4}He), chemistry (photochemical reactions, Briggs-Rauscher reaction) and biology (population dynamics, genetics, nerve membranes). Also, experimental evidence for noise-induced transitions is reported and new experiments are proposed. In Chaps. 8 and 9 we leave the limit case of white noise and extend the formalism to the case of environments with nonzero memory, i.e., noise with a nonvanishing correlation time. We model such situations by so-called colored noise. Chapter 8 treats in particular colored noise of the Ornstein-Uhlenbeck type, i.e., a Gaussian Markov process, whereas Chap. 9 is devoted to the dichotomous Markov noise, a two state process also called random telegraph signal.

The theory of noise-induced transitions, and even more so their experimental study, is at its beginnings. It is our hope that this book will lead to further advances in this field and that it will convince the reader of the nonintuitive role which noise can play in nonlinear systems. Furthermore, we hope that the reader will come to share our enthusiasm for this subject and will have occasion to study the fascinating phenomenon of noise-induced transitions in his own field of research.

It is a pleasure to acknowledge the many fruitful discussions we have had on the subject of noise-induced transitions with our colleagues in the Service de Chimie Physique II over the last few years, in particular with P. Allen, P. Borckmans, L. Brenig, J. L. Deneubourg, G. Dewel, M. Malek Mansour, G. Nicolis, J. W. Turner, C. Van den Broeck and D. Walgraef. We would especially like to thank Prof. I. Prigogine for his constant encouragement and support of our work. We benefited greatly from the collaboration with Prof. L. Arnold, Universität Bremen, Dr. P. De Kepper, Centre de Recherche Paul Pascal Bordeaux, Mr. C. R. Doering, University of Texas at Austin, Prof. K. Kitahara, Shizuoka University, Dr. J. C. Micheau, Université de Toulouse, and Prof. J. W. Stucki, Universität Bern, and we would like to express our gratitude to all of them. Furthermore, we wish to thank Dr. A. S. Chi, University of Beijing, Prof. W. Ebeling, Universität Rostock, Prof. R. Graham, Universität Essen,

Dr. M. O. Hongler, Université de Genève, Prof. S. Kabashima, Tokyo Institute of Technology, Prof. T. Kurtz, University of Wisconsin-Madison, Prof. F. Moss, University of Missouri-St. Louis, Dr. J. C. Roux, Centre de Recherche Paul Pascal Bordeaux, Profs. J. M. Sancho and M. San Miguel, Universidad de Barcelona, Prof. H. L. Swinney, University of Texas at Austin, and Prof. V. Wihstutz, Universität Bremen, for helpful discussions and suggestions.

Very special thanks are due to Prof. H. Haken for inviting us to write this monograph and having it included in the Springer Series in Synergetics.

We thank Dr. H. Lotsch and the staff of the Springer-Verlag for the care and attention they brought to this manuscript.

We are grateful to Ms. D. Hanquez and Ms. S. Dereumaux for typing the manuscript and to Mr. P. Kinet for drawing the figures and technical help in preparing the manuscript.

This research has been supported by the Studienstiftung des deutschen Volkes, by the Belgian Government, Actions de Recherche Concertée no 76/81II3, by the Instituts Internationaux de Physique et de Chimie, fondés par E. Solvay, by the U.S. Department of Energy, DE-AS05-81ER10947, by the NATO research grant no. 12582 and by the National Foundation for Cancer Research.

Austin, Brussels, August 1983 *W. Horsthemke · R. Lefever*

List of Abbreviations

a.s.:	almost surely
t.p.d.:	transition probability density
w.r.t.:	with respect to
FPE:	Fokker Planck equation
KBE:	Kolmogorov backward equation
SDE:	stochastic differential equation
Ito-SDE:	Ito-stochastic differential equation
S-SDE:	Stratonovich differential equation
D-noise:	Dichotomous noise
O-U-noise:	Ornstein-Uhlenbeck noise
O-U-process:	Ornstein-Uhlenbeck process
F-boundary:	boundary classified according to Feller's convention
GS-boundary:	boundary classified according to Gihman-Skorohod convention

Contents

1. Introduction

1.1 Deterministic and Random Aspects of Macroscopic Order

Science is often understood as a quest for uncovering marvelous regularities which nature hides in its inexhaustible diversity. There is indeed a profound belief that fundamental regularities are present in the midst of the variability and complexity of natural phenomena; regularities which once understood enlighten and simplify everything. This conviction has been beautifully expressed by Johannes Kepler when he wrote that, "The diversity of the phenomena of Nature is so great, and the treasures hidden in the heavens so rich, precisely in order that the human mind shall never be lacking in fresh nourishment" [1.1].

Order is here intimately connected with a profoundly deterministic conception of the universe. Laws are sought that make exactly predictable events which at first sight were undeterminable, like eclipses or the path followed by comets in the sky. Newton's dynamics crowned this quest with tremendous success and as a result a deterministic conception of the laws of nature became the basis of scientific methodology. This determinism assumes the materialistic origin of natural phenomena: it postulates that behind any phenomenon as mysterious as could be, purely materialistic causes are acting which sooner or later will be identified. This concept of determinism also assumes that nature is precisely organized, i.e., it obeys laws in which there is no place for such "imperfections" as elements of chance or randomness. With the discovery of the precision and beauty of celestial mechanics all doubts were brushed away, even for the most complex systems, that random processes play a significant role in the natural world and need to be taken into account.

This conception rapidly found further nourishment in the early progress of the natural sciences. Such illuminating laws as Dalton's law of definite proportions for the combinations of chemical species or Boyle-Mariotte's law for the expansion of gases strongly supported the view that no arbitrariness lies behind the complexity of physico-chemical processes. Like celestial mechanics, physics and chemistry in turn were found to obey strict deterministic principles.

The idea that such principles are in all fields the keys to understanding and progress dominated the development of science in the entire nineteenth century. The achievements of the natural sciences were viewed as a universal model for any field still subjected to empiricism. In the case of medicine this conviction is for example expressed in the response of Laplace to somebody who was astonished that he had proposed to admit medical doctors to the Academy of Sciences, since medicine at that time was not considered as a science. It is, Laplace said simply, in order that they be among scientists. Some forty years later, medicine had clearly advanced in the direction indicated by Laplace when Claude Bernard noted in [Ref. 1.2, p.116] that, "Il y a un déterminisme absolu dans tout phénomène vital; dès lors il y a une science biologique". With a few decades of delay, the same ideas appear in the social sciences. For *Durkheim* [1.3] determinism and science were one and the same thing; the idea that societies are submitted to natural laws and constitute "un règne naturel" is equivalent to admitting that they are governed by "the principle of determinism" which at that time was so firmly established in the so-called exact sciences.

Yet it is now clear that the complete rejection of randomness prevailing in the nineteenth century was unwarranted. Classical determinism experienced a strong blow when the atomic and subatomic properties of matter revealed irreducible indeterminacies and required the formulation of a new mechanics: quantum mechanics. But independently even of these microscopic findings, the strength of deterministic positions has also been eroded for other reasons related to problems posed by the properties of matter at the very macroscopic level. In macroscopic physics the oldest and still actual of these problems concerns the meaning of entropy and of irreversibility which entropy purports to describe. Taking a step which shocked some of his most illustrious contemporaries, Boltzmann proposed an interpretation of entropy which supplemented the laws of dynamics with stochastic considerations. The latter have generally been regarded as added approximations made unavoidable by the practical impossibility to treat exactly the dynamics of a many-body problem. This point of view is likely to be revised. Indeed the idea gains ground that irreversibility is already rooted in dynamics and is not an illusion due to approximations. It has been shown that in the evolution of *classical* dynamical systems an *intrinsic* randomness coexists with the perfectly deterministic laws of dynamics [1.4–7] (for a review see [1.8–10]). More precisely, whereas the motion of individual particles follows trajectories which are deterministic in the fullest sense of the word, the motion of regions of phase space, i. e., of bundles of trajectories, acquires stochastic features.

Also related with irreversibility another problem arises in which the role of stochastic elements can hardly be neglected, namely the constructive role of irreversible processes in the formation of large-scale supramolecular organisations commonly called dissipative structures. What is striking in this problem, posing a most pressing question, is the profound dichotomy which appears between the behavior of matter at the macroscopic level and its behavior at the microscopic level. How is the space-time coherence of chemical dissipative structures, of laser beams or Bénard rolls possible; how can such a long-range macroscopic order spontaneously appear and maintain itself in spite of molecular chaos and internal fluctuations? [1] The same dichotomy is found in the processes of self-organization taking place in biology. Metabolic processes are essentially chemical transformations. The element of chance in these transformations is seemingly quite large, due to the fact that in living cells the number of molecules involved in

[1] In the following we shall call *internal fluctuations* all elements of randomness which directly derive from the many degrees of freedom involved in the processes and interactions at the microscopic level of the system.

these transformations is often very small. Yet metabolism is extraordinarily precise. It produces, for instance, with an astonishing degree of dependability, protein molecules whose sequence of amino acids and spatial structure are so particular that their probability of occurrence by pure chance is practically zero.

As another facet of the constructive role of irreversible processes and of the dichotomy between order and randomness which is involved let us consider for a moment the mechanism of biological evolution. Since Darwin it is admitted that the biosphere is quite unlike the *static*, harmoniously deterministic world that Kepler envisioned in comtemplating the heavens. Biological species and even prebiotic macromolecular compounds [1.11, 12] are self-organizing systems. They are in a perpetual state of becoming which depends in an essential manner on events of chance. At random and independently of the direction of evolution, a large pool of hereditary genetic variations develops. This pool is the indispensable raw material for evolution. In it, evolution finds the favorable variations whose frequency in the population it subsequently amplifies and stabilizes via the precise well-defined rules of heredity transmission. Thus the distinguishing characteristic of evolution theory, which clearly had no analog in the physical sciences at the time when evolution theory was formulated, is that it gives an unusually important role to random events. Mutations are the random triggers of progress. However their effects are even more far reaching and decisive than that; these events of chance may decide at random between different possible roads of evolution. It is now generally admitted that the outcome of the biosphere is not uniquely determined. If life has evolved on another planet under exactly the same environmental conditions as on earth, we nevertheless are prepared to encounter very different forms of life, perhaps even based on a different chemistry. There is a consensus that given the right conditions the emergence of life is inevitable. In this sense, it is a physical, materialistic, deterministic phenomenon. But this does not mean that it is predictable. Quite to the contrary, using more modern language one could say that in the course of its unfolding, life continuously chooses stochastically among many, perhaps infinitely many, possible scenarios. In one given realization of the process, the scenario which will be followed cannot be predicted with certainty.

Still today some researchers feel that randomness is given too much importance in Darwin's theory [1.13]. They look for a stronger coupling between biological variability and environmental conditions. Be that as it may, mutations and other random aspects of evolution are generally considered as so deeply rooted in the specificities of the living state of matter as to be entirely particular to it. More often than not the differences between the evolution of the biosphere and that of the physical world have been attributed to it. This situation however is changing. In recent years the mechanisms of self-organization in the physical sciences have become much better understood and a new appreciation of the role of chance in natural phenomena has emerged. We shall review more in detail in Sect. 1.2 some of the essential advances which have been made, but we should like to note already here the turn in ideas to which these advances led. They suggest that the macroscopic world is far less deterministic, i.e., predictable in the classical sense, than we ever thought. In fact, completely new aspects of randomness have come to light which call for a profound reappraisal of the role and importance of random phenomena in nature.

First, it has been found that the mechanisms of self-organization become much more complex in strongly dissipative systems than in conservative, equilibrium-type systems. In the vicinity of a stable thermodynamic equilibrium state, the behavior of a dissipative system can easily be predicted, given that in this domain it possesses a unique attractor, namely the thermodynamic branch. Far from thermodynamic equilibrium on the contrary, the same system may possess an amazingly complex network of bifurcations. The importance of elements of chance such as internal fluctuations then inevitably increases. Their influence becomes crucial in the choices which the system makes in the course of its evolution between the numerous basins of attraction, or dissipative structures, to which bifurcations give rise [1.14, 15]. When an external parameter is changing, somewhat as in biological evolution, different scenarios can unfold: some attractors will be visited, others will not, depending only on the random fluctuations which occur at each instant of time. Remarkably, this sensitivity to fluctuations already appears in the simplest self-organizing hydrodynamical systems. It is known, for example, that a Bénard system whose parameters are controlled with the best possible experimental accuracy nevertheless in two identical experiments evolve unpredictably according to different scenarios [1.16].

The second blow to conventional ideas regarding the properties of the macroscopic world comes from the facility with which the scenarios govering the evolution of deterministic macroscopic systems, e.g., systems described by ordinary differential equations, generate irregular aperiodic solutions called chaotic or turbulent. These results obtained in parallel with the development of nonequilibrium stability theory created a shock in the physical and biological sciences. They deviate drastically from the scenario proposed by Landau to explain hydrodynamical turbulence, namely the excitation of an infinite number of frequency modes in a continuous system. Indeed, the first alternative scenario, proposed by *Ruelle* and *Takens* [1.17], involves only three frequencies. The "noisy behavior" is here associated with a strange attractor which appears after three successive Hopf bifurcations. A characteristic feature of a strange attractor is the sensitive dependence on initial conditions; nearby trajectories separate exponentially in time [1.18 – 21]. Astonishingly, a strange attractor, implying turbulent behavior, can occur already in low-dimensional systems, namely in systems described by as little as three first-order differential equations.

Furthermore, not only do simple deterministic systems, contrary to naive expectations, easily become *intrinsically noisy*, but also it has been found that other routes to chaos are possible rather than via a sequence of Hopf bifurcations. At least two other major scenarios have been described, namely the intermittent transition to turbulence [1.22] and the period doubling scenario [1.23 – 28]. (See [1.29, 30] for recent reviews.) It is also to be noted that when a control parameter of a dissipative system is changed in a continuous systematic way chaos is not necessarily the "ultimate" kind of behavior appearing after the possibilities of more "coherent" regimes in bifurcation diagrams have been exhausted. For instance, in the Lorenz model which furnishes an approximation of the Bénard instability [1.31], chaotic domains alternate with temporally organized regimes [1.32]. The same type of behavior has recently been proven to exist for the periodically forced van der Pol oscillator [1.33]. This property can also be seen in the Belousov-Zhabotinsky reaction [1.34–36]. Experimental evidence supporting the relevance of those scenarios in the interpretation of hydrodynamical [1.37–42] and chemical turbulence [1.35, 43–45] is rapidly accumulating.

The investigation of self-organization in nonequilibrium systems which are coupled to fluctuating environments has brought forth the third major impetus to reappraise the role of randomness and constitutes the subject of this book. A naive intuitive belief, which commonly accompanies our inclination to look at nature with "deterministic eyes", holds that the influence of environmental fluctuations, generally understood to mean *rapid* random variations, is trivial. It is argued that (i) rapid noise is averaged out and thus a macroscopic system essentially adjusts its state to the average environmental conditions; (ii) there will be a spreading or smearing out of the system's state around that average state due to the stochastic variability of the surroundings. Fluctuations are a nuisance, a disorganizing influence, but after all play only a secondary role. These expectations are borne out for a certain special type of coupling between the system and its environment. Surprisingly, however, more often than not the behavior of a nonlinear system in a noisy environment does not conform to the common intuitive expectations. Over the past few years, systematic theoretical and experimental studies have demonstrated that in general the behavior is stupendously different from the aforementioned simple picture. In a large class of natural phenomena environmental randomness can, despite its apparently disorganizing influence, induce a much richer variety of behaviors than that possible under corresponding deterministic conditions. Astonishingly, an increase in environmental variability can lead to a structuring of nonlinear systems which has no deterministic analog. Perhaps even more remarkably, these transition phenomena display features similar to equilibrium phase transitions and to transition phenomena encountered in nonequilibrium systems under deterministic external constraints as, for instance, the Bénard instability and the laser transition. The notion of phase transition was extended to the latter about a decade ago, since certain properties which characterize equilibrium phase transition are also found in these phenomena [1.14, 46–52]; for reviews see [1.53–56]. As we emphasize in this book, it is possible to go even one step further and to extend the concept of phase transition to the new class of nonequilibrium transition phenomena which are induced by environmental randomness. We have thus called them noise-induced nonequilibrium phase transitions or, for short, noise-induced transitions. This choice of name is intended to express that this new class of transition phenomena is close kin to the classical equilibrium phase transitions and to the more recent class of nonequilibrium phase transitions. However, it is not meant to imply, and it should not be expected, that noise-induced transitions display exactly the same features as equilibrium transitions. Deterministic nonequilibrium conditions already lead to a richer transition behavior with such new possibilities as the transition to sustained periodic behavior known as limit cycle. More importantly, for the new class of transition phenomena to which this monograph is devoted, one cannot of course overlook the fact that the new states, to which noise-induced transitions give rise, carry a permanent mark of their turbulent birth. They are creatures of noise and as such at first sight foreign to our deeply ingrained deterministic conceptions of order.[2]

In fact, the phenomenon of transitions induced by external noise belongs to a whole stream of ideas which really upsets our classical conceptions concerning the relation between determinate and random behavior. These ideas constitute a refutation of our gut feeling about the role of fluctuations. Though for noise-induced transitions the situation is not as neat as it is for classical equilibrium and nonequilibrium phase transitions, it is far from unpredictable and lawless. The notions and concepts, developed for classical transition phenomena and essentially rooted in a deterministic conception of nature, can be extended and adapted to deal with situations where noise plays an important role. A theoretical investigation is thus made possible. More important even, the situation is accessible to experimental investigation. Transitions induced by external noise are an observable physical phenomenon as documented by various experiments in physico-chemical systems. Noise-induced transitions are thus more than a mere theoretical figment and their existence has profound consequences for our understanding of self-organization in macroscopic systems. As stated above, they force us to reappraise the role of randomness in natural phenomena.

The organization of this monograph is as follows. To place the class of noise-induced transition phenomena in its proper context, in the next sections we shall briefly discuss disorder-order transitions under deterministic environmental conditions and the effect of internal fluctuations on them. Then we shall present the phenomenon of noise-induced transitions and have a first go at the question of modeling macroscopic systems subjected to a fluctuating environment. To make this work self-contained we shall present in a concise, but we hope nevertheless clear way the mathematical tools needed for an unambiguous treatment of nonlinear systems driven by external noise. This will be followed by a precise and operational definition of noise-induced transitions. Their properties will be investigated in detail for Gaussian white-noise environments as well as for two types of colored noise. Three experiments, in which noise-induced transitions have been observed, will be described in detail. Furthermore, new experiments in physics, chemistry and biology will be proposed and the significance of noise-induced transitions for natural systems will be discussed with concrete examples.

1.2 From Crystals to Dissipative Structures

Order-disorder transitions have always been discussed in the physical sciences under environmental conditions which are deterministic, i.e., constant in time or

[2] Outside the physical sciences also, though still somewhat confused, the idea is gaining ground that random factors which do not enter into consideration in the usual Darwinian theory play an important role in evolution. Writing about the possible causes of the Permian extinction which some 225 million years ago wiped out more than 80 percent of all species living at that time, *Gould* [Ref. 1.57, p. 349] concludes in a manner which sounds like a response to Johannes Kepler that, "Perhaps randomness is not merely an adequate description for complex causes that we cannot specify. Perhaps the world really works this way, and many happenings are uncaused in any conventional sense of the word. Perhaps our gut feeling that it cannot be so reflects only our hopes and prejudices, our desperate striving to make sense of a complex and confusing world, and not the ways of nature."

time periodic. These conditions have been universally adopted in the development of thermodynamics and statistical mechanics. They correspond to a simplified approach which in all logic had to be pursued first and which fitted a general inclination to play down the importance of environmental randomness. Randomness being synonymous with imperfection, it had to be eliminated by all means in experiments; in theories it could only be a source of unnecessary complications obscuring the fundamental beauty of the processes of self-organization of matter.

Therefore deterministic stationary environmental conditions have so far always been considered as self-evident in macroscopic physics. We recall in this section some important results which have been established within this framework. Later on this will permit us to situate more easily the novelties which are brought in by noise-induced transitions.

1.2.1 Macroscopic Description of Self-Organization in a Constant Environment

Many of the commonly encountered macroscopic systems can be described in terms of a set of state variables $\{X_i\}$ obeying evolution equations of the form

$$\partial_t X(r,t) = f_\lambda(X(r,t)) , \tag{1.1}$$

where $X(r,t)$ and $f_\lambda(X(r,t))$ are vectors whose components are respectively the state variables X_i and the functional relations expressing the local evolution of the X_i's in time t and in space r. The state variables X_i may denote for example the temperature, the electrical potential, the velocity field or the chemical composition. The functionals in general contain partial derivatives with respect to space and are non-linear, due, for example, to the occurrence of chemical reactions inside the system or transport phenomena like convection in hydrodynamics. They also depend on a set of control parameters λ (kinetic constants, diffusion coefficients, fixed concentrations of some compounds, etc.).

In order that (1.1) constitutes as well-posed problem, the value of these parameters together with the boundary conditions which are maintained on the surface S of the system must be known. The latter usually are either Dirichlet conditions which fix the values $\{X_i^S\}$ of the state variables on the surface or Neumann conditions which fix the values of the fluxes on the surface $\{\boldsymbol{n} \nabla X_i^S\}$ ($\boldsymbol{n}$ vector normal to the surface).

The control parameters λ (or a subset thereof) and the boundary condition acting on the system constitute the constraints imposed by the external world. As mentioned above, it has been common up to now in the study of self-organization phenomena to make the simplification that the environment is constant in time.[3] Obviously this requires that the value of all the control parameters λ and the boundary conditions be constant. The problem of the onset of an ordered

[3] Henceforth in this chapter we shall always refer to the simplest case of a constant environment when speaking of a deterministic environment. Time-periodic environments have been much less studied [1.58, 59].

behavior in the system can then be investigated as follows. Since the environment is constant we may suppose that there exists at least one time-independent solution $\{\bar{X}_i\}$ of (1.1), i.e., satisfying

$$f_\lambda(\bar{X}) = 0 , \tag{1.2}$$

which we can take as a reference state and which corresponds to a "banal" unorganized state. The onset of ordered behavior is then associated with the idea of instability and symmetry breaking: self-organization occurs when the banal solution $\bar{X}$ becomes unstable and is replaced by a new solution of (1.1) whose symmetry properties are lower. The simplest way to investigate this possibility is to test first the stability of the reference state with respect to small perturbations. One sets

$$X(r, t) = \bar{X} + x(r, t) \quad \text{with} \quad |x_i/\bar{X}_i| \ll 1 \tag{1.3}$$

and replaces it in (1.1). The time evolution of the perturbation $x(r, t)$ is then given by the solution of the system of equations

$$\partial_t x_i = \sum_j A_{ij} x_j \tag{1.4}$$

obtained by linearization of (1.1). The elements A_{ij} are time independent, since we have linearized around a time-independent reference state, and thus (1.4) admits solutions of the form

$$x_i(r, t) = x_i^k(r) \exp(\omega_k t) . \tag{1.5}$$

The x^k must satisfy the boundary conditions imposed on the system but may have lower symmetry properties than the reference state. In fact they are simply the eigenvectors of the eigenvalue problem (k refers to the wave numbers possible)

$$(\omega_k I - A) x^k(r) = 0 . \tag{1.6}$$

The values of $\mathrm{Re}\{\omega_k\}$ determine the rate at which the perturbations of the state variables evolve. Typically the life-time of disturbances in the system is of the order of

$$\tau_{\text{macro}} = |1/\mathrm{Re}\{\omega_k\}| . \tag{1.7}$$

Therefore τ_{macro} can be called the macroscopic time scale of evolution of the system. Obviously if the reference state $\bar{X}$ is asymptotically stable, all $\mathrm{Re}\{\omega_k\}$ must be negative. The onset of a transition can then be found simply by studying the behavior of $\mathrm{Re}\{\omega_k\}$ as a function of the values of the control parameters λ and of the boundary conditions imposed on the system. To be specific, let us assume that we explore the properties of the system by manipulating a single control parameter λ. At that point $\lambda = \lambda_c$ at which at least one $\mathrm{Re}\{\omega_k\}$ changes

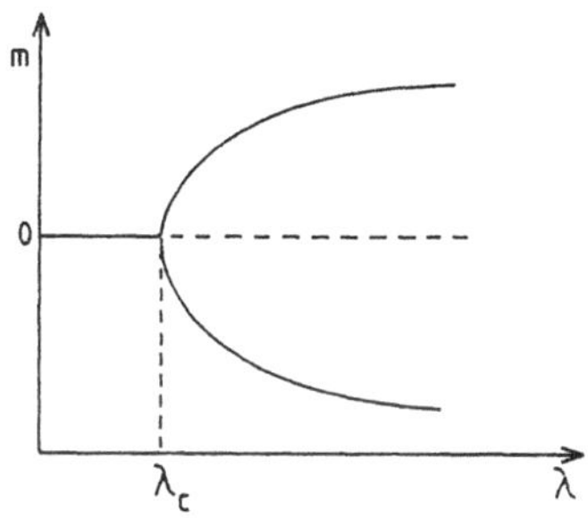

Fig. 1.1. Bifurcation diagram of a second-order phase transition. Some order parameter m is plotted versus the external constraint λ. At λ_c, the reference state becomes unstable (broken line) and two new stable branches of solution branch off supercritically

from negative to positive, the lifetime of fluctuations tends in first approximation to become infinite. In other words there is a *slowing down* of the rate of relaxation of the fluctuations. The value λ_c is called a *point of bifurcation*: it is a point at which one or several new solutions of the equations (1.1) coalesce with the reference state $\bar{X}$ considered.[4] It is customary to associate with a new solution of (1.1) a quantity m, or *order parameter*, which vanishes at λ_c and which measures the deviation from the reference state, e.g., the difference between the concentration of a compound on the new branch of solutions and the value of its concentration on the reference state, the amplitude of a spatial or temporal oscillatory mode, etc. Plotting these quantities as a function of λ yields what one calls a *bifurcation diagram*. An example of a bifurcation diagram for a second-order phase transition is sketched in Fig. 1.1. Below λ_c there exists a unique asymptotic solution which is stable and corresponds to the regime having the highest symmetry compatible with the constraints imposed on the system. At λ_c this solution becomes unstable and simultaneously new branches of solution of lower symmetry bifurcate supercritically. This bifurcation is encountered in the classical hydrodynamical instabilities described by Bénard or Taylor.[5] In these systems, it describes the onset of a coherent spatial pattern of convection cells in an initially unstructured fluid phase when the temperature gradient or the angular velocity gradient imposed across the systems passes through a threshold value. This bifurcation is also frequently encountered with the onset of temporal and/or spatial oscillations in chemical and enzymatic reaction systems; well-known systems are the Belousov-Zhabotinsky reaction and the reaction of the glycolytic enzyme phosphofructokinase. Typically in second-order phase transitions, the order parameter grows for $\lambda > \lambda_c$ like

$$m = \text{const}\,(\lambda - \lambda_c)^{1/2} \tag{1.8}$$

and the relaxation time of fluctuations behaves in the vicinity of λ_c like, compare with (1.7),

$$\tau_{\text{macro}} = \text{const}/\,|\lambda - \lambda_c|\,. \tag{1.9}$$

---

4 By new solution we refer here to the stable or unstable asymptotic regimes which the system may approach respectively for $t \to +\infty$ or $t \to -\infty$. We are not interested in transient behaviors.

5 For reviews see [1.40, 60].

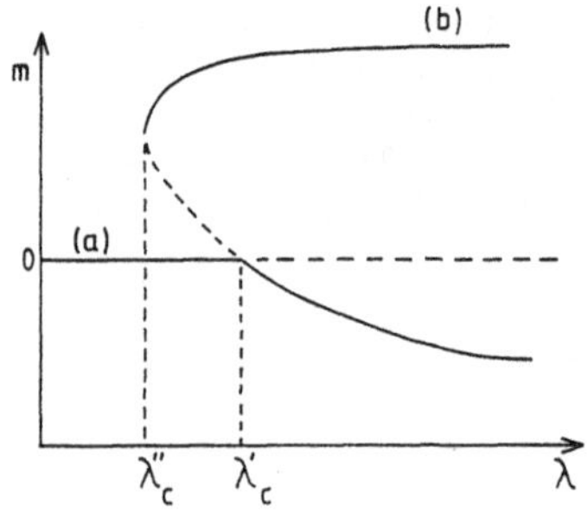

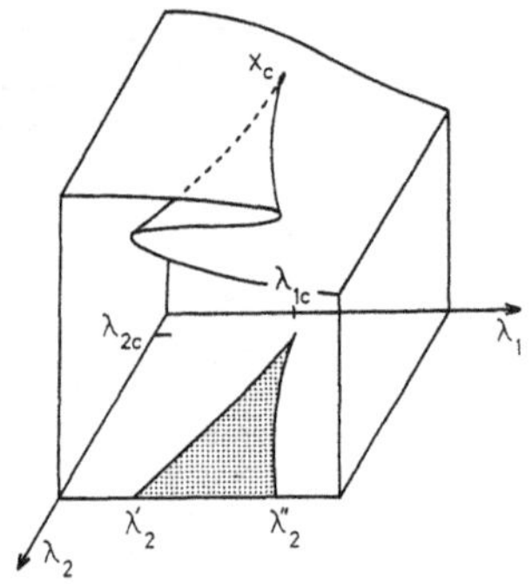

Fig. 1.2. Bifurcation diagram of a first-order phase transition. At λ_c' a new branch of solutions bifurcates subcritically. For $\lambda_c'' < \lambda < \lambda_c'$ two locally stable states, namely the reference state (*curve a*) and a new branch of solutions (*curve b*) coexist, separated by an unstable threshold state

Fig. 1.3. Cusp catastrophe

In addition to the behavior sketched in Fig. 1.1, one finds in all branches of the physical and biological sciences an overabundance of discontinuous transition phenomena similar to first-order phase transitions. They are characterized by the existence of a branch of solutions which bifurcates subcritically and is part of an *hysteresis* loop as represented in Fig. 1.2. When the external parameter λ increases in a continuous fashion from zero, the state variables X may jump at $\lambda = \lambda_c'$ from the lower branch of steady states (a) to the upper branch (b). If λ is then decreased the down jump from (b) to (a) takes place at a different value λ_c''.

In this monograph, we shall often encounter the simplest form of first-order transition possible in spatially homogeneous (well-mixed) systems. In the language of catastrophe theory, it corresponds to a cusp catastrophe for the homogeneous steady-state solutions of (1.2): when plotted in terms of two appropriate control parameters λ_1, λ_2 these states lie on a surface which typically exhibits a fold (Fig. 1.3). The coordinates (λ_{1c}, λ_{2c}, X_c) where the fold has its origin constitute the *critical point* of the system.

It would be a task out of proportion to the scope of this book to try to present here a complete panorama of what is known of the bifurcation and self-organization phenomena observed even in the simplest natural or laboratory systems, especially since in the last ten years giant advances have been accomplished in many widely diverse fields. A number of books can be consulted where these advances are reviewed extensively and where further references to original papers can be found [1.14, 15, 40, 44, 54, 61 – 68]. However, there are two aspects of the organization of macroscopic systems in a constant environment which are of fundamental importance and which we need to recall explicitly. Keeping these aspects in mind permits us to situate more exactly the framework in which noise-induced transitions have to be discussed. We shall thus devote the last part of this section to the question of the thermodynamic interpretation of bifurcations, with Sect. 1.2.2 as a brief summary of the influence of internal fluctuations on bifurcation diagrams.

So far we have introduced the mechanisms of self-organization and bifurcation without considering the dependence of these phenomena on the strength of the external constraints imposed by the environment. We also mentioned the simplest examples of bifurcation taking place when a "banal" reference state becomes unstable. However, since equations (1.1) are in general highly nonlinear, their exploration in parameter space usually reveals a whole network of further instabilities. This network is responsible for the complex behaviors and the multiplicity of scenarios mentioned in Sect. 1.1. As we already emphasized there, the richness of dynamical behaviors in a macroscopic system is specific to the domain far from thermodynamic equilibrium. Conversely, in the parameter space there exists a domain close to thermodynamic equilibrium where the nonlinearities present in (1.1) cease to play a role whatever system is investigated. The dynamical properties of *any* macroscopic system then become fairly simple and can be apprehended in a model-independent fashion. We should like to recall these thermodynamic results here because they demonstrate the clear-cut distinction which exists between two types of order in a constant environment.

First let us recall the situation at thermodynamic equilibrium. If a system is in contact with a constant environment which furthermore is at equilibrium, i.e., which imposes no constraints in the form of fluxes of energy or matter on the system, then only the class of coherent organization known as equilibrium structures can appear. The standard example is the crystal. The formation of these structures obeys a universal mechanism which, at least qualitatively, is well understood. This mechanism immediately follows from the second law of thermodynamics. At constant temperature and volume, it amounts to looking for the type of molecular organization that minimizes the system's free energy

$$F = E - TS,$$

where E is energy, T temperature and S entropy. Self-organization is then the result of competition between the energy and entropy. At low temperatures, the system adopts the lowest energetic configuration even if its entropy is small as in the case of the crystal.

During the last fifty years physicists have striven to understand how far the basic simplicity and beauty of the laws which govern self-organization under constant equilibrium conditions can be transposed to self-organization phenomena taking place in systems which are subjected to a constant environmental stress and which therefore cannot approach a state of thermodynamic equilibrium. The motivations in this direction are very strong, since clearly many of the most organized systems of nature like biological systems are submitted to an environment far from thermodynamic equilibrium.

The natural way to try to extend the ideas explaining the formation of equilibrium structures to nonequilibrium situations is to look for the conditions under which the dynamical properties of macroscopic systems can be expressed in terms of a potential function which plays the role of the free energy. A first answer to this question was found in the development of the linear thermodynamic theory of irreversible processes. This theory applies to systems where the environmental constraints are small so that the thermodynamic forces induced by these constraints only slightly depart from their zero equilibrium value. The rates of the irreversible processes are then linearly related with the thermodynamic forces. Furthermore, the phenomenological proportionality coefficients expressing this linear dependence are constants satisfying symmetry requirements well known as Onsager's reciprocity relations. This guarantees the existence of a state function, the entropy production P, which is nonnegative everywhere in the space of the $\{X_i\}$,

$$P(X) \geqslant 0 \tag{1.10}$$

and which has the properties of a potential [1.69]. Inequality (1.10) derives immediately from the second law; it simply expresses that irreversible processes necessarily dissipate energy when a system departs from its equilibrium state for which by definition $P = 0$. If now the environment keeps the system away from a stable equilibrium state, the steady states which are its continuation in the nonequilibrium domain, i.e., which form the thermodynamic branch [1.70] correspond to a minimum of P. Indeed one has

$$d_t P(X) \leqslant 0, \tag{1.11}$$

where the equality sign is satisfied only for steady states [1.69, 71, 72]. According to Lyapounov's theorem [1.73], inequalities (1.10, 11) ensure that the steady states of the thermodynamic branch are asymptotically stable. Furthermore, as a byproduct of the symmetry properties which near equilibrium guarantee that $P(X)$ is a potential function, it can be proven that the approach of the thermodynamic branch is *monotone* in time, i.e., even damped oscillations are excluded [1.74]. In summary, the dynamical as well as the steady-state properties of macroscopic systems submitted to small constant environmental constraints are qualitatively identical to the properties of equilibrium states: no new type of order is possible.

For bifurcations from the thermodynamic branch to become possible, the properties of the potential expressed by (1.10, 11) must be lost. This can take place only at a finite distance from thermodynamic equilibrium and has given rise to the notion of a thermodynamic threshold for self-organization. A theorem proposed by *Glansdorff* and *Prigogine* [1.14] summarizes this fundamental result.

Theorem. Consider a single phase, open, nonlinear system, subject to time-independent nonequilibrium boundary conditions. Then steady states belonging to a finite neighborhood of the state of thermodynamic equilibrium are asymptotically stable and their approach is monotonous in time. Beyond a critical distance from equilibrium, they *may* become unstable.

Thus the thermodynamic threshold for self-organization is reached when the thermodynamic branch undergoes for the first time a primary bifurcation. At this point, the dynamics of the system are driven by its nonlinearities. The onset of a coherent behavior of large populations of atoms or molecules becomes possible and may lead to the formation of *dissipative structures*. One can also say that beyond the thermodynamic threshold for self-organization one enters the field of synergetics: there is a tremendous reduction of the enormous number of degrees of freedom of macroscopic systems. Typically billions or more of molecules are, using *Haken*'s terminology [1.54] "slaved" to a few modes.

1.2.2 Internal Fluctuations

In the mechanisms of self-organization presented above fluctuations were not included. The approach is entirely deterministic: external fluctuations are not taken into account since the environment is considered strictly constant; internal fluctuations, though they are unavoidable, are supposed to be negligible. In this monograph, we want to relax the assumption of constancy of the environment and to discuss specifically the transition phenomena which are induced by a randomly fluctuating environment. Obviously, it is desirable to keep the approach simple and thus, if possible, to include external fluctuations in the description without also taking into account the spontaneous internal fluctuations of the system. Since (1.1) form our starting point, their validity needs to be questioned; the more so that they present instabilities where by definition the sensitivity to internal fluctuations increases. The latter are an intrinsic part of the kinetic processes by which the system evolves. Therefore it is essential to assess if these fluctuations modify the outcome of the deterministic description, in particular the bifurcation diagrams.

Under the influence of its internal fluctuations a system can no longer stay in a definite state. Instead it will effect a random walk in the state space leading to a distribution of values for the state variables. Hence the appropriate quantity to describe the system is the probability that the variables take a certain value. To find this probability and its temporal evolution essentially three different methods are used in the literature (see [1.75] where further references can be found). It is not necessary in the context of this monograph to present these methods in detail, especially since they all yield the same result, namely internal fluctuations *do not* change the local stability properties of the system. In particular the position of transition points is in no way modified by the presence of these fluctuations. [6] Furthermore, all three methods are in complete qualitative agreement as to the behavior of internal fluctuations: around a stable macroscopic state the magnitude of fluctuations in concentration like variables scales as one over the volume V of the system. At a critical point these fluctuations are enhanced, they are of the order of $V^{-1/2}$ [1.76]. Thus in the limit of a macroscopically large system (thermodynamic limit $V \to \infty$) they again become negligible as in the case of a stable reference state. The most important upshot of all this is that the enhancement of internal fluctuations in the neighborhood of the critical point does not affect the position of this point and does not compromise the deterministic description. Furthermore, it turns out that the extrema of the probability density, which is sharply peaked, are in general in the

[6] We refer here only to spatially homogeneous systems which are the kind of systems we mainly are interested in this work.

immediate vicinity of the solution of the deterministic system and coincide with it in the limit of a large system, i.e., for $V \to \infty$. Therefore the bifurcation diagrams, obtained by the deterministic description, remain valid in the sense that they basically describe the behavior of the extrema of the probability density, which correspond to the macroscopic states of the system.

1.3 External Noise

Contrary to internal fluctuations which can safely be neglected for macroscopically large systems, this is not true for fluctuations due to environmental randomness. The main distinction between internal fluctuations and external noise is that the intensity of the latter does in general *not* scale with an inverse power of the system size. In view of the essential role which the environment plays in the behavior of nonequilibrium systems, it should come as no surprise that the influence of environmental fluctuations can under some conditions be far from negligible. Strangely enough it is only during these last twenty-five years that from time to time some results in this direction have been reported, first from a theoretical point of view and coming from widely dispersed fields. Even stranger, these results attracted only scarce attention, perhaps largely due to the way in which they were presented and to the fact that they come from outside the main realm of physics. To our knowledge the first description of the non-negligible effect of external noise was given by *Kuznetsov* et al. [1.77] in a paper on the valve oscillator, which is a circuit of interest in radio engineering (see also [1.78]). The authors note that as the strength of the external noise is changed there are essentially two regions of operations (Fig. 1.4). If the external noise level is high, the amplitude of the oscillations is chiefly zero. When the intensity of the noise is decreased below a certain threshold, the amplitude is chiefly found near the nonzero deterministic value. The only comment of the authors on this phenomenon is that the latter case is the more interesting one. A similar result is noted by *Stratonovich* and *Landa* [1.79] in their article on a different type of oscillator as well as in *Stratonovich*'s book [1.80]. However the phenomenon is barely commented upon. This is understandable since the framework of nonequilibrium phase transitions within which the importance of these phenomena could have been better appreciated did not yet exist.

These phenomena were rediscovered years later in the completely different context of ecological systems. In dealing with the simple problem of the logistic

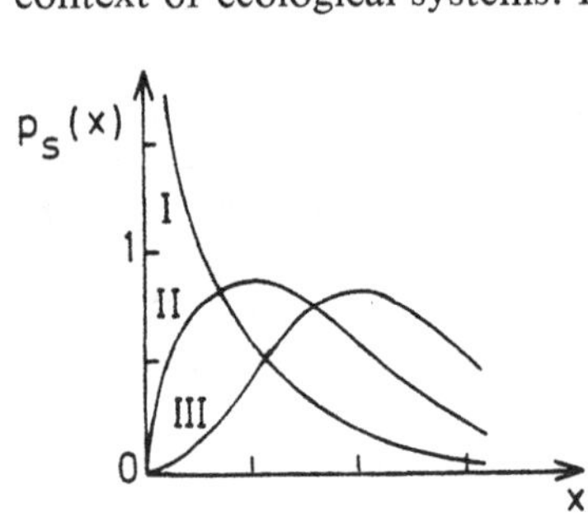

Fig. 1.4. Sketch of the stationary probability density $p_s(x)$ of the amplitude of the valve oscillator for three values of the external fluctuations intensity. At strong intensities $p_s(x)$ diverges near $x = 0$ (*curve I*) while at low intensities it is centered around the deterministic state (*curves II and III*)

growth of a population, *May* [1.81] emphasized that the population would become extinct in a fluctuating environment whatever the mean Malthusian growth parameter, provided the fluctuations of this parameter are strong enough. In other words, the transition point from survival of the population to extinction, which in this simple ecological system occurs under deterministic conditions when the death rate exactly balances the birth rate, becomes dependent on the additional parameter corresponding to the variance of the environmental fluctuations. This results in a shift of this transition point. A similar phenomenon was described at about the same time by *Hahn* et al. [1.82] for the onset of a limit cycle in an enzymatic oscillator system.

The one feature that is common to all the above-described systems, coming from widely different fields, is that the effect of external noise depends on the state of the system. It is not simply additive noise as in the common Langevin treatment of random phenomena in physics. Noise of the above type, now generally known as multiplicative noise, was first considered by *Kubo* [1.83] in his stochastic theory of lineshape, i.e., the motional narrowing in magnetic-resonance phenomena. Kubo considered a linear system, a randomly modulated oscillator, a physical realization of which is the spin precession in a magnetic field containing a random component. However, due to the linearity of the problem, no transition phenomena, similar to the aforementioned ones, are found.

1.4 Noise-Induced Nonequilibrium Phase Transitions

The above miscellaneous collection of systems suggests that external noise can, in contrast to internal fluctuations, modify the local stability properties of macroscopically large systems. The transition point is shifted, depending on the intensity of the external noise. We shall establish in this monograph that this is a general phenomenon for nonlinear systems subjected to multiplicative external noise. The shift of the bifurcation diagram is probably not too surprising once one thinks about it. After all it is very plausible that fluctuations which are of order V^0 and not V^{-1} play an important role in the vicinity of transition points.

The main thrust of this work will, however, be to bring to light even deeper and far less intuitive modifications which external noise can induce in the macroscopic behavior of nonlinear systems. Nonequilibrium systems are, by their very nature, closely dependent on their environment, a point stressed in Sect. 1.2. This fact gives rise to the following question: could not the interplay of the *nonequilibrium* of the system and of the *environmental randomness* lead to drastic changes in the macroscopic behavior of the system, *even* outside the neighborhood of a deterministic instability point? In other words, could not the external noise modify the bifurcation diagrams in a much more profound way than just by a shift in parameter space? Yet a different way to ask this question: do nonlinear systems, coupled to a rapidly fluctuating environment, always adjust their macroscopic behavior to the average properties of the environment or can one find situations in which the system responds in a certain, more active way to the randomness of the environment, displaying for instance behavior forbidden under deterministic external conditions? The answer to these questions is indeed positive. It has been established that even extremely rapid totally random external noise can deeply alter the macroscopic behavior of nonlinear systems: it can induce new transition phenomena which are quite unexpected from the usual phenomenological description. We shall present here a thorough discussion of the theoretical methods used to analyze these noise-induced phenomena, study various aspects of their phenomenology and discuss their consequences for some representative model systems chosen from physics, chemistry and biology. The next section will be devoted to a first discussion of our modeling procedure.

1.5 Modeling Environmental Fluctuations

Due to the omnipresence of external noise, the need to investigate its effects arises in numerous, vastly different situations. To give only a few examples: the propagation of waves through random media, stochastic particle acceleration, signal detection, optimal control with fluctuating constraints, etc. As already mentioned above, our aim here is to describe a new class of nonequilibrium phase transitions, namely changes in the macroscopic behavior of nonlinear systems induced by external noise. To be able to do so in a clear and transparent way, bringing out the essential features of noise-induced phenomena without getting bogged down in the mire of particularities and unwarranted complexities, we shall restrict our attention to those kinds of systems and environments in which noise-induced phenomena are not obscured by other complicating factors. This motivates the following choice of systems:

i) We shall consider systems that are spatially homogeneous. [7] This is a satisfactory approximation in a broad class of applications, namely either if the transport is fast compared to the "reaction" kinetics or if the system is artificially kept homogeneous, e.g., by stirring in the case of chemical reactions. Such systems are commonly called zero-dimension systems.

ii) We shall deal with macroscopically large systems and assume that the thermodynamic limit, system size $V \to \infty$, has been taken. In this way, complications from other noise sources in the system are avoided. Indeed, as discussed in Sect. 1.2.2, internal fluctuations can be safely neglected in the description of macroscopically large systems. It was pointed out that although internal fluctuations are enhanced at the critical point, the deterministic description is not compromised. [8] Accordingly, in a constant environment macroscopic state variables, such as concentration, temperature, etc., can adequately be described by deterministic phenomenological (rate) equations of the form (1.1). These equations form a

---

[7] The exception will be transition phenomena in liquid crystals in Sect. 8.7.

[8] The spatial dimensionality of the system is of importance in this respect. Taking as starting point a zero-dimension deterministic description, we consistently assume that fluctuations do not break the homogeneity of the system. It is known, however, that when inhomogeneous fluctuations need to be considered, they may have some influence on the nature or on the position of transition points [1.84, 85].

valid basis on which to build a phenomenological treatment of the influence of external noise in macroscopic systems.

iii) We shall consider systems which can satisfactorily be described by *one* intensive variable. This is motivated by the fact that exact analytical results are in general only available for one-variable systems. It is desirable, in the context of noise-induced phenomena, to avoid the use of approximation procedures as far as possible, since modifications of the behavior of the system are often very drastic and rather unintuitive. It seems to us preferable, at least in the first stage of the investigation, to establish the recently discovered effects of external noise with the help of rigorous procedures. Otherwise, the results would be tainted with doubts as to the validity of the approximation schemes. Certain physical assumptions and idealizations are of course unavoidable, as in any description of the real world. We shall, however, use only those which are well accepted in the literature and we shall carefully discuss their consequences for the results obtained. Ultimately of course, the results have to be subjected to an experimental verification.

Having defined the class of systems we intend to deal with, let us now turn our attention to the environment, especially the manner in which its variability can be modeled. In contradistinction to the internal fluctuations, the random variability of the environment is in general not of microscopic origin. The external noise is often the expression of a turbulent or chaotic state of the surroundings or reflects the fact that the external parameters depend on numerous interfering environmental factors. This has the consequence that environmental fluctuations are not scaled by an inverse power of the system size. For this reason they will not disappear on the macroscopic level of description of the system. In the laboratory the experimenter has obviously a certain control over the strength of the external noise. By careful experimental procedure the noise level can be reduced, however it is impossible to get rid of all external noise completely. On the other hand, the intensity of environmental fluctuations can of course be increased in a controlled way to investigate its influence on the behavior of the system. This is another feature in which external noise differs from internal fluctuations, and the experimenter has a far greater control over it.

In any case, for natural as well as for laboratory systems external noise is never strictly zero. Therefore the need arises to refine the phenomenological description in order to include the effects of environmental randomness. In the following, we shall consider that there is no feedback from the system onto the environment and that the environment undergoes no systematic temporal evolution. The first is a standard assumption and requires essentially that the environment is much larger than the system. The second assumption is fulfilled in most applications, at least over the time spans one is interested in. It is made in order to separate clearly the effects of the environmental fluctuations, the topic of this work, from those effects due to a systematic evolution of the surroundings, as for instance the influence of periodic seasonal variations on natural systems.

The influence of the environment on the macroscopic properties of the system is described on the level of the phenomenological equation via the external parameters λ. If the system is coupled to a fluctuating environment, then these parameters become in turn stochastic quantities. Under the two assumptions made above, they can be represented by stationary stochastic processes λ_t. For the following it is convenient to decompose λ_t into two parts, i.e., $\lambda + \zeta_t$, where λ corresponds to the average state of the environment, and ζ_t describes the fluctuations around it. Obviously we have $E\{\zeta_t\} = 0$.[9] Including external noise in the phenomenological description leads to the following stochastic differential equation (SDE):

$$\dot{X}_t = f_{\lambda_t}(X_t) . \tag{1.12}$$

In a very large class of phenomenological equations, encountered in applications, the external parameter appears in a linear way. (The nonlinear case is discussed in Sect. 8.7 and Chap. 9.) We consider for the time being only one fluctuating external parameter. Equation (1.12) is then of the form

$$\dot{X}_t = h(X_t) + \lambda g(X_t) + \zeta_t g(X_t) . \tag{1.13}$$

The next step in the modeling procedure is to specify explicitly the probabilistic characteristics of the random process in terms of the physical properties of the environment. While a detailed discussion of this point has to be postponed till we have introduced the necessary tools from the theory of probability in more precise terms, it is worthwhile to address this problem already at this stage in a somewhat heuristic fashion, relying on the common intuitive comprehension of the probabilistic terms employed. In some cases the mechanism, giving rise to the environmental randomness, can be precisely identified. This is of course trivially true for experiments in the laboratory, carried out to study the influence of fluctuating external constraints and where the experimenter can control the external noise. In most instances, specially for natural systems, the situation is generally so complex that the variations of the external parameters cannot be attributed to a single well-defined cause. One has to be satisfied with the experimental observation that the system perceives its surroundings as a noise source. It turns out, however, that in these situations it is not necessary to inquire into the exact origin of the environmental fluctuations to specify the stochastic process ζ_t. Indeed, consider the following two important situations which cover most applications.

i) *Continuously Varying External Parameters.* It is an experimental observation that in an astonishingly large class of situations the values of the external parameter are distributed according to a curve that is satisfactorily described by the familiar bell-shaped curve of the Gaussian distribution, also known as the normal distribution. This fact can be understood as the consequence of a profound and fundamental theorem of probability theory, known as the central limit theorem. In most situations, fluctuations in the external parameters are the cumulative effect of numerous environmental factors. Whatever the probability distributions of these factors, provided they are not too dissimilar and not too strongly correlated, the central limit theorem assures that the fluctuations in the

[9] $E\{\}$ denotes the mean value of random variable (Chap. 2).

external parameter are Gaussian distributed. A precise formulation of this fundamental theorem of probability theory as well as its conditions of applicability can be found in any standard textbook [1.86, 87]. In light of this theorem the ubiquitous appearance of the Gaussian distribution in applications is hardly surprising any more. The temporal properties of this kind of external noise will be discussed below.

ii) *Shot Noise.* A second important class of external fluctuations consists in the occurrence of well-defined discrete events in the environment at random times. An appropriate way to model this type of noise is often given by the Poisson process, the characteristics of which are discussed in detail in [1.88]. This kind of noise resembles that which arises from the shot effect in vacuum tubes and hence is generally designated as shot noise.

These two classes of noise cover most of the situations encountered in natural systems. Clearly, the detailed mechanism of variations in the environment is not needed to arrive at a satisfactory model of external noise in most situations. According to the fundamental limit theorems of probability theory, external noise displays under most circumstances a kind of universal behavior. Thus the models for environmental fluctuations can be chosen among the most simple and basic classes of stochastic processes, namely Gaussian and Poisson processes.

Let us now briefly discuss the temporal properties of external noise. It turns out that in a large class of applications there exists a very clear-cut separation between the time scale of the macroscopic evolution of the system and the time scale of the environmental fluctuations. The external noise varies on a much faster time scale than the system. The environment forgets, so to speak, very quickly what state it was in a short time ago. Since the memory of the environment is so extremely short from the viewpoint of the system, it is usual in the physical literature to pass to the idealization of a memoryless environment. This sounds rather harmless but it is actually at this point that dangerous territory is entered, which contains hidden pitfalls and traps to ensnare the unwary theoretician. The passage to the limit of an environment without memory is rather subtle and beset with danger if it is not carefully effected. This limiting procedure will be treated with due respect in Sect. 3.2 and the various pitfalls will be explained. If one succeeds in avoiding the various traps, either by luck, intuition or whatever else, one captures a treasure which might be bane or boon: the white noise. This is a random process which forgets immediately the value it took the instant before. In more technical terms, it has independent values at every instant of time. Such processes are known as completely random. It is obvious that they are extremely irregular, jumping wildly around. To make matters worse, it turns out, if the limit procedure is properly effected, that white noise has infinite intensity, since it jumps between minus and plus infinity. On the one hand, it is intuitively obvious that a completely random process with finite intensity would have no effect whatsoever on the system. On the other hand, this means that white noise is a very strange object, hardly an ordinary random process. Indeed, white noise is for random processes what the Dirac delta-function is for deterministic functions. It is a generalized random process (Appendix A) and as such has to be treated rather carefully. For instance, it is meaningless to carry out a nonlinear operation on a delta function, e.g., squaring it.

This makes one wonder if any sense can be given to a differential equation like (1.13) with a white noise on the right-hand side. The extreme irregularity of the white noise implies that the time derivative of X_t is certainly not defined in any ordinary sense. However, it turns out that these properties of the white noise can often be ignored, if $g(X)$ is a constant, i.e., if the influence of the external noise does not depend on the state of the system. Then (1.13) can be handled, in most circumstances, as if it were an ordinary differential equation and meaningful results are obtained. In general however, the influence of the external noise depends on the state of the system. Then the white noise is multiplied by a function, i.e., $g(X_t)$, which is also quite irregular and one has to investigate carefully if a meaning can still be given to (1.13). There is some ground for optimism: if a sense can be given to (1.13), then X_t is one order less irregular than the white noise. It would have been obtained by the integration of the white noise input and hence would have been smoothed. The way by which one succeeds in giving a well-defined meaning to (1.13) proceeds via the equivalent integral equation[10]:

$$X_t = X_0 + \int_0^t f_\lambda(X_s)\, ds + \int_0^t g(X_s)\, \xi_s\, ds\,. \tag{1.14}$$

Formulated in this manner, the question is now how to arrive at a consistent definition of the stochastic integral $\int g(X_s)\, \xi_s\, ds$. For the sake of concreteness, let us for the time being consider only the case of continuously varying external parameters, which will in any case occupy a central role in this monograph. Then at each instant of time the fluctuations should be Gaussian distributed. This frequently encountered case of white noise is known as Gaussian white noise. Obviously the question arises in what sense a quantity that jumps between minus and plus infinity can be Gaussian distributed. The answer to this question as well as to the other questions raised by the strange properties of white noise will be presented in detail in the following chapters. For the moment, in order to give readers not familiar with the notion of white noise a true feeling for its unusual properties, we should like to present it in a handwaving way.

White noise is an immensely useful concept, if treated with the proper respect for the subtleties it involves. A naive approach to white noise which assimilates it to an ordinary random process is rather dangerous and can lead to meaningless results. Our insistence on these particular features of white noise is not mere mathematical hairsplitting. It is necessary to avoid the confusion and controversy that has plagued the treatment of systems with multiplicative noise for almost two decades now. The main source of confusion is linked to the definition of the stochastic integral $\int g(X_s)\, \xi_s\, ds$. The problem is that though a sense can be given to this integral and thus to the SDE (1.13), in spite of the extremely irregular nature of the white noise, there is *no unique* way to define it, precisely because white noise is so irregular. This has nothing to do with the different definitions of ordinary integrals by Riemann and Lebesgue. After all, for the class of functions for which the Rieman integral as well as the Lebesgue integral can be defined,

[10] White noise in the following will always be designated by the symbol ξ while ζ as in (1.13) will be reserved for other kinds of noise.

both integrals yields the same answer. The difference between the two definitions for the above stochastic integral, connected with the names of Ito and Stratonovich, is much deeper; they give different results. To make this statement more comprehensible and as a sneak preview, let us briefly describe in qualitative terms, how Ito and Stratonovich define the integral $\int g(X_s)\,\xi_s ds$. Both definitions are based on the heuristic relation that integration of Gaussian white noise yields Brownian motion, which we shall denote by W_t (see Chap. 2 for details). Therefore the above integral can be written

$$\int g(X_s)\,\xi_s ds = \int g(X_s)\,dW_s\,.$$

The integral on the right-hand side is then defined, as in the case of an ordinary integral, by the limit of the approximating sums. The Ito definition corresponds, roughly speaking, to

$$\int g(W_s)\,dW_s = \lim \sum g(W_{t_{i-1}})(W_{t_i} - W_{t_{i-1}})\,,$$

whereas the Stratonovich integral is given by

$$\int g(W_s)\,dW_s = \lim \sum g\left(\frac{W_{t_{i-1}} + W_{t_i}}{2}\right)(W_{t_i} - W_{t_{i-1}})\,.$$

So the only difference is the choice of the evaluation point. Ito chooses the left-hand point $W_{t_{i-1}}$ in the partition of the time interval, whereas Stratonovich opts for the middle point $(W_{t_{i-1}} + W_{t_i})/2$. For an ordinary (deterministic) integral,

$$\int U(X)\,dX = \lim \sum U(\tilde{X}_i)(X_i - X_{i-1})\,,$$

any evaluation point $\tilde{X}_i$, as long as $\tilde{X}_i \in [X_{i-1}, X_i)$, can be chosen; the limit is independent of it. Due to the extremely wild behavior of the Gaussian white noise, this is no longer true for the stochastic integral. The limit of the approximating sums depends on the evaluation point; Ito and Stratonovich yield different answers for the same integral (Chap. 5). For instance,

Ito: $\int_0^t W_s dW_s = \frac{1}{2}(W_t^2 - W_0^2) - t/2\,,$

Stratonovich: $\int_0^t W_s dW_s = (W_t^2 - W_0^2)/2\,.$

It should go without saying that both the Ito and Stratonovich definitions are mathematically correct and can serve as the basis for a consistent calculus. Nevertheless, the fact that no unique definition for the stochastic integral with white noise exists has puzzled and bewildered quite a few scientists. In 1969 *Mortensen* found: "Although this sujet" (the Ito and the Stratonovich calculus) "has been discussed in several papers in the last two or three years, reading some of these papers can leave one more bewildered than before one started" [1.89, p. 272]. Amazingly, these words are still true today, perhaps even more so. In spite of clear and carefully written papers like Mortensen's and others, the confusion continues. There are still attempts to prove one integral definition wrong and establish the other as the only legitimate one in scientific applications. The bewilderment persists because too many workers in this field fail to treat white noise as the strange, but powerful object that it is. They blind themselves to the fact that it is a generalized random process, with features totally different from ordinary processes. The attitude is still widespread that most of the subtleties connected with white noise are mere mathematical hairsplitting and of no importance for practical applications. This leads to confusion when finally situations are encountered in which what is supposed to be nothing more than highbrow mathematics has solid practical consequences. This is the case in the treatment of systems coupled to a fluctuating environment in a multiplicative way. Actually only a minimal level of mathematical rigor and proper respect for the characteristic features of white noise is needed to dissipate the confusion rapidly.

Therefore to keep the door closed to many ambiguities, imprecisions and controversies which have plagued the treatment of noise phenomena in the past, it is necessary to present in the following chapters some basic notions of probability theory, of the theory of Markovian diffusion processes and of the theory of stochastic differential equations. We shall, however, always have our minds firmly set on the practical concepts. Furthermore, we shall ignore any mathematical subtleties which are only of marginal importance for our approach.

2. Elements of Probability Theory

Probability theory is the adequate mathematical framework within which to tackle the problem of the effect of external noise on nonlinear systems. To be able to discuss the modeling procedure of these phenomena in a clear way, it is necessary to recall in this chapter, with a certain degree of mathematical precision, some basic notions of probability theory. At the same time it will serve to establish our notations. We have made every effort to make this monograph as far as possible self-contained, especially the mathematical aspects of our approach. If the reader is totally unfamiliar with probability theory and feels the need to read more about the subject we suggest that he consults [1.86, 87, 2.1] or other standard textbooks.

2.1 Probability Triple and Random Variables

The basic notion of probability theory is the probability triple $(\Omega, \mathscr{A}, P)$ consisting of the sample space Ω, a field of events $\mathscr{A}$ and a probability measure P. The first two elements of this triple constitute the only ingredients used in the definition of a random variable.

2.1.1 The Sample Space Ω and the Field of Events $\mathscr{A}$

Consider an experiment in which the outcome cannot be predicted with certainty, e.g., picking a particular molecule in a gas container or choosing in an unbiased way a certain petri dish in an experiment on bacterial growth. In order to keep track of all the different possible outcomes, let us give a label ω to each individual outcome that can occur in the experiment in question. In these examples, we would "place a tag" on each molecule and on each petri dish. So the experiment can be characterized, roughly speaking, by the set of all (labeled) individual outcomes. Abstracting from the particular examples, this motivates the following definition: A sample space Ω is the ensemble of elementary outcomes, labeled $\omega: \omega \in \Omega$. The number of elementary outcomes may be finite, as in the above examples, countably or uncountably infinite.

The second element of the probability triple $\mathscr{A}$ is the σ field (or σ algebra) of events. This is, contrary to its possibly intimidating name, a quite simple and easily understood concept. Consider a collection of elementary outcomes that is meaningful, or of interest, for a particular experiment. In the above examples this could be the set of all molecules with a speed lower than $(kT/m)^{1/2}$ or the set of all petri dishes containing populations of more than N individuals. Such a subset A of Ω, $A \subset \Omega$, is called an event. This event occurs if the elementary outcome ω belongs to A, i.e., if a molecule with $|v| < (kT/m)^{1/2}$ or a petri dish with a number of bacteria greater than N is picked. The σ field $\mathscr{A}$ is the ensemble of all events, i.e., $A \in \mathscr{A}$. This set $\mathscr{A}$ of events is necessarily smaller or equal to the set of all subsets of Ω, the so-called power set of Ω and denoted by $\mathscr{P}(\Omega)$. If the sample space Ω is finite and contains, say, M elements, then $\mathscr{A}$ also contains only a finite number of elements, which is less or equal to 2^M, the number of elements in $\mathscr{P}(\Omega)$. The set $\mathscr{A}$ of events quite naturally possesses the following properties:

1) It contains the certain event:

$$\Omega \in \mathscr{A}. \tag{2.1}$$

(If the experiment is performed, one of all the possible outcomes will obviously occur.)

2) It contains the impossible event, namely the empty set ϕ:

$$\phi \in \mathscr{A}. \tag{2.2}$$

(If the experiment is performed, it is impossible that no outcome results.)

3) If A is an event, then the completement $\bar{A} = \Omega - A$ is an event. Thus

$$A \in \mathscr{A} \quad \text{implies} \quad \bar{A} \in \mathscr{A}. \tag{2.3}$$

4) If A and B are events, so is their union and intersection

$$A, B \in \mathscr{A} \Rightarrow \begin{cases} A \cup B \in \mathscr{A} \\ A \cap B \in \mathscr{A} \end{cases}. \tag{2.4}$$

Any set $\mathscr{A}$ of subsets of Ω that fulfills conditions (2.1 – 4) is called a field. For practical purposes, it is convenient that also the union of countably many events is an element of $\mathscr{A}$, i.e.,

$$A_n \in \mathscr{A} \quad n = 1, 2, \ldots \Rightarrow \bigcup_{n=1}^{\infty} A_n \in \mathscr{A}. \tag{2.5}$$

This condition is of course trivially satisfied for sample spaces consisting only of a finite number M of elementary outcomes. Then $\mathscr{A}$ also contains only a finite number of elements. Otherwise condition (2.5) has to be imposed and a field of events with this additional property is precisely what is understood by a σ field. In the above example of bacterial growth, the number of petri dishes is finite and the most natural choice of σ field is the set of all subsets.

It is often necessary to consider a collection of events $\{A_i | i \in J\}$. The smallest σ algebra $\mathscr{C}$ that contains this set of events, i.e., $A_i \in \mathscr{C}$ for all i, is said to be the σ field generated by the events $\{A_i\}$. Essentially it is the σ algebra that is obtained by repeatedly applying operations (2.3 – 5) to the set of events $\{A_i \mid i \in J\}$. The σ field $\mathscr{C}$ will be denoted by $\sigma\{A_i\}$ or $\mathscr{A}\{A_i\}$.

2.1.2 Random Variables

A random variable X is a function from the sample space Ω into some state space. In the following it will be the set of real numbers or some subset thereof, $X: \Omega \to \mathbb{R}$. However, not any real-valued function qualifies as a random variable. It is furthermore required that this function has the following property:

$$A = \{\omega \mid X(\omega) \leqslant x\} \in \mathscr{A}, \quad \forall x \in \mathbb{R}. \tag{2.6}$$

Introducing the notion of the inverse image X^{-1}, defined by

$$X^{-1}(B) = \{\omega \mid X(\omega) \in B\}, \quad B \subset \mathbb{R} \tag{2.7}$$

this condition is often written as

$$A = X^{-1}((-\infty, x]) \in \mathscr{A}, \quad \forall x \in \mathbb{R}. \tag{2.8}$$

A real valued function X: $(\Omega, \mathscr{A}) \to \mathbb{R}$, which fulfills the requirement (2.6), is also said to be *measurable* with respect to the σ field $\mathscr{A}$. In words this means that the subset A of the sample space, which consists of all those elementary outcomes ω for which $X(\omega)$ is smaller than some arbitrary real number x, is an event. Care should be taken to distinguish between the random variable, denoted by X, and the value it takes for a particular sample, denoted by x, which is an element of the state space. If any subset of Ω is an event, namely $\mathscr{A}$ is the power set $\mathscr{P}(\Omega)$, then any real valued function is a random variable; (2.8) is trivially fulfilled. This situation occurs frequently when the sample space is finite as in the above examples. If the sample space has uncountably many elements, then it is in general not meaningful to consider each subset of Ω as an event. In a certain sense Ω is too large; it contains rather bizarre subsets which are not meaningful as an event. To anticipate a little bit, it would be impossible to define a probability for each element of $\mathscr{P}(\Omega)$, i.e., for each subset of Ω. For a discussion of this more subtle point see [1.86, 87].

The motivation to make (2.8) a defining feature of a random variable is based on the following considerations: suppose that the underlying sample space is not accessible to any direct observation but only via the function X: $\Omega \to \mathbb{R}$. This is a common situation in many applications; X can be thought of as some measuring device. Then it is legitimate to consider the state space $\mathbb{R}$ as a new sample space for the quantity X. The events in $\mathbb{R}$ are naturally given by the intervals: $B = [x, y]$, $x, y \in \mathbb{R}$. Using the operations union, intersection and complement these intervals generate a σ field, denoted $\mathscr{B}$ and which goes by the name of Borel σ field. This is the field that is naturally associated with the state space $\mathbb{R}$.[1]

---

[1] The Borel σ field $\mathscr{B}$ is an example for a σ field which is strictly smaller than the power set.

To generate the Borel σ field, it is sufficient to use only intervals of the type $(-\infty, x]$. It is of course reasonable to require that an event of X in $\mathbb{R}$ corresponds to an event in the underlying sample space Ω. Otherwise the "measuring device" X would yield spurious information. This is what the condition (2.8) is all about. It ensures that for any event $B \in \mathscr{B}$ in the state space, there corresponds an event in the underlying sample space, i.e., the inverse image of B is an event. It is sufficient to consider in (2.8) only events of the form $B_x = (-\infty, x]$, $x \in \mathbb{R}$ since $\mathscr{B}$ is generated by them, as mentioned above. Hence, if (2.8) holds, then $X^{-1}(B) \in \mathscr{A}$ for all $B \in \mathscr{B}$. Of course, some information might be lost. Not every event in the underlying sample space might be "detected" by the "measuring device X". In more mathematical terms, in general not every event $A \in \mathscr{A}$ is of the form $A = X^{-1}(B)$, $B \in \mathscr{B}$. This is expressed by the fact that the σ field generated by all sets of the form $X^{-1}((-\infty, x])$, $x \in \mathbb{R}$, and denoted by $\mathscr{A}(X)$ is in general only a subfield of $\mathscr{A}$, $\mathscr{A}(X) \subset \mathscr{A}$.

A simple random variable in the case of the gas container mentioned above is the speed $|v|$ of a gas molecule at a given instant of time, if $\mathscr{A} = \mathscr{P}(\Omega)$. Since the number of gas molecules, i.e., the number of elementary outcomes, is finite, the σ field of events can be chosen to be the power set and is indeed the natural choice as remarked above. Condition (2.8) is trivially satisfied in this case. Consider now a slightly different situation. Let the σ field of events this time not be equal to the power set. Instead let us choose the following, admittedly somewhat artificial, σ field $\tilde{\mathscr{A}} = \{\phi, A_l, A_r, \Omega\}$, where A_r corresponds to the set of all molecules for which $v \cdot \hat{x}$ is positive. Here A_l is the complement, i.e., the set of molecules for which $v \cdot \hat{x}$ is negative or zero and $\hat{x}$ is the unit vector in some arbitrary direction. It is now easy to see that $|v|$ is not a random variable in this case. Indeed, the set of molecules whose speed is lower than some value u, $\{\omega \,||\, v\,|(\omega) \leqslant u\}$ coincides in general neither with ϕ, nor with A_r, nor A_l, nor Ω.

A similar example for the bacterial growth experiment is the following. In a large number of petri dishes E. coli bacteria are incubated, half of them contain the strain B, the other half the strain C. The sample space is the set of petri dishes. An elementary outcome occurs when the experimenter chooses in an unbiased way, i.e., at random, a particular petri dish. Since the number of elementary outcomes is again finite, consider the natural choice for the σ field of events, i.e., $\mathscr{A} = \mathscr{P}(\Omega)$. Let N be the number of bacteria at a given instant of time, in a given petri dish, regardless of the strain to which it belongs. It is obvious that condition (2.8) is fulfilled in this case and N is a random variable. On the contrary, if the following σ field $\mathscr{A} = \{\phi, A_B, A_C, \Omega\}$ had been chosen, where A_B corresponds to the ensemble of all petri dishes containing bacterial strain B and A_C corresponds to the set of remaining petri dishes containing strain C, then the number of bacteria in a petri dish is not a random variable. Indeed $\{\omega \,|\, N(\omega) \leqslant x\}$ coincides in general neither with ϕ, nor with A_B, nor A_C, nor Ω.

This underlines that the sample space Ω as well as the σ field $\mathscr{A}$ are essential ingredients in the definition of a random variable. Let us close this section by the slightly paradoxical sounding remark that there is nothing random with a random variable. As has been nicely formulated by *Chung* [1.86, p. 75]: "What might be said to have an element of randomness in $X(\omega)$ is the sample point ω, which is picked "at random" [...]. Once ω is picked, $X(\omega)$ is thereby determined and there is nothing vague, indeterminate or chancy about it anymore".

2.1.3 The Probability Measure *P*

Let us now turn our attention to the third element P of the probability triple $(\Omega, \mathscr{A}, P)$ which we neglected so far because it plays no role in the definition of a random variable. Here P is a measure for the frequency of the occurrence of the event A and this is of course what we intuitively understand by a probability. To be precise, P is a function from the σ *field* $\mathscr{A}$ onto the interval $[0, 1]$, $P\colon \mathscr{A} \to [0, 1]$. Note that P is a function defined on the σ field $\mathscr{A}$ and not on the sample space Ω, that is, the probability is defined for events A and not only for the elementary outcomes ω. The distinction is essential if Ω has uncountably many elements. In this case the probability that precisely the elementary outcome ω occurs is often strictly zero. A meaningful definition of probability then has to use as its basis the σ field of events. By definition P has the following properties:

$$P(\phi) = 0\,, \tag{2.9}$$

$$P(\Omega) = 1\,, \tag{2.10}$$

$$A_n \in \mathscr{A}, A_n \bigcap_{n \neq m} A_m = \phi \Rightarrow P\left(\bigcup_{n=1}^{\infty} A_n\right) = \sum_{n=1}^{\infty} P(A_n)\,. \tag{2.11}$$

Such a set function P is called a probability measure. The above three conditions reflect in mathematical terms our intuitive notions about probabilities. These are namely that the impossible event ϕ has probability zero, and the certain event Ω probability one. Furthermore, the probability of a union of mutually exclusive events is of course intuitively obtained by summing up the probabilities of the single exclusive events. The third condition requires this to remain true for a countable union of exclusive events. In concrete applications, the probability measure P of the basic probability triple is obtained either by a priori reasoning or is deduced from a long series of experiments in which the relative frequency of the event A is determined. (There are some mathematical subtleties involved in the frequency interpretation of probability theory on which we do not intend to dwell here).

In any case, to quote *Arnold* [Ref. 2.2, p. 4], "The object of probability theory is not the determination of the probability measure P of the underlying sample space Ω, but it is concerned with the calculation of new probabilities from given ones." In particular, the probability of events of the Borel σ field associated with the state space $\mathbb{R}$ of some random variable $X(\omega)$ can immediately be expressed in terms of the probability P of events of the σ field $\mathscr{A}$ of the probability triple $(\Omega, \mathscr{A}, P)$. Indeed, consider the random variable $X\colon (\Omega, \mathscr{A}, P) \to (\mathbb{R}, \mathscr{B})$. The function X transports, so to speak, the probability measure P from the underlying sample space Ω onto the state space $\mathbb{R}$. If B is an observable event in the state space, $B \in \mathscr{B}$, then we can define its probability by

$$P_X(B) = P(X^{-1}(B)) = P(\{\omega | X(\omega) \in B\})\,. \tag{2.12}$$

In this way, a probability measure is defined on the state space. This definition is meaningful since it is a defining property of a random variable that $X^{-1}(B)$ is an event in the underlying probability triple. Thus the probability is defined for this subset $X^{-1}(B)$ of the sample space Ω. This throws additional light on requirement (2.8). Consider a real-valued function $Z: \Omega \to \mathbb{R}$ which is not endowed with property (2.8). Then there exists an event $B' \in \mathscr{B}$, such that $Z^{-1}(B') \notin \mathscr{A}$. In other words, on the level of the measuring device Z an event has occurred for which however no probability can be determined, since it does not correspond to an event in the underlying sample space. Since $Z^{-1}(B') \notin \mathscr{A}$, $P(Z^{-1}(B'))$ is undefined. Obviously, this is an undesirable situation; Z is not a suitable measurable quantity. This examplifies that requirement (2.8) is absolutely essential for a meaningful notion of a random variable. Otherwise, no viable theory can be built up.

2.1.4 The Distribution Function

The probability measure P_X on $\mathscr{B}$ is induced by the random variable and is called the distribution of the random variable. Due to the structure of $\mathscr{B}$ it is of course sufficient, as in (2.8), to consider only events of type $B_x = (-\infty, x]$. For a real-valued random variable the distribution P_X is completely defined by the distribution function $F_X(x)$:

$$F_X(x) = P(\{\omega | X(\omega) \leqslant x\}) = P(X \leqslant x), \quad x \in \mathbb{R}. \tag{2.13}$$

(In the following we will suppress the subscript X, indicating the random variable F refers to, if no confusion can arise.) It is easily seen from definition (2.13) that a distribution function has the following properties:

$$\begin{aligned} &\text{i)}\ F(-\infty) = 0 \\ &\text{ii)}\ F(+\infty) = 1 \\ &\text{iii)}\ F(x) \text{ is an increasing right continuous}^{2} \text{ function}. \end{aligned} \tag{2.14}$$

The inverse also holds. Every function fulfilling the three conditions (2.14) can act as the distribution function of some random variable.

Using the distribution (function), the state space with its associated σ field $\mathscr{B}$ can be completed to form a probability triple $(\mathbb{R}, \mathscr{B}, P_X)$. If we are interested only in the quantity X, then we could forget the underlying probability triple $(\Omega, \mathscr{A}, P)$ and take $(\mathbb{R}, \mathscr{B}, P_X)$ as the new sample space and interpret the random variable as the identity:

$$X: (\mathbb{R}, \mathscr{B}, P_X) \to (\mathbb{R}, \mathscr{B}, P_X), \quad X(x) = 1(x) = x. \tag{2.15}$$

Loosely speaking, we could then say that a real-valued random variable is a quantity characterized by a probability law, namely its distribution function.

---

[2] $\lim_{\varepsilon \downarrow 0} F(x+\varepsilon) = F(x)$; $\varepsilon \downarrow 0$ means that ε is positive and tends to zero.

This is the situation commonly encountered in practical applications, where in general the underlying probability triple is unknown or inaccessible. However, from a theoretical point of view the use of an underlying probability triple is very advantageous, especially when dealing with more than one random variable X_i, $i = 1, \ldots n$, e.g., for questions of convergence. Indeed, theoretical considerations become considerably more transparent in such cases by the use of one and the same probability measure. The different distributions P_{X_i} are then just transformations of P by the various random variables. Therefore even in practical applications it is often useful to resort to the unknown underlying probability triple $(\Omega, \mathscr{A}, P)$ to disentangle the lines of reasoning.

In application two types of random variables are of particular importance. First, those corresponding to physical quantities that by their very nature take only discrete values, as e.g., the number of bacteria in a petri dish. Second, random variables which correspond to continuously varying physical quantities, as e.g., concentrations. Discrete random variables are characterized by the fact that the distribution function increases only at countably many places by jumps and is constant between them. In the applications we shall be interested in, we shall mostly encounter continuous random variables, for which the probability density $p(x)$, defined by

$$F(x) = \int_{-\infty}^{x} p(x')\,dx' \tag{2.16}$$

or equivalently $p(x) = \partial_x F(x)$, exists. Roughly speaking, $p(x)\,dx$ is the probability that $X(\omega)$ takes a value in the infinitesimal neighborhood of x:

$$p(x)\,dx = P(\{\omega | X(\omega) \in (x, x+dx)\}) = P(X \in (x, x+dx)) = P(d\omega)\,. \tag{2.17}$$

In the following, most definitions and expressions will be stated in the form they take for random variables with probability densities. Their transcription to the case of discrete variables is straightforward.

2.1.5 Moments and Extrema

The distribution function or probability density gives a complete characterization, from a probabilistic point of view, of the random variable X. However, this complete information is not always needed and certain numerical characteristics of the random variable are often sufficient. Among the most important ones are the so-called moments. The first moment, commonly known as the mean value or mathematical expectation, is the sum of the states of the random variable, weighted by the probability with which the former occur. Note, however, that a random variable X and the probability P are defined on two different objects. The first is a function on the sample space Ω, the second is defined on the σ field $\mathscr{A}$. It is therefore necessary to say a few words on integration over a probability triple $(\Omega, \mathscr{A}, P)$. We begin by considering random variables of a particularly simple form, namely so-called random step functions

$$X(\omega) = \sum_{i=1}^{n} x_i I_{A_i}(\omega)\,. \tag{2.18}$$

Here I_A is the indicator function of the set A, i.e.,

$$I_A(\omega) = \begin{cases} 1 & \text{if} \quad \omega \in A \\ 0 & \text{if} \quad \omega \notin A \end{cases}. \tag{2.19}$$

The events A_i, $A_i \in \mathscr{A}$, are mutually disjunct, $A_i \cap A_j = \phi$, $i \neq j$, and cover the whole sample space: $\bigcup_{i=1}^{n} A_i = \Omega$. A step function is simple, since it takes only finitely many different values. Introducing the notation $E\{X\}$ for the mathematical expectation, it is natural to define

$$E\{X\} = \sum_{i=1}^{n} x_i P(A_i)\,. \tag{2.20}$$

We shall also write this in the following form, which turns out to be useful later on, as

$$E\{X\} = \int_{\Omega} X\,dP = \int_{\Omega} X(\omega)\,dP(\omega) = \int_{\Omega} X(\omega) P(d\omega)\,. \tag{2.21}$$

As the next step, consider now a random variable $X \geqslant 0$. There exists an increasing sequence X_n of random step functions with

$$\lim_{n\to\infty} X_n(\omega) = X(\omega)\,, \quad \forall\, \omega \in \Omega \tag{2.22}$$

and

$$\lim_{n\to\infty} \int_{\Omega} X_n\,dP = m \leqslant \infty\,, \tag{2.23}$$

where m stands here for mean value and is obviously in general a real number. The limiting value m is independent of the particular sequence X_n and it is thus legitimate to define

$$E\{X\} = \int_{\Omega} X\,dP = m = \lim_{n\to\infty} \int_{\Omega} X_n\,dP\,. \tag{2.24}$$

If X is an arbitrary random variable, we decompose it into its positive and negative parts

$$X = X^+ - X^-\,, \quad X^+ = X I_{[X \geqslant 0]}\,, \quad X^- = -X I_{[X<0]}$$

and define

$$E\{X\} = \int_{\Omega} X\,dP = \int_{\Omega} X^+\,dP - \int_{\Omega} X^-\,dP\,. \tag{2.25}$$

For real-valued random variables, the expectation can easily be calculated using the induced distribution:[3]

$$m \equiv E\{X\} = \int_{\mathbb{R}} x \, dF(x) \,, \tag{2.26}$$

and if the probability density exists

$$E\{X\} = \int_{\mathbb{R}} x p(x) \, dx \,. \tag{2.27}$$

Higher moments are defined in an analogous way. The rth moment (r positive integer, $r \in \mathbb{N}$) is

$$\begin{aligned} E\{X^r\} &= \int_{\Omega} X^r(\omega) \, dP(\omega) \\ &= \int_{\mathbb{R}} x^r dF(x) \qquad (2.28) \\ &= \int_{\mathbb{R}} x^r p(x) \, dx \,. \qquad (2.29) \end{aligned}$$

Again the latter equality holds only if $p(x)$ exists. As far as the moments are concerned, let us remark that contrary to the often encountered belief, the knowledge of all moments $E\{X^r\}$ is not necessarily sufficient to determine uniquely $p(x)$ [Ref. 2.3, p. 166].

In general we have the following transformation theorem: Let $Y = h(X)$ be a random variable, where $h(x)$ is a non-random function ($h(x)$ does not depend on ω), then

$$\begin{aligned} E\{Y\} &= \int_{\Omega} h(X(\omega)) \, dP(\omega) \qquad (2.30) \\ &= \int_{\mathbb{R}} h(x) \, dF(x) = \int_{\mathbb{R}} h(x) p(x) \, dx \,. \qquad (2.31) \end{aligned}$$

Whereas the zeroth moment always exists, namely

$$\begin{aligned} 1 = P(\Omega) &= \int_{\Omega} dP(\omega) \\ &= \int_{\mathbb{R}} dF(x) \\ &= \int_{\mathbb{R}} p(x) \, dx \,, \qquad (2.32) \end{aligned}$$

the higher moments are not necessarily finite. A measure for the strength of fluctuations, so to speak for the "randomness" of the random variable, is the width of its probability density, or the average size of the deviations from its mean value. The mean square deviation or variance of a random variable is given by

---

[3] $\int_{\mathbb{R}} dx = \int_{-\infty}^{+\infty} dx$.

$$\sigma^2 = E\{(\delta X)^2\} = E\{(X-m)^2\}$$
$$= \int_{\mathbb{R}} (x-m)^2 dF(x)\,, \tag{2.33}$$

i.e., the second central moment. The general definition of the central moments is

$$E\{(\delta X)^r\} = E\{(X-m)^r\}\,. \tag{2.34}$$

While σ^2 is a measure for the width of the probability density, the third central moment is related to the skewness or asymmetry of the random variable. If fluctuations are symmetric around the mean value, then

$$E\{(\delta X)^{2k+1}\} = 0 \quad \text{with} \quad k = 1, 2, \ldots\,. \tag{2.35}$$

Other important characteristics of the probability density (or the probability for discrete variables) are the number and location of its extrema x_m:

$$\partial_x p(x)\,|_{x=x_m} = 0\,. \tag{2.36}$$

[For discrete variables: $p(\nu_m) = p(\nu_m - 1)$]. The maxima

$$\partial_{xx} p(x)\,|_{x=x_m} < 0 \tag{2.37}$$

are the so-called most probable states. These are the states that are preferentially observed in a series of experiments. Note that according to definitions (2.36, 37) the property of being a most probable state is a *local* property. These states are more probable than the states in a certain neighborhood around them. No global statement referring to the largest value of the probability density on the state space is implied.

Let us consider a few instructive examples.

The Gaussian Distribution. A random variable is said be Gaussian or normally distributed if its probability density is given by

$$p(x) = [(2\pi)^{1/2}\sigma]^{-1} \exp\left(-\frac{(x-m)^2}{2\sigma^2}\right), \tag{2.38}$$

often denoted by: X is $N(m, \sigma^2)$. As already implied by the notation, it is easily verified that

$$E\{X\} = m\,, \tag{2.39}$$
$$E\{(\delta X)^2\} = \sigma^2\,. \tag{2.40}$$

It follows immediately from (2.38) that a Gaussian variable is completely characterized by its first two moments. Indeed we have for the central moments

$$E\{(\delta X)^r\} = \begin{cases} 0 & r \geqslant 1 \quad \text{odd} \\ (r-1)!!\,\sigma^r & r \geqslant 2 \quad \text{even}^4 \end{cases} . \tag{2.41}$$

The Gaussian probability density is symmetric around its mean value. All moments exist, cf. (2.41), since the exponential function decreases faster than any power of x increases. The Gaussian or normal distribution plays an important role in applications, as already mentioned in Chap. 1, and occupies a central place in probability theory as manifested by the central limit theorem.

The Poisson Distribution. A random variable X that takes as values only the nonnegative integers is said to be Poisson distributed, if

$$P[X=\nu] = p(\nu) = \frac{m^\nu}{\nu!}\exp(-m), \quad \nu = 0, 1, 2, \ldots \, . \tag{2.42}$$

As is implied by the notation,

$$E\{X\} = \sum_{\nu=0}^{\infty} \nu \frac{m^\nu}{\nu!}\exp(-m) = m \, . \tag{2.43}$$

Whereas the Gaussian distribution is characterized by two parameters, mean m and variance σ^2, the Poisson distribution is already completely determined by its mean value as is clear from (2.42). The variance of a Poisson variable is equal to its mean value:

$$\sigma^2 = E\{(\delta X)^2\} = \sum_{\nu=0}^{\infty} (\nu - m)^2 \frac{m^\nu}{\nu!}\exp(-m) = m \, . \tag{2.44}$$

For the higher moments we have

$$E\{X^r\} = \sum_{k=0}^{r} S(r,k)\, m^k \, , \tag{2.45}$$

where the $S(r,k)$ are the Stirling numbers of the second kind. The Poisson distribution plays the same central role for discrete variables that the Gaussian distribution plays for continuous variables.

In the examples presented above, the probability (density) has only one extremum, a maximum, and it coincides with the mean value. This is a rather special situation. If the probability is asymmetric, then certainly mean value and most-probable value do not coincide. Furthermore, the random variable can, of course, possess more than one most probable state. This is a very interesting situation from an applied point of view since it reflects a certain structuring in the fluctuations of the quantity X. The importance of the extrema is due to the fact that they are the appropriate order parameter for equilibrium and nonequilibrium phase transitions in the presence of fluctuations. We shall come back to this topic in more detail in Chap. 6.

---

[4] $(r-1)!! = 1 \cdot 3 \cdot 5 \cdot \cdots \cdot (r-1)$.

2.1.6 Joint Random Variables

Let us now consider a pair of random variables X, Y,

$$\begin{aligned} X &: (\Omega, \mathscr{A}, P) \rightarrow (\mathbb{R}, \mathscr{B}) \\ Y &: (\Omega, \mathscr{A}, P) \rightarrow (\mathbb{R}, \mathscr{B}) . \end{aligned} \tag{2.46}$$

Their joint distribution function $F_{XY}(x, y)$ is given by

$$F_{XY}(x, y) = P[X \leqslant x, \ Y \leqslant y] . \tag{2.47}$$

This definition makes sense since obviously

$$\begin{aligned} A &= \{\omega \,|\, X(\omega) \leqslant x, \ Y(\omega) \leqslant y\} \\ &= \{\omega \,|\, X(\omega) \leqslant x\} \cap \{\omega \,|\, Y(\omega) \leqslant y\} \\ &= A_x \cap A_y \end{aligned}$$

is an observable event, due to the fact that by definition A_x and A_y are elements of the σ field $\mathscr{A}$ and thus, according to (2.4), $A \in \mathscr{A}$. The function $F_{XY}(x, \infty)$ is called the marginal distribution of X, and similarly $F_{XY}(\infty, y)$, the marginal distribution of Y. It is easily verified that $F_{XY}(x, \infty) = F_X(x)$ and $F_{XY}(\infty, y) = F_Y(y)$. In order not to burden the following discussion, we shall consider only random variables for which the joint probability density $p(x, y)$ exists:

$$F_{XY}(x, y) = \int_{-\infty}^{x} \int_{-\infty}^{y} p_{XY}(x', y') \, dx' \, dy' , \tag{2.48}$$

or

$$p_{XY}(x, y) = \partial_x \partial_y F_{XY}(x, y) .$$

The random variables X and Y are said to be *independent* if their joint probability density factorizes

$$p_{XY}(x, y) = p_X(x) \, p_Y(y) . \tag{2.49}$$

The covariance of two random variables is the product moment

$$\begin{aligned} \sigma_{XY} &= E\{(X - m_X)(Y - m_Y)\} \\ &= \int_{\mathbb{R}} \int_{\mathbb{R}} (x - m_X)(y - m_Y) \, p_{XY}(x, y) \, dx \, dy . \end{aligned} \tag{2.50}$$

Obviously if X and Y are independent, then

$$\sigma_{XY} = 0 . \tag{2.51}$$

The inverse is, however, in general not true; $\sigma_{XY}=0$ does not imply that $p_{XY}(x,y)=p_X(x)p_Y(y)$. A pair of variables that fulfills (2.51), which as just noted is a weaker condition than (2.49), is called *uncorrelated*. All these notions can be straightforwardly extended to n random variables: $X_1, \ldots, X_n$.

2.1.7 Conditional Probabilities

An interesting question that is often encountered in applications is to determine the probability that an event A will take place, knowing with certainty that another event B occurred. This is the so-called conditional probability and is denoted by $P(A\,|\,B)$. If the two events are independent, which in analogy to (2.49) is defined as

$$P(A\cap B)=P(A)P(B), \tag{2.52}$$

then, of course, the equality, $P(A\,|\,B)=P(A)$ should hold.[5] In the elementary case, the conditional probability $P(A\,|\,B)$ is defined as

$$P(A|B)=\frac{P(A\cap B)}{P(B)}. \tag{2.54}$$

Note that this definition is meaningful only if $P(B)\neq 0$. It is obvious that for fixed B, $P(A\,|\,B)$ is a probability measure on the σ field $\mathscr{A}$. Thus the conditional expectation of the random variable X, denoted by $E\{X\,|\,B\}$, is defined as

$$E\{X|B\}=\int_\Omega X(\omega)P(d\omega|B). \tag{2.55}$$

Taking into account the above definition this can be written as

$$E\{X|B\}P(B)=\int_B X(\omega)P(d\omega). \tag{2.56}$$

This elementary definition will not always be sufficient for our needs. Indeed, the conditional probability and conditional expectation are only defined with respect to events that occur with nonzero probability. Often, however, conditions have to be considered which have probability zero, as for instance the event that a continuous random variable takes precisely a given value $\tilde{x}$: $P(X=\tilde{x})$. If a probability density exists, we have

$$P(X\in[\tilde{x}-\varepsilon,\tilde{x}+\varepsilon])=\int_{\tilde{x}-\varepsilon}^{\tilde{x}+\varepsilon}p(x)\,dx\simeq p(\tilde{x})2\varepsilon \tag{2.57}$$

implying $P(X=\tilde{x})=0$. There is, however, a second and perhaps more important reason to aim for a more general concept of conditional probabilities. In the

[5] Let us note here as a side remark that two σ fields, $\mathscr{A}_1$ and $\mathscr{A}_2$, are said to be independent, if

$$P(A_1\cap A_2)=P(A_1)P(A_2)\quad \forall A_1\in\mathscr{A}_1 \quad\text{and}\quad \forall A_2\in\mathscr{A}_2. \tag{2.53}$$

elementary case, the conditional probability is defined with respect to *one* given event. Often this to too restrictive and it is necessary to condition with respect to a collection of events, which reflect, for instance, our knowledge gained in past experiments. Anticipating a bit, to formalize the important concept of Markov processes in a satisfactory way necessitates the consideration of conditioning with respect to the history of the process. This history is given in a natural way by the set of events that occurred in the past. To illustrate the concept of conditioning with respect to a set of events on a simple situation, consider a certain experiment, characterized by a probability triple $(\Omega, \mathscr{A}, P)$ and a random variable X. For instance, let us take the experiment on bacterial growth and the size of the population or the gas vessel and the speed of the molecules. Suppose further that a certain number of measurements or observations have been performed, corresponding to the occurrence of a set of events $\{A_i\}$. The sub-σ-field $\mathscr{C}$, generated by $\{A_i\}$, $\mathscr{C} \subset \mathscr{A}$ is precisely the information that can be inferred from them and that the experimenter has to make estimates on the character of the random variables X. The experimenter will use this available information to achieve the best possible prediction of the actual random variable X. In other words, he will approximate the random variable X by a random variable, say Y, that takes fully into account the information so far obtained. What are the properties the random variable Y should have? First, it should be a random variable with respect to the known σ field $\mathscr{C}$, i.e., Y has to be $\mathscr{C}$ measurable:

$$Y^{-1}(B) \in \mathscr{C} \quad \text{for all} \quad B \in \mathscr{B} . \tag{2.58}$$

Note that in general X is of course not a random variable with respect to the sub-σ-field $\mathscr{C}$, but only with respect to the full σ field $\mathscr{A}$. In other words, there exist $B \in \mathscr{B}$, such that $X^{-1}(B) \in \mathscr{A}$ but $X^{-1}(B) \notin \mathscr{C}$. This only reflects the fact that the actual random variable X contains generally more information on the system than the random variable Y, inferred from the measurements so far. Second, the random variable Y, the estimate of X, certainly has to yield the same information as the actual variable restricted to the observed collection of events. In precise mathematical terms

$$\int_{A_i} Y(\omega) P(d\omega) = \int_{A_i} X(\omega) P(d\omega) \quad \forall A_i \subset \mathscr{C} . \tag{2.59}$$

An important theorem in measure theory, the so-called Radon-Nikodym theorem (whose exact formulation is however not needed in the following and will therefore be skipped here) assures the existence of a random variable Y with the above two properties (2.58 and 59). Furthermore, the random variable Y is almost surely unique. This means that for any other random variable $\tilde{Y}$ with the above two properties, we have

$$P(\{\omega | Y(\omega) = \tilde{Y}(\omega)\}) = 1 . \tag{2.60}$$

The terms "almost surely" (a.s.) and "with probability one" will be used synonymously throughout this book. This random variable Y is called the conditional expectation of X under the condition $\mathscr{C}$ and denoted by

$$Y = E\{X|\mathscr{C}\}. \tag{2.61}$$

Similarly to (2.55), one expects to be able to write this as

$$Y = E\{X|\mathscr{C}\} = \int_{\Omega} X(\omega) P(d\omega|\mathscr{C}). \tag{2.62}$$

While in the elementary case $P(A|B)$ is a probability on $\mathscr{A}$ for fixed B, this point is more delicate in the general case and the validity of (2.62) needs careful discussion. This point will be taken up again after a brief summary of the essential features of conditional expectations.

Conditional probabilities are a special case of conditional expectations. Indeed, in the elementary case, we can obviously write

$$P(A|B) = E\{I_A|B\}. \tag{2.63}$$

Accordingly we have for the general situation

$$P(A|\mathscr{C}) = E\{I_A|\mathscr{C}\}. \tag{2.64}$$

Note that for fixed A the conditional probability is a *random* variable with respect to $\mathscr{C}$. It satisfies, as follows straightforwardly from (2.59),

$$\int_{A_i} P(A|\mathscr{C}) P(d\omega) = P(A \cap A_i) \quad \forall A_i \in \mathscr{C}. \tag{2.65}$$

Let us now summarize some important properties of conditional expectations. All the following statements hold almost surely. They are

i) $E\{E\{X|\mathscr{C}\}\} = E\{X\}$, (2.66)

ii) $X \geqslant 0 \Rightarrow E\{X|\mathscr{C}\} \geqslant 0$, (2.67)

iii) X measurable w.r.t. $\mathscr{C} \Rightarrow E\{X|\mathscr{C}\} = X$. (2.68)

In words this says that if the random variable contains no more information than that inferred from the experiments, then obviously the best estimate of X is the random variable X itself.

iv) Linearity: $E\{X\} < \infty, E\{Y\} < \infty$

$$\Rightarrow E\{aX + bY|\mathscr{C}\} = aE\{X|\mathscr{C}\} + bE\{Y|\mathscr{C}\} \quad (a, b \text{ constants}). \tag{2.69}$$

v) If X and $\mathscr{C}$ are independent, i.e., $\sigma(X)$ and $\mathscr{C}$ are independent, then

$$E\{X|\mathscr{C}\} = E\{X\}. \tag{2.70}$$

vi) If $\mathscr{C}_1$ and $\mathscr{C}_2$ are sub-σ-fields of $\mathscr{A}$, such that $\mathscr{C}_1 \subset \mathscr{C}_2 \subset \mathscr{A}$, then

$$E\{E\{X|\mathscr{C}_2\}|\mathscr{C}_1\} = E\{E\{X|\mathscr{C}_1\}|\mathscr{C}_2\} = E\{X|\mathscr{C}_1\}. \tag{2.71}$$

The conditioning with respect to $\mathscr{C}_1$ prevails since the σ field $\mathscr{C}_1$ is the smaller and imposes the more stringent conditions.

Let us now turn our attention again to conditional probabilities. Note that $P(A \mid \mathscr{C})$ is a random variable $P(A \mid \mathscr{C})(\omega)$ and is defined for each $A \subset \mathscr{A}$ only almost surely, i.e., up to an event of probability zero. Since this exceptional event can depend on A, one can in general *not* consider $P(\cdot \mid \mathscr{C})(\omega)$[6] as a probability on the σ field $\mathscr{A}$ for fixed $\omega \subset \Omega$. However, in a number of cases this interpretation is valid. Fortunately these are the only ones we have to deal with. The following theorem holds.

Theorem [2.2, 4]. Let X be a random variable and consider the conditional probability:

$$P(X \in B \mid \mathscr{C}) = P(\{\omega \mid X(\omega) \in B\} \mid \mathscr{C}), \quad B \in \mathscr{B}.$$

Then there exists a function, which we write suggestively as $P_X(B \mid \mathscr{C})(\omega)$, with the following properties:

i) for a fixed $\omega \in \Omega$, $P_X(\cdot \mid \mathscr{C})(\omega)$ is a probability on $\mathscr{B}$
ii) for a fixed event B, $P_X(B \mid \mathscr{C})(\cdot)$ is a version of $P(X \in B \mid \mathscr{C})$.

This means $P_X(B \mid \mathscr{C})(\cdot)$ is a random variable with respect to $\mathscr{C}$ and, compare with (2.65),

$$\int_{A_i} P_X(B \mid \mathscr{C})(\omega) P(d\omega) = P(A_i \cap [X \in B]), \quad \forall A_i \in \mathscr{C}. \tag{2.72}$$

This function $P_X(B \mid \mathscr{C})(\omega)$ is uniquely defined, up to an event with probability zero in $\mathscr{C}$ and this event is *independent* of B. Here $P_X(B \mid \mathscr{C})$ is known as the conditional probability distribution of X, given $\mathscr{C}$. If

$$E\{g(X)\} < \infty, \quad \text{then} \quad E\{g(X) \mid \mathscr{C}\} = \int_{\mathbb{R}} g(x) P_X(dx \mid \mathscr{C})(\omega). \tag{2.73}$$

Remember, as again indicated by writing the explicit dependence on ω, that the conditional expectation $E\{g(X) \mid \mathscr{C}\}$ is a random variable with respect to $\mathscr{C}$.

Let us now consider in more detail a particular class of conditions that will be of special importance in the following, namely conditioning with respect to a random variable. In fact, as expounded in Chap. 1, it is the rule in applications that the state of a system or environment is described by a random variable and not by a precise value for the macroscopic variables. We shall often encounter the problem of making estimates and inferring something about a system knowing that certain quantities depending on the environment are described by a given random variable. That is why we shall now consider the special case where the σ field $\mathscr{C}$ is generated by a random variable, say Z, i.e.,

[6] The dot · in place of the respective argument indicates that the argument varies freely over the appropriate space.

$$\mathscr{C} = \sigma\{A \mid A = Z^{-1}(B), \quad B \in \mathscr{B}\} = \sigma(Z)$$

and we write

$$E\{X \mid \mathscr{C}\} = E\{X \mid Z\}. \tag{2.74}$$

This random variable, i.e., the conditional expectation with respect to a given random variable Z takes a particularly simple form. Recall that by definition $E\{X|Z\}$ is a random variable with respect to $\sigma(Z)$, meaning that $E\{X|Z\}$ is $\sigma(Z)$ measurable, i.e., $E\{X|Z\}^{-1}(B) \in \sigma(Z)$ for all $B \in \mathscr{B}$. This property allows us to exploit a nice theorem [2.4] which facilitates the evaluation of $E\{X|Z\}$ and reads as follows: If a random variable is measurable with respect to $\sigma(Z)$, then it can be written as a measurable function of Z. Applied to the conditional expectation $E\{X \mid Z\}$, this theorem says that $E\{X \mid Z\} = h(Z)$. In other words, the value of $E\{X \mid Z\}$ is fixed by the value of Z at ω. This function h is a.s. uniquely defined. It is unique except for a set N of images of Z which occurs with zero probability $P[Z \in N] = 0$. Here h is a real-valued function, defined on the real line $h\colon \mathbb{R} \to \mathbb{R}$, and we write

$$E\{X|Z = z\} = h(z). \tag{2.75}$$

This gives a precise meaning to the expectation of X under the condition that the random variable Z takes on the value z. It is a typical example of a condition that occurs for continuous random variables with probability zero in general. Applying these results to the conditional probability $P(X \in B|Z)$, the conditional distribution can be expressed via an almost surely unique function Q

$$P_X(B \mid \mathscr{C})(\omega) = Q_X(B|Z(\omega)). \tag{2.76}$$

This allows us to write

$$P(X \in B|Z = z) = Q_X(B|z), \tag{2.77}$$

which is for fixed z a probability on $\mathscr{B}$. If $Q_X(\cdot\,|z)$ has a density, i.e.,

$$Q_X(B|z) = \int_B p(x|z)\,dx, \tag{2.78}$$

this density is called the conditional probability density of X, given that $Z = z$. If the marginal probability density

$$p_Z(z) = \int_{\mathbb{R}} p_{XZ}(x, z)\,dx$$

is positive, then $p(x \mid z)$ is simply given by

$$p(x|z) = \frac{p_{XZ}(x, z)}{p_Z(z)}. \tag{2.79}$$

2.2 Stochastic Processes

As discussed in Chap. 1, real environments vary randomly in the course of time. This can be modeled by using a random variable to describe the state of the environment at each instant of time. We thus obtain a family of random variables indexed by the parameter time t. The fluctuations of the environment induce in their turn random variations in the state of the system. Here again we can describe the temporal evolution of the system by a family of random variables that at each instant of time represents the state of the system.

2.2.1 Definitions

A family of random variables indexed by the parameter time t is known as a random (or stochastic) process. More precisely, a set $\{X_t;\ t\in\theta\}$ of real-valued random variables, i.e., X_t: $(\Omega, \mathscr{A}, P)\rightarrow(\mathbb{R}, \mathscr{B})$, is called a random process (or random function) with the index set θ and state space $\mathbb{R}$. In the following, the index parameter will be the time and the index set θ is either the real line $\mathbb{R}$ or the non-negative half line, if the process is started at $t=0$. As far as notation is concerned, stochastic processes will be denoted by X_t, whereas deterministic time-dependent functions will be written as $X(t)$. Note, as recalled above, that a random variable is a function from the sample space into the real numbers. So a random process can be considered to depend on two arguments: the index t and the elementary outcome ω: $X_t(\omega)$. If we keep the first argument, i.e., time fixed, and let ω vary over the sample space – this is denoted by a point in place of the second argument – then $X_t(\cdot)$ is by definition a random variable. If we keep ω fixed, i.e., pick a particular sample point which corresponds to considering a single observation of the stochastic process, and let t vary over the set θ, then $X.(\omega)$ is a real-valued function on the time axis (or some portion thereof). $X.(\omega)$ is called a realization or sample path of the process X_t. To illustrate these notions let us consider again the experiment on bacterial growth. The size X_t of the bacterial colonies is a random process. If the experimenter fixes one particular instant t' and determines the number of bacteria in the different petri dishes, then we clearly have the first situation; $X_{t'}(\cdot)$ is a random variable. If on the other hand, the experimenter picks at random one particular petri dish, i.e., a $\omega\in\Omega$, and monitors the evolution of the bacterial population, then we have the second case. In this way he records a realization or sample path of the random process X_t.

As we have seen in Sect. 2.1.4, a random variable is characterized by its distribution functions, a pair of random variables by their joint distribution function. A random process, being a family of random variables, is characterized by a hierarchy of distribution functions, namely:

$$F(x,t)=P[X_t\leqslant x]\,,$$
$$F(x_1,t_1;x_2,t_2)=P[X_{t_1}\leqslant x_1;X_{t_2}\leqslant x_2]\,,\ \ldots \tag{2.80}$$

with t, $t_i \in \theta$ and x, $x_i \in \mathbb{R}$. This system of distribution functions has naturally the following two properties:

i) Symmetry: If $i_1, \ldots, i_n$ is a permutation of $1, \ldots n$, then

$$F(x_{i_1}, t_{i_1}; \ldots; x_{i_n}, t_{i_n}) = F(x_1, t_1; \ldots; x_n, t_n); \tag{2.81}$$

ii) Compatibility: For $m < n$, and $t_1, \ldots t_m, t_{m+1} \ldots t_n \in \theta$ we have

$$F(x_1, t_1; \ldots; x_m, t_m; \infty, t_{m+1}; \ldots; \infty, t_n) = F(x_1, t_1; \ldots; x_m\, t_m), \tag{2.82}$$

i.e., the lower members of the hierarchy can be obtained from the higher ones. For the probability densities, which are defined by

$$F(x_1, t_1; \ldots; x_n, t_n) = \int_{-\infty}^{x_1} dx_1' \ldots \int_{-\infty}^{x_n} dx_n' p(x_1', t_1; \ldots; x_n', t_n), \tag{2.83}$$

this corresponds to integrating out the variables $\{x_{m+1}, \ldots, x_n\}$

$$p(x_1, t_1; \ldots; x_m, t_m) = \int_{\mathbb{R}} dx_{m+1}' \ldots \int_{\mathbb{R}} dx_n' p(x_1, t_1; \ldots; x_m, t_m; \ldots; x_n', t_n). \tag{2.84}$$

The fundamental theorem of Kolmogorov establishes that the inverse is also true. For every hierarchy of distribution functions, satisfying the symmetry condition (2.81) and the compatibility condition (2.82), there exists a probability triple $(\Omega, \mathscr{A}, P)$ and a stochastic process X_t defined on it that possesses the given distribution functions. This allows us to adopt the following convention: if no underlying probability triple is explicitly specified, which will be the common situation henceforth, but only a hierarchy of distribution functions is given, then it will always be understood that the so-called canonical choice has been made, i.e.,

$\Omega = \mathbb{R}^\theta$: set of all real-valued functions $\omega(\cdot)$ defined on the interval θ.

Since any real-valued function qualifies as an element of Ω, the elements of the sample space can be quite erratic. It is necessary to choose the sample space so large, in order to be able to accommodate all real-valued stochastic processes, which may have quite irregular realizations. The class of real-valued functions which does not occur as sample paths of the process under consideration will be excluded by assigning zero probability to them.

The Borel σ field corresponding to $\mathbb{R}^\theta$ is $\mathscr{A} = (\mathscr{B})^\theta$; it is generated by $B_1 \times B_2 \times \ldots \times B_n$, where $B_i \in \mathscr{B}$ and n arbitrary.

$X_t(\omega) = \omega(t) =$ value of the function ω at the point t.

The canonical choice is the extension of (2.15) to stochastic processes. It is characterized by the fact that the elementary outcomes coincide with the sample paths.

2.2.2 Separability

Two stochastic processes X_t and $\tilde{X}_t$ are said to be equivalent, if for every $t \in \theta$: $P[X_t = \tilde{X}_t] = 1$. In this case $\tilde{X}_t$ is called a version of X_t, and vice versa, and the two processes have an identical hierarchy of distribution functions. This does *not* however imply that the realizations of the two processes are identical. Especially their analytical properties, such as continuity and differentiability, can be quite different. This is due to the fact that the realizations of X_t can be arbitrarily modified on a set of time points, which has probability zero, without destroying the property $P[X_t = \tilde{X}_t] = 1$. Consider the following simple standard example [Ref. 1.88, 2.2], though it is of a rather artificial nature. Let τ be a random variable with state space equal to the interval [0, 1] and a uniform distribution on it, i.e.,

$$P[\tau(\omega) \in (a, b)] = b - a \quad \text{with} \quad 0 \leqslant a \leqslant b \leqslant 1 . \tag{2.85}$$

Define the random processes X_t and $\tilde{X}_t$ on the same probability triple by

$$X_t(\omega) \equiv 0 \quad t \geqslant 0 \tag{2.86}$$

and

$$\tilde{X}_t(\omega) = \begin{cases} 1 & t = \tau(\omega) \\ 0 & \text{otherwise} \end{cases} .$$

In words, while the process X_t is zero for all instants of time, the process $\tilde{X}_t$ is obtained from it by modifying it at one particular random point in time, namely by setting the sample path, labeled ω, equal to one for $t = \tau(\omega)$. Since the random time variable τ is uniformly distributed over the interval [0, 1], the probability that it takes the particular value $t \in [0, 1]$ is of course equal to zero, $P[\tau = t] = 0$. This obviously implies that $P[X_t = \tilde{X}_t] = 1$, i.e., the two processes are equivalent. However, whereas X_t has continuous sample paths, this is not true for $\tilde{X}_t$. In the following, we shall always choose a "nice" version of a stochastic process that avoids such undesirable features as $\tilde{X}_t$ in the above example. Mathematicians call this "to choose a separable version", which always exists. It is however not very useful for the following to present the abstract mathematical definition of separability. It suffices to remark that "nice" versions of stochastic processes always exist and that we expect, of course, those to be the ones describing real systems.

2.2.3 Continuity

A more interesting question is whether the knowledge of the hierarchy of distribution functions allows us to decide if a (separable) version with almost surely continuous sample paths exists. A stochastic process is said to have almost surely continuous sample paths, if

$$P(\{\omega | X_{.}(\omega) \text{ is a continuous function on } \theta\}) = 1 . \tag{2.87}$$

The answer is furnished by Kolmogorov's criterion which ensures the existence of such a version if for $s, t \in \theta$

$$E\{|X_t - X_s|^\alpha\} \leqslant \gamma |t-s|^{1+\beta} \tag{2.88}$$

for some positive number α, β and γ.

In this context let us briefly enumerate the various ways in which a stochastic process can be considered to be continuous:

i) X_t is called continuous in mean square if for every instant of time t

$$\lim_{s \to t} E\{|X_s - X_t|^2\} = 0\,; \tag{2.89}$$

ii) X_t is continuous in probability if for every t and positive ε

$$\lim_{s \to t} P[\,|X_s - X_t| > \varepsilon] = 0\,; \tag{2.90}$$

iii) X_t is continuous almost surely, if for every t

$$P(\{\omega \,|\, \lim_{s \to t} X_s(\omega) = X_t(\omega)\}) = 1\,. \tag{2.91}$$

Note that all three conditions, even the last one, in no way imply the continuity of the sample paths of the process. The difference between (2.87) and (2.91) is that in the latter condition, the subset A_0 of Ω defined by

$$A_0 = \{\omega \,|\, \lim_{s \to t} X_s(\omega) \neq X_t(\omega)\}\,,$$

which has probability zero, $P(A_0) = 0$, though it is not necessarily empty, can be different from one instant of time to the other. As an example, consider the process defined by (2.86). It fulfills all three conditions (2.89 – 91), but obviously

$$P\{\omega \,|\, \tilde{X}_{\cdot}(\omega) \text{ is discontinuous}\} = 1\,.$$

Indeed, the subset A_0 obviously depends on t, namely $A_0 = \{\omega \,|\, \tau(\omega) = t\}$. In general, a process will be continuous in mean square, in probability or continuous almost surely, if the probability that a discontinuity occurs in its sample paths at a precise instant of time is equal to zero, as in (2.86).

2.2.4 Stationarity

A stochastic process is called stationary (in the strict sense) if all its finite-dimensional probability densities are invariant against time shifts, i.e.,

$$p(x_1, t_1; \ldots; x_n, t_n) = p(x_1, t_1 + t; \ldots; x_n, t_n + t)\,. \tag{2.92}$$

In particular, this implies that the one-dimensional probability density does not at all depend on time:

$$p(x,t)=p_s(x)\,. \tag{2.93}$$

Therefore, the expectation value (if it exists) of a stationary stochastic process is constant:

$$\begin{aligned} E\{X_t\} &= \int_{\mathbb{R}} x\,p(x,t)\,dx \\ &= \int_{\mathbb{R}} x\,p_s(x)\,dx = m\,. \end{aligned} \tag{2.94}$$

Furthermore, the two-dimensional probability density $p(x_1,t_1;x_2,t_2)$ depends only on the time difference t_2-t_1:

$$p(x_1,t_1;x_2,t_2)=p(x_1,x_2;t_2-t_1)\,. \tag{2.95}$$

This has the consequence that the covariance, which in the case of stochastic processes is often called correlation function, $C_X(t,s)$, defined according to (2.50) by

$$E\{\delta X_{t_1}\delta X_{t_2}\}=\int_{\mathbb{R}}\int_{\mathbb{R}}(x_1-m)(x_2-m)\,p(x_1,t_1;x_2,t_2)\,dx_1dx_2\,, \tag{2.96}$$

also depends only on t_2-t_1:

$$C_X(t_1,t_2)=C_X(|t_2-t_1|)\quad \text{(if it exists)}\,. \tag{2.97}$$

A stochastic process that fulfills the properties (2.94) and (2.97) with $E\{X_t^2\}<\infty$ but not necessarily (2.92) is called stationary in the wide sense. (Note that the class of wide sense stationary processes cannot be said to be larger than the class of strictly stationary processes, since for the latter the mean value does not have to be finite).

The stage is now set to introduce three simple stochastic processes that play a fundamental role in probability theory and which will also occupy a central place in our modeling of fluctuating environments.

2.3 Brownian Motion: The Wiener Process

Brownian motion has played a central role in the theory of random phenomena in physics as well as mathematics. It is the rapid, perpetual, highly irregular motion of a small particle suspended in a fluid. The main features of Brownian motion, as established by experiments in the last century, are:

i) smaller particles move more rapidly;
ii) lowering the viscosity of the fluid also leads to more rapid motion;
iii) motion becomes more active when the fluid is heated;

iv) the motion is ceaseless and the trajectories are so irregular, their details are so fine, that they seem to have no tangent, i.e., the velocity of a Brownian particle is undefined.

Quite a few explanations were proposed for this strange phenomenon before the true cause of this perpetual motion was understood and the first theoretical treatment was given by Einstein. The chaotic motion of the suspended particle is maintained by the collisions with the molecules of the surrounding medium.

Due to the thermal motion of the surrounding fluid molecules, the Brownian particle suffers in a short time interval an enormous number of collisions, typically 10^{21} per second [2.5]. Since the particle is much heavier than the fluid molecules, the effect of each individual collision alone is negligible. However, due to the large number of continuously occurring collisions an effective motion results that can be observed under the microscope. Furthermore, it should be noted that each collision is independent of the others. Taking into account these facts, one arrives at a mathematical model of Brownian motion which is generally known as the Wiener process. This basic stochastic process will now be presented in detail. We shall consider here the motion of a Brownian particle in only one spatial dimension, i.e., on a line. Since the spatial components of the motion are independent, the extension to n-dimensional Brownian motion is straightforward.

Let W_t denote the displacement of the Brownian particle from some arbitrary starting point at time $t=0$. The standard Wiener process has thus the initial condition:

$$W_0 = 0\,. \tag{2.98}$$

The hierarchy of probability densities is given by the following formulae:

$$p(x, t) = (2\pi t)^{-1/2} \exp(-x^2/2t) \equiv n(x, t) \tag{2.99}$$

$$\vdots$$

$$p(x_1, t_1; \ldots; x_m, t_m) = n(x_1, t_1)\, n(x_2 - x_1, t_2 - t_1) \ldots n(x_m - x_{m-1}, t_m - t_{m-1})\,. \tag{2.100}$$

The Wiener process is a Gaussian process. It fulfills the defining condition that all finite dimensional probability densities are Gaussian, i.e., of the form

$$p_G(z) = [(2\pi)^n \det C]^{-1/2} \exp[-\tfrac{1}{2}(z-m)^{tr} C^{-1}(z-m)]\,, \tag{2.101}$$

where $z^{tr} = (z_1, \ldots, z_n)$, $m^{tr} = (m_1, \ldots, m_n)$ and C a positive-definite $n \times n$ matrix.[7]

The Wiener process has stationary *independent increments,* one of its most important properties. A stochastic process X_t is said to have independent increments if the random variables $(t_1 < \ldots < t_n)$ $X_{t_1}, X_{t_2} - X_{t_1}, \ldots, X_{t_n} - X_{t_{n-1}}$ are independent. Obviously, the probability density of the increment $W_t - W_s$ of the Wiener process is $n(\Delta x, t-s)$, i.e., $N(0, t-s)$. Since it depends only on $t-s$, it is

[7] $E\{z^{tr}\} = m^{tr}$, $E\{(z-m)(z-m)^{tr}\} = C$.

invariant with respect to time shifts, implying the increments are stationary. The joint probability density of $W_{t_1}, W_{t_2-t_1}, \ldots, W_{t_n-t_{n-1}}$ factorizes, due to the particular form of (2.100), and thus according to (2.49) the random variables are independent.

These properties of the Wiener process, namely to be Gaussian distributed and to have independent increments, reflect closely the characteristic features of Brownian motion. The stochastic process W_t is thus a satisfactory mathematical model of the latter. Indeed, as is clear from our above discussion concerning the source of the ceaseless erratic motion of a Brownian particle, the displacement of a Brownian particle is the sum of a very large number of independent infinitesimally small displacements due to the collisions. Invoking the central limit theorem we therefore expect the change in position of the Brownian particle to be Gaussian distributed. Furthermore, the displacements occurring over nonoverlapping time intervals should be stochastically independent, since they are due to collisions which are independent of each other. The stationarity of the displacements reflects the fact the the fluid is in equilibrium.

The Wiener process itself is obviously not a stationary process since $p(x, t+u) \neq p(x, t)$, as follows immediately from (2.99). The expectation value and the correlation function are easily calculated to be

$$E\{W_t\} = 0\,, \tag{2.102}$$

$$E\{W_t W_s\} = \min(t, s)\,. \tag{2.103}$$

The mean square displacement of a Brownian particle

$$E\{W_t^2\} = t \tag{2.104}$$

increases thus only linearly in time. This is characteristic of a spatial diffusion and was first theoretically derived by Einstein. Thus the Wiener process W_t is also not stationary in the wide sense. It has almost surely continuous sample functions, a feature that a model for the motion of a Brownian particle should possess. (To be precise, there exists a "nice", i.e., separable version that has a.s. continuous sample paths. It is obvious that only this version is an acceptable model for Brownian motion. This illustrates our remark of choosing a "nice", physically meaningful version). We have immediately, using (2.41), that

$$E\{(W_t - W_s)^4\} = 3(t-s)^2\,. \tag{2.105}$$

This means that the Kolomogorov criterion (2.88) is fulfilled with $\alpha = 4$, $\gamma = 3$, $\beta = 1$.

Though the sample paths of the Wiener process are with probability one continuous functions, the Wiener process is, as befits a model of Brownian motion, quite "irregular". With probability one, *the sample functions are nowhere differentiable,* i.e., the velocity of a Brownian particle is undefined, and *they have infinite length on any finite time interval* ([Ref. 2.1, p. 238] and [Ref. 2.6, p. 52]). (These features are of course idealizations, they reflect the fact that the

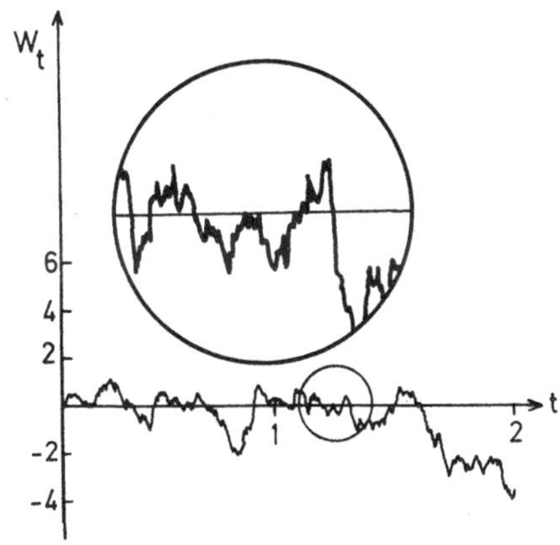

Fig. 2.1. A typical path of the Wiener process obtained for $\sigma^2 = 1$. The values on abscissa and ordinate have been divided respectively by 10^2 and 10^5. Though the sample paths are continuous, they are nowhere differentiable and have an infinite length on a finite time interval

details of the trajectories of a real Brownian particle are incredibly fine and cannot be followed in detail; the sample path represented in Fig. 2.1 gives an approximate idea of these properties.) We shall not give a rigorous proof of the first statement, but only observe that for a specified instant of time the probability density of the difference quotient $(W_{t+h} - W_t)/h$ is $N(0, 1/h)$ as follows from (2.100). Obviously, in the limit $h \to 0$ this Gaussian distribution diverges. In other words, the difference quotient does not converge with positive probability to a finite random variable at a specified instant of time.

Since the second property of the sample paths of W_t will play an important role later on, we shall prove explicitly that the sample paths of W_t have almost surely infinite length on any time interval $[s, t]$. To do this we first establish the following result, which is of interest in its own right, namely:

$$\operatorname*{qm-lim}_{n\to\infty} \sum_{k=1}^{n} (W_{t_k} - W_{t_{k-1}})^2 = t - s\,. \tag{2.106}$$

Here $s = t_0^{(n)} < t_1^{(n)} < \cdots < t_n^n = t$ is a sequence of partitions of the interval $[s, t]$, such that the mesh size $\delta_n = \max(t_k^{(n)} - t_{k-1}^{(n)})$ tends to zero for $n \to \infty$. Limit in the mean square is denoted by qm – lim and is defined by

$$X = \operatorname*{qm-lim}_{n\to\infty} X_n \quad \text{if} \quad \lim_{n\to\infty} E\{(X_n - X)^2\} \to 0\,. \tag{2.107}$$

To establish the result (2.106) we have to consider

$$\lim_{n\to\infty} E\left\{\left[\sum_{k=1}^{n} (W_{t_k} - W_{t_{k-1}})^2 - (t-s)\right]^2\right\} \equiv \lim_{n\to\infty} V_n$$

and to show that it converges towards zero.

$$\begin{aligned}
V_n &= E\left\{\sum_{k=1}^{n}\sum_{l=1}^{n}(W_{t_k}-W_{t_{k-1}})^2(W_{t_l}-W_{t_{l-1}})^2\right\} \\
&\quad -2(t-s)E\left\{\sum_{k=1}^{n}(W_{t_k}-W_{t_{k-1}})^2\right\}+(t-s)^2 \\
&= \sum_{k=1}^{n}E\{(W_{t_k}-W_{t_{k-1}})^4\}+\sum_{\substack{k=1\\k\neq l}}^{n}\sum_{l=1}^{n}E\{(W_{t_k}-W_{t_{k-1}})^2\}E\{(W_{t_l}-W_{t_{l-1}})^2\} \\
&\quad -2(t-s)\sum_{k=1}^{n}E\{(W_{t_k}-W_{t_{k-1}})^2\}+(t-s)^2,
\end{aligned}$$

where we have exploited the fact that W_t has independent increments. With

$$\begin{aligned}
&E\{(W_{t_k}-W_{t_{k-1}})^2\}=t_k-t_{k-1} \quad \text{and} \\
&E\{(W_{t_k}-W_{t_{k-1}})^4\}=3(t_k-t_{k-1})^2
\end{aligned}$$

according to (2.104 and 105) we obtain

$$\begin{aligned}
V_n &= \sum_{k=1}^{n}3(t_k-t_{k-1})^2+\sum_{\substack{k=1\\k\neq l}}^{n}\sum_{l=1}^{n}(t_k-t_{k-1})(t_l-t_{l-1}) \\
&\quad -2(t-s)\sum_{k=1}^{n}(t_k-t_{k-1})+(t-s)^2 \\
&= 2\sum_{k=1}^{n}(t_k-t_{k-1})^2+\sum_{k=1}^{n}\sum_{l=1}^{n}(t_k-t_{k-1})(t_l-t_{l-1})-2(t-s)^2+(t-s)^2 \\
&= 2\sum_{k=1}^{n}(t_k-t_{k-1})^2+\left(\sum_{k=1}^{n}(t_k-t_{k-1})\right)^2-(t-s)^2 \\
&= 2\sum_{k=1}^{n}(t_k-t_{k-1})^2\leqslant 2\sum_{k=1}^{n}(t_k-t_{k-1})\delta_n\leqslant 2(t-s)\delta_n.
\end{aligned}$$

This indeed converges to zero for n tending to infinity and thus proves (2.106). If the sequence of partitions is chosen in such a way that $\sum\delta_n<\infty$, then the convergence in (2.106) is almost sure or, what is the same, with probability 1. X is the almost sure limit of X_n if

$$P(\{\omega|\lim_{n\to\infty}X_n(\omega)\neq X(\omega)\})=0. \tag{2.108}$$

For $X_n\to X$ almost surely, it is sufficient that there exists some $q>0$ such that

$$\sum_{n=1}^{\infty}E\{|X_n-X|^q\}<\infty. \tag{2.109}$$

As is clear from the proof

$$E\left\{\left[\sum_{k=1}^{n}(W_{t_k}-W_{t_{k-1}})^2-(t-s)\right]^2\right\}\leqslant 2(t-s)\,\delta_n\,.$$

Hence, if $\sum\delta_n<\infty$, we have (2.109) with $q=2$, i.e.,

$$\lim_{n\to\infty}\sum_{k=1}^{n}[W_{t_k}(\omega)-W_{t_{k-1}}(\omega)]^2=t-s\quad\text{almost surely}\,. \tag{2.110}$$

Consider now the partition with points $t_k^{(n)}=s+(t-s)k/2^n$, $k=0,1,2,\ldots,2^n$. Since $\delta_n=(t-s)/2^n$ and therefore $\sum\delta_n<\infty$, (2.110) holds. Then the left-hand side of the inequality

$$\sum_{k=1}^{2^n}[W_{t_k}(\omega)-W_{t_{k-1}}(\omega)]^2\leqslant\max_{k=1,\ldots,2^n}|W_{t_k}(\omega)-W_{t_{k-1}}(\omega)|\cdot\sum_{k=1}^{2^n}|W_{t_k}(\omega)-W_{t_{k-1}}(\omega)| \tag{2.111}$$

converges for almost every ω, i.e., almost every sample path, to the finite value $(t-s)$, according to (2.110). Since the Wiener process has almost surely continuous sample paths, we have for almost every ω, i.e., with probability one,

$$\lim_{h\to 0}[W_{t+h}(\omega)-W_t(\omega)]=0\,.$$

It follows that the first factor of the rhs converges towards zero with probability one. Thus the second factor of the rhs necessarily diverges with probability one. Otherwise we would end up with $P[t-s\leqslant 0\cdot a=0]>0$ which is an obvious contradiction. In other words, almost every sample path of the Wiener process W_t has infinite length on an (arbitrary) time interval $[s,t]$. In this mathematical model of Brownian motion, the Brownian particle travels a path of infinite length on *any* finite time interval. As already mentioned, this feature is due to mathematical idealization, a price that has to be paid if the temporal evolution of the displacement of a Brownian particle is to be described by a mathematical model with simple appealing features. Nevertheless, this price is justified, since experiments show that the model approximates the reality very well.

2.4 Brownian Motion: The Ornstein-Uhlenbeck Process

If the Wiener process is used to model Brownian motion, i.e., is chosen as a stochastic process to represent the position of the Brownian particle, then the instantaneous velocity is not defined in this model. It is infinite, since the sample paths of W_t are nowhere differentiable. This can be avoided by considering the velocity of the Brownian particle instead as the main random quantity as done by *Uhlenbeck* and *Ornstein* [2.7]. This stochastic process is therefore known as the

Ornstein-Uhlenbeck process. The position is then obtained by integration and not given any more by the Wiener process. The starting point for this model of Brownian motion is the decomposition of the force acting on the suspended particle into a systematic part, the friction $-\alpha v$, and into a random part F_t, due to the perpetual kicks of the surrounding fluid molecules:

$$m\dot{v}_t = -\alpha v_t + F_t . \tag{2.112}$$

Since the average effect of the fluid molecules is already taken into account in the friction term, it is plausible to assume that F_t has mean value zero. As discussed above, F_t is the sum of a large number of independent infinitesimally small contributions. We assume it therefore to be Gaussian distributed in the light of the central limit theorem. Furthermore, since F_t is due to the numerous collisions of the light fluid molecules with the much heavier Brownian particles, it is plausible to suppose that F_t varies on a much faster time scale than v_t. Dividing (2.112) by m, we write

$$\dot{v}_t = -\gamma v_t + \zeta_t . \tag{2.113}$$

If we suppose that ζ_t has a short but nonvanishing memory, (2.113) can be interpreted as an ensemble of ordinary differential equations for the sample paths and thus be solved realizationwise. This point will be commented upon in much more detail in Chap. 8. In any case a naive approach to (2.113) is not likely to be dangerous since it is a linear problem. The solution of (2.113) is given by

$$v_t = v_0 \mathrm{e}^{-\gamma t} + \int_0^t \zeta_s \mathrm{e}^{\gamma(s-t)} ds . \tag{2.114}$$

A linear operator transforms a Gaussian process into a Gaussian process. Thus the integral over a Gaussian process is Gaussian [2.1, p. 161 f.]. This implies that v_t will be a Gaussian process if the initial condition v_0 is a Gaussian random variable independent of F_t (or a constant). Remember, Gaussian processes are completely characterized by the mean value function and the correlation function. We have:

$$E\{v_t\} = E\{v_0\}\mathrm{e}^{-\gamma t} + \int_0^t E\{\zeta_s\}\mathrm{e}^{\gamma(s-t)} ds = E\{v_0\}\mathrm{e}^{-\gamma t} \tag{2.115}$$

$$E\{v_t v_{t+\tau}\} = \mathrm{e}^{-\gamma(2t+\tau)}E\{v_0^2\} + \int_0^t \int_0^{t+\tau} \mathrm{e}^{\gamma(s-t)}\mathrm{e}^{\gamma(s'-t-\tau)}E\{\zeta_s\zeta_{s'}\} ds\, ds' . \tag{2.116}$$

Obviously, in agreement with the physics of the problem, ζ_t should be a stationary process, i.e.,

$$E\{\zeta_s\zeta_{s'}\} = \varphi(s-s') . \tag{2.117}$$

Since furthermore ζ_t varies on a much faster time scale than v_t, the function φ should be sharply peaked around zero and decrease rapidly to zero for $|s-s'| > 0$. With $\int_{\mathbb{R}} \varphi(\tau)\, d\tau \equiv \sigma^2$, we obtain for the covariance of the velocity

$$E\{(v_t - E\{v_t\})(v_{t+\tau} - E\{v_{t+\tau}\})\} = e^{-\gamma(2t+\tau)} E\{(\delta v_0)^2\} + (\sigma^2/2\gamma)(e^{2\gamma t} - 1) e^{-\gamma(2t+\tau)} . \tag{2.118}$$

If v_0 is $N(0, \sigma^2/2\gamma)$, then v_t is (at least) a wide sense stationary Gaussian process with $E\{v_t\} = 0$ and

$$E\{v_t v_s\} = (\sigma^2/2\gamma) \exp(-\gamma|t-s|) .$$

In fact we shall now define the Ornstein-Uhlenbeck process X_t as the process given by the following hierarchy of probability densities:

$$p(x, t) = \left(\frac{2\pi\sigma^2}{2\gamma}\right)^{-1/2} \exp\left(-\frac{1}{2}\frac{x^2}{\sigma^2/2\gamma}\right) \tag{2.119}$$

$$p(x_1, t_1; \dots; x_n, t_n) = p(x_1) p(x_2, x_1; t_2 - t_1) \dots p(x_n, x_{n-1}; t_n - t_{n-1}) \tag{2.120}$$

where

$$p(y, x; \Delta t) = [2\pi(\sigma^2/2\gamma)(1 - e^{-2\gamma\Delta t})]^{-1/2} \exp\left[-\frac{1}{2}\frac{(y - x e^{-\gamma\Delta t})^2}{(\sigma^2/2\gamma)(1 - e^{-2\gamma\Delta t})}\right] . \tag{2.121}$$

The so-defined Ornstein-Uhlenbeck process X_t solves (2.113) if ζ_t is a Gaussian white noise. At this point, we do not yet have available all the necessary mathematical ingredients to establish this fact. We shall come back to this problem in Chap. 5. By definition, the Ornstein-Uhlenbeck process (O-U-process) is like the Wiener process a Gaussian process. Also it shares with the Wiener process the property that it has almost surely continuous sample paths. Since we shall obtain this result in Chap. 4 as a byproduct in a more general context, we shall not go to any lengths now to prove it using Kolmogorov's criterion. The Ornstein-Uhlenbeck process differs from the Wiener process by two important features:

i) The Ornstein-Uhlenbeck process is, as a simple inspection of (2.119, 120) shows, a stationary process (in the strict sense). Its mathematical expectation vanishes:

$$E\{X_t\} = 0 , \tag{2.122}$$

and its correlation function decreases exponentially:

$$E\{X_t X_s\} = (\sigma^2/2\gamma) \exp(-\gamma|t-s|) . \tag{2.123}$$

ii) The Ornstein-Uhlenbeck process does *not* have independent increments; indeed they are not even uncorrelated. To see this, let $t > s > u > v$ and consider

$$\begin{aligned} E\{(X_t - X_s)(X_u - X_v)\} &= E\{X_t X_u\} + E\{X_s X_v\} - E\{X_t X_v\} - E\{X_s X_u\} \\ &= (\sigma^2/2\gamma)(e^{-\gamma|t-u|} + e^{-\gamma|s-v|} - e^{-\gamma|t-v|} - e^{-\gamma|s-u|}) . \end{aligned} \tag{2.124}$$

This expression is in general nonzero. Thus the changes in the velocity of the Brownian particle are correlated, and a fortiori stochastically dependent, over nonoverlapping time intervals.

Let us now consider the integrated Ornstein-Uhlenbeck process

$$Y_t = \int_0^t X_s ds , \tag{2.125}$$

which represents the position of the Brownian particle, started at time zero at the origin: $Y_0 = 0$. (The integration is understood realizationwise, i.e.,

$$Y_t(\omega) = \int_0^t X_s(\omega)\, ds . \qquad \text{a.s.}$$

It is well defined since $X_s(\omega)$ is, with probability one, a continuous function of s). Obviously we have

$$E\{Y_t\} = E\left\{\int_0^t X_s\, ds\right\} = \int_0^t E\{X_s\} ds = 0 . \tag{2.126}$$

For the correlation function we obtain

$$\begin{aligned}
E\{Y_t Y_s\} &= E \int_0^t \int_0^s X_u X_v du\, dv \\
&= \int_0^t \int_0^s E\{X_u X_v\} du\, dv \\
&= \int_0^t \int_0^s \frac{\sigma^2}{2\gamma} \exp(-\gamma |u-v|)\, du\, dv \qquad \text{[cf. (2.123)]} \\
&= \frac{\sigma^2}{2\gamma}\left[\int_0^t du \int_0^u dv\ \mathrm{e}^{-\gamma|u-v|} + \int_0^t du \int_u^s dv\ \mathrm{e}^{-\gamma|v-u|}\right], \quad t<s \\
&= \frac{\sigma^2}{2\gamma}\left[\int_0^t du \frac{1}{\gamma} \mathrm{e}^{-\gamma u}(\mathrm{e}^{\gamma u}-1) - \int_0^t du \frac{\mathrm{e}^{\gamma u}}{\gamma}(\mathrm{e}^{-\gamma s} - \mathrm{e}^{-\gamma u})\right] \\
&= \frac{\sigma^2}{2\gamma^2}\left[\int_0^t du (1-\mathrm{e}^{-\gamma u}) - \int_0^t du (\mathrm{e}^{-\gamma s} \cdot \mathrm{e}^{\gamma u} - 1)\right] \\
&= \frac{\sigma^2}{2\gamma^2}\left[t + \frac{1}{\gamma}(\mathrm{e}^{-\gamma t}-1) - \frac{\mathrm{e}^{-\gamma s}}{\gamma}(\mathrm{e}^{\gamma t}-1) + t\right] \\
&= \frac{\sigma^2}{\gamma^2} t + \frac{\sigma^2}{2\gamma^3}[\mathrm{e}^{-\gamma t} - 1 + \mathrm{e}^{-\gamma s} - \mathrm{e}^{-\gamma(s-t)}] .
\end{aligned} \tag{2.127}$$

As already mentioned above, the integral over a Gaussian process is again a Gaussian process. Thus Y_t is completely characterized by its mean (2.126) and its correlation function (2.127). We see that the Wiener process as a model for Brownian motion in position space is recovered in the limit $\sigma \to \infty$, $\gamma \to \infty$ such that $\sigma^2/\gamma^2 = \text{const}$ (=1 for the standard Wiener processes). The Wiener process is thus an adequate description of Brownian motion in the high friction and high noise intensity limit.

2.5 The Poisson Process

Wiener and Ornstein-Uhlenbeck processes are two examples of processes that model physical quantities, which vary continuously, and have thus fittingly almost surely continuous realizations. Let us now turn our attention to the other extreme, i.e., quantities that vary only by discrete steps. The basic example for this class of processes and in a certain sense the discrete counterpart to the Wiener process is the Poisson process. It measures the number of times a specified event occurs during the time period from 0 to t. Thus each sample function is a nondecreasing step function, i.e., the Poisson process does *not* have almost surely continuous sample paths. Obviously, by definition, all sample paths start at zero at time $t = 0$:

$$\nu_0 = 0 \,. \tag{2.128}$$

The Poisson process ν_t is used to model the disintegration of radioactive particles, the number of chromosome breakages under harmful irradiation, the number of incoming telephone calls, the arrival of customers for service, etc. In all these examples, the number of events occurring in an interval $[s, t]$ does not depend, to a good approximation, on how many events occurred in a preceding nonoverlapping interval $[u, v]$. The Poisson process has, as does its continuous counterpart the Wiener process W_t, independent increments. To be precise, its hierarchy of probabilities is given by

$$\begin{aligned} &P(i, t) = [(\lambda t)^i/i!]\, \mathrm{e}^{-\lambda t} \qquad i = 0, 1, \ldots \\ &P(j, t) = 0 \qquad\qquad\qquad\quad j = -1, -2, \ldots^{8} \end{aligned} \tag{2.129}$$

$$P(i_1\, t_1; \ldots; i_n, t_n) = p(i_1, t_1)\, p(i_2 - i_1, t_2 - t_1) \ldots p(i_n - i_{n-1}, t_n - t_{n-1}) \,. \tag{2.130}$$

The structure of this hierarchy coincides, as a simple glance at (2.99, 100) reveals, completely with that of the Wiener process, except that the Gauss distribution has been replaced by the Poisson distribution. The independent increments of the Poisson process are also stationary, whereas in complete analogy to the Wiener process, the Poisson process itself is not even stationary in the wide sense, since

[8] To define $P(j, t) = 0$ for negative integers allows us to write the n-fold probability in a compact way as done in (2.130). Otherwise we would have to use a string of inequalities like $i_2 > i_1, \ldots, i_n > i_{n-1}$.

$$E\{\nu_t\} = \lambda t \tag{2.131}$$

$$E\{\delta\nu_t \delta\nu_s\} = \lambda \min(t, s) . \tag{2.132}$$

As follows from (2.129) and the fact that the increments are independent, the probability that during the interval $(t, t+h)$ one or more jumps have taken place is

$$P_{\text{jump}}(h) = 1 - \exp(-\lambda h) . \tag{2.133}$$

This obviously tends to zero, if h tends to zero. In other words, the probability that at a specified instant of time t the Poisson process undergoes at least one jump is zero. In accordance with (2.91), the Poisson process is almost surely continuous. (As is easily seen (2.89, 90) are also fulfilled.) Let us, however, stress again that in contrast to the Wiener process the Poisson process does not have almost surely continuous sample paths. A fortiori, the sample paths of the Poisson process are not almost surely differentiable functions.

3. Stochastic Models of Environmental Fluctuations

Three main elements characterize stochastic processes: the nature of the state space, the index parameter set θ and the dependence relations among the random variables $\{X_t\}$. So far we have essentially discussed the first two elements for the stochastic processes which will serve as models for environmental fluctuations. The parameter set θ is trivially in all cases the time axis. As to the state space, we distinguished between continuously varying external parameters and discrete external parameters. Invoking the central limit theorem we concluded that the former can be modeled by a process with a Gaussian probability law. As basic examples for Gaussian stochastic processes we discussed the two different ways to describe the motion of a Brownian particle in the previous chapter, namely the Wiener and Ornstein-Uhlenbeck processes.

Let us now turn our attention to the third element, the dependence relation between the random variables making up the stochastic process used to model environmental fluctuations. This point was addressed in a brief and rather heuristic way already in Sect. 1.5. It was observed that in a broad class of applications a clear cut separation of time scales exists, namely that the environmental state varies much faster than the macroscopic state of the system. This led us to consider stochastic processes with extremely short memories and in a rather natural way the notion of white noise, a completely random process with independent values at every instant of time, arose. However, this discussion had to remain inevitably rather qualitative and handwaving, since it turned out that some rather subtle concepts were involved. It was stressed that a naive approach is beset with dangers and that white noise has to be handled with due respect for its unusual features. In this chapter, we shall start furnishing those mathematical tools necessary to avoid the pitfalls mentioned in Sect. 1.5. In particular, we shall have a close look at the passage from a real noise with a short memory to the idealization of white noise with zero memory. The properties of white noise will be presented in detail and its great value in the theory of external noise will be explained.

3.1 Correlation Function and Noise Spectrum

As a first step towards a clear and precise formulation of our way to model environmental fluctuations, we have to quantify the notion of rapid external noise. Let us therefore look for the characteristics defining the time scale of the system and that of its environment [3.1]. Recall that the systems we shall deal with here are governed by a phenomenological equation of the type

$$\dot{X}(t) = h(X(t)) + \lambda g(X(t)) = f_\lambda(X(t)) \,. \tag{3.1}$$

We denote by τ_{macro} the time that is typical for the macroscopic temporal evolution of the system. As in Sect. 1.2.1, we shall in general identify τ_{macro} with the relaxation time of the system towards a reference steady state $\bar{X}$ found under the average environmental conditions. To be precise $\bar{X}$ is defined by

$$h(\bar{X}) + E\{\lambda_t\} g(\bar{X}) = 0 \,, \tag{3.2}$$

and we determine τ_{macro} via the usual linear stability analysis (1.4 – 7). This yields here

$$\omega(\bar{X}) = \partial_X f_\lambda(X)\,|_{X=\bar{X}} \tag{3.3}$$

and hence the characteristic macroscopic time is the relaxation time of the system

$$\tau_{\text{macro}} = |1/\omega(\bar{X})| \,. \tag{3.4}$$

In some applications it might be appropriate to use a more refined definition of τ_{macro} including nonlinear effects.

A measure for the rapidity of the random environmental fluctuations is the correlation time τ_{cor}. It is, so to speak, the memory time of the stochastic process and it is defined for a stationary (or wide sense stationary) process as

$$\tau_{\text{cor}} = \frac{1}{C(0)} \int_0^\infty C(\tau)\, d\tau \,. \tag{3.5}$$

The rationale of this definition is easily understood. The right-hand side is the area beneath the normalized correlation function $\tilde{C}(\tau) = E\{\delta X_t \delta X_{t+\tau}\}/E\{\delta X_t^2\}$; $\tilde{C}(\tau) \leqslant 1$ and $\tilde{C}(0) = 1$. Intuitively one would say that the process has a long memory, if $C(\tau)$ or $\tilde{C}(\tau)$ decreases only slowly, implying a large area beneath $\tilde{C}(\tau)$. On the other hand, for a process with a short memory, $C(\tau)$ or $\tilde{C}(\tau)$ decreases rapidly, thus giving rise to a small area beneath $\tilde{C}(\tau)$. The normalized correlation function is used in order to be able to compare processes with different values for the variance.

For the frequently occurring case of an exponentially decreasing correlation function, as for the Ornstein-Uhlenbeck process (2.123),

$$C(\tau) = C(0) \exp(-\gamma\tau) \,, \tag{3.6}$$

it is easy to see that definition (3.5) yields the familiar result

$$\tau_{\text{cor}} = \gamma^{-1} \,. \tag{3.7}$$

A rapidly fluctuating environment can be characterized by the property that the correlation time τ_{cor} of the stochastic process λ_t is much smaller than the typical macroscopic time τ_{macro} of the system:

$$\tau_{\text{cor}} \ll \tau_{\text{macro}} . \tag{3.8}$$

We shall now consider an alternative way, different from the correlation function, to characterize the dependence relation between the random variables $\{X_t\}$. This is based on the fact, which we quote without proof, that any stationary process can be written as a superposition of oscillations with frequency ν, with random amplitude and phase. The so-called frequency spectrum $S(\nu)$ is then a measure for the mean square power with which an oscillation of frequency ν contributes to the process X_t. We shall see that $S(\nu)$ is just the Fourier transform of the correlation function and hence contains the same information on the process. Depending on the concrete application, one or the other is more readily obtainable or more convenient to characterize the temporal dependence relation of $\{X_t\}$.

Let us now repeat these statements in precise mathematical form [2.3, 3.2a, 3.3]. Let X_t be a (wide sense) stationary stochastic process with mean value zero: $E\{X_t\}=0$. (If $E\{X_t\}=m \neq 0$, consider $\delta X_t = X_t - m$.) Then X_t is the sum of harmonic functions $\exp(\mathrm{i}\nu t)$ with random amplitude and phase, given by the differential dZ_ν of a complex-valued stochastic process Z_ν:

$$X_t = \int_{\mathbb{R}} e^{\mathrm{i}\nu t} dZ_\nu . \tag{3.9}$$

In other words, X_t can be written as a Fourier integral with random coefficients. This theorem is known as the spectral decomposition. The process Z_ν is a stochastic process with the index parameter set $\theta = \mathbb{R}$ and its state space is the space of complex numbers $\mathbb{C}$. Furthermore, it has uncorrelated increments of zero mean value, i.e.:

$$\text{i)} \quad E\{Z_\nu - Z_{\nu'}\} = 0 , \tag{3.10}$$

$$\text{ii)} \quad E\{(Z_\nu - Z_{\nu'})\overline{(Z_\mu - Z_{\mu'})}\} = 0 \tag{3.11}$$

for nonoverlapping $[\nu, \nu']$ and $[\mu, \mu']$,

$$\text{iii)} \quad E\{(Z_\nu - Z_{\nu'})\overline{(Z_\nu - Z_{\nu'})}\} = \int_{\nu'}^{\nu} S(\bar{\nu})\, d\bar{\nu} \tag{3.12}$$

(the bar denotes complex conjugation).

The integral appearing in (3.9) is a special case of the more general situation $\int_{\mu'}^{\mu} \varphi(\nu)\, dZ_\nu$, where $\varphi(\nu)$ is a *nonrandom* function on $\mathbb{R}$. This integral is defined as

$$\int_{\mu'}^{\mu} \varphi(\nu)\, dZ_\nu = \text{qm-}\lim_{n\to\infty} \sum_{i=1}^{n} \varphi(\nu_{i-1}^{(n)})(Z_{\nu_i^{(n)}} - Z_{\nu_{i-1}^{(n)}}) , \tag{3.13}$$

where $\{\nu_i^{(n)}\}$ is a partition of $[\mu', \mu]$ with mesh size tending to zero. It is easily verified that it possesses the following properties:

$$\text{i)}\quad E\left\{\int_{\mu'}^{\mu} \varphi(\nu)\,dZ_\nu\right\} = 0 \tag{3.14}$$

$$\text{ii)}\quad E\left\{\int_{\mu'}^{\mu} \varphi(\nu)\,dZ_\nu \cdot \overline{\int_{\mu'}^{\mu} \psi(\nu)\,dZ_\nu}\right\} = \int_{\mu'}^{\mu} \varphi(\nu)\,\overline{\psi(\nu)}\,S(\nu)\,d\nu\,. \tag{3.15}$$

Using these properties it is now easy to show that $C(\tau)$ and $S(\nu)$ are Fourier transforms of each other. Indeed:

$$\begin{aligned} C(\tau) &= E\{X_\tau X_0\} \\ &= E\{X_\tau \bar{X}_0\} \qquad \text{(since } X_t \text{ is a real-valued stochastic process)} \\ &= E\left\{\int_{\mathbb{R}} \mathrm{e}^{\mathrm{i}\nu\tau} dZ_\nu \int_{\mathbb{R}} \overline{dZ_\nu}\right\} \\ &= \int_{\mathbb{R}} \mathrm{e}^{\mathrm{i}\nu\tau} S(\nu)\,d\nu \end{aligned} \tag{3.16}$$

according to (3.15), which proves the assertion. Due to a well-known property of the Fourier transform, a narrow frequency spectrum $S(\nu)$ corresponds to a slowly decreasing broad correlation function $C(\tau)$. And vice versa, a broad frequency spectrum is associated with a rapidly decreasing correlation function. This implies that rapid external fluctuations $\tau_{\text{cor}} \ll \tau_{\text{macro}}$ having a narrow correlation function possess a broad frequency spectrum with an effective band width ν_{b}, defined in analogy to (3.5) as

$$\nu_{\text{b}} = \frac{1}{S(0)} \int_0^\infty S(\nu)\,d\nu\,, \tag{3.17}$$

which is very large compared to the typical frequency of the system:

$$\nu_{\text{b}} \gg \omega(\bar{X})\,. \tag{3.18}$$

Typically, the environments of natural systems fulfill condition (3.18). This feature is easily understood: as discussed above, external noise can be an expression of a turbulent or chaotic state, a defining property of which is a broad-band spectrum, or the external parameter depends on a multitude of interfering environmental factors, implying that a large number of harmonic modes are "excited" and intervene in its temporal behavior. Therefore in a large class of applications, the environmental fluctuations are very rapid in the sense of (3.8) or (3.18). Furthermore, it will turn out that this case of broad-band external noise is also the most tractable from a mathematical viewpoint. It is thus appropriate that we start our analysis of the effects of external noise on nonlinear systems with the limiting case of extremely rapid environmental fluctuations.

3.2 The White-Noise Process

If $\tau_{\mathrm{cor}} \ll \tau_{\mathrm{macro}}$, one is tempted to pass to the limit $\tau_{\mathrm{cor}}=0$. The rationale to adopt this idealization is the following. The memory of the environment is extremely short compared to that of the system. It is therefore reasonable to expect that any effects related to it are barely perceptible in the macroscopic system. Hence, no qualitative change in the macroscopic behavior should occur if we set the non-vanishing but extremely short correlations equal to zero. This means that the environment can be adequately described by a process with *independent values* at each instant of time, i.e., a so-called completely random process. Some circumspection, however, has to be exerted in passing to the limit $\tau_{\mathrm{cor}}=0$ as already emphasized in Sect. 1.5. If we approach the limiting process, having independent values, simply by letting the correlation time go to zero, we shall not only neglect memory effects but at the same time get rid of *any* effect of the environmental fluctuations. To see this, let us consider a Gaussian process with an exponentially decreasing correlation function, namely the Ornstein-Uhlenbeck (O-U) process.[1] According to (3.16), the frequency spectrum of the O-U process is given by

$$\begin{aligned} S(\nu) &= \frac{1}{2\pi}\int_{\mathbb{R}} \mathrm{e}^{-\mathrm{i}\nu\tau} C(\tau)\,d\tau \\ &= \frac{1}{2\pi}\int_{\mathbb{R}} \mathrm{e}^{-\mathrm{i}\nu\tau}(\sigma^2/2\gamma)\,\mathrm{e}^{-\gamma|\tau|}\,d\tau \\ &= (\sigma^2/2\pi)(\nu^2+\gamma^2)^{-1}\,. \end{aligned} \tag{3.19}$$

The correlation time of the O-U process is

$$\tau_{\mathrm{cor}} = \gamma^{-1} \tag{3.20}$$

as follows from (3.7). Hence the limit $\tau_{\mathrm{cor}} \to 0$ corresponds to $\gamma \to \infty$. It is easily seen by inspecting (3.19) that in this limit the mean square power with which oscillations of frequency ν contribute to the O-U process X_t vanishes, i.e.,

$$\lim_{\gamma\to\infty} S(\nu) = 0\,, \qquad \nu \in \mathbb{R}\,. \tag{3.21}$$

This implies that a decrease in the correlation time τ_{cor}, without changing the other characteristics of the process, e.g., the variance $C(0)$, leads eventually to a situation in which the random variations of the environment have no impact at

[1] For the sake of clarity and concreteness we shall consider in the following explicitly only the case of a normally distributed external parameter. For the reasons exposed above, this situation is most frequently encountered in applications and it will therefore occupy a central place in this work. Exponentially decreasing correlation functions are almost as ubiquitous, as far as correlations are concerned, as the normal distribution is among probabilities. Again there is a deep reason for this, namely a fundamental theorem in the theory of stochastic processes, the so-called Doob theorem, which we shall present in Chaps. 4, 8.

all on the system, simply because in this limit the total input power $S = 2\int_0^\infty S(\nu)d\nu = \sigma^2/2\gamma = C(0)$ is spread uniformly over infinitely many frequencies. The limit $\tau_{cor} \to 0$ is obviously too simplistic. It implies more than just neglecting the memory of the noise, in fact it is a *noiseless* limit. This has sometimes been overlooked and has led some authors to the conclusion that for very rapid fluctuations ($\tau_{cor} \to 0$) the system does not feel environmental fluctuations and follows the deterministic evolution. This is rather trivial: there is almost no randomness left in the surroundings for the system to respond to.

If external noise with a short memory is to be replaced by an equivalent idealized noise with zero memory, then in the light of the above discussion the appropriate limiting procedure is to couple the decrease in τ_{cor} with an adequate increase in the strength of fluctuations. From (3.19) it follows that a finite limit is obtained, if concomitant with $\tau_{cor} \to 0$, σ^2 goes to infinity, such that σ^2/γ^2 is a constant and not the variance $\sigma^2/2\gamma$. Note that this is just the limit in which the integrated O-U process converges to the Wiener process, cf. (2.127)! The particular importance of this observation will become clear later on. In the limit $\tau_{cor} \to 0$, $\sigma \to \infty$ such that $\sigma^2/\gamma^2 = \text{const} = \bar{\sigma}^2$ the frequency spectrum of the O-U process converges to

$$S(\nu) = \bar{\sigma}^2/2\pi\,, \tag{3.22}$$

i.e., a completely flat spectrum. For its correlation function we obtain in this limit

$$C(\tau) = \bar{\sigma}^2\delta(\tau)\,. \tag{3.23}$$

Here $\delta(\tau)$ denotes the Dirac delta function. It is not a function in the ordinary sense. It belongs to the class of generalized functions[2] and is defined only by its properties under the integral sign, namely

$$\int_{\mathbb{R}} \varphi(\tau)\delta(\tau)d\tau = \varphi(0) \tag{3.24}$$

for any continuous function $\varphi(\tau)$. Roughly speaking, the δ-function can be characterized by saying that it is zero everywhere except at $\tau = 0$ where it is infinitely high such that

$$\int_{\mathbb{R}} \delta(\tau)d\tau = 1\,. \tag{3.25}$$

The δ-function can be visualized as the limit of Gaussian densities whose variance tends to zero:

$$\delta(\tau - s) = \lim_{\sigma^2 \to 0} (\sqrt{2\pi}\,\sigma)^{-1} \exp\left(-\frac{1}{2}\,\frac{(\tau - s)^2}{\sigma^2}\right). \tag{3.26}$$

[2] These are also known as distributions in the literature. However, to avoid confusion with probability distributions we shall not use this name.

As is clear from (3.22, 23), a δ-correlated process has a flat spectrum. This property is at the origin of the name *white noise* for such processes; all frequencies are present with equal power as in white light. The O-U process is a Gaussian process, a property which is conserved in the limiting procedure. For this reason, the limiting process for $\tau_{\text{cor}} \to 0$ of the O-U process is known as Gaussian white noise and is in the following denoted $\bar{\sigma}\xi_t$ [where ξ_t is the standard Gaussian white noise with $E\{\xi_t\} = 0$ and $E\{\xi_t \xi_{t+\tau}\} = \delta(\tau)$]. Gaussian white noise is an extremely irregular process. It jumps widely around; its realizations are nowhere continuous.

Naturally the question arises, are there other kinds of white noise besides the Gaussian white noise? The answer is yes and the class of white noises is completely known [3.2b]. In fact it is not at all difficult to characterize a white-noise process. Its defining feature is that it is a completely random process, i.e., it has independent values at every instant of time, and has an infinite variance. In other words, any process whose correlation function is proportional to a Dirac delta function qualifies as white noise. Furthermore, it is easy to obtain all possible white noises. Consider a process V_t with stationary independent increments, as for instance, the Wiener process W_t or the Poisson process ν_t. Then the random variables $(V_{t+h} - V_t)/h$ and $(V_{s+h} - V_s)/h$ are independent random variables for h sufficiently small and $t > s$. This property would also be conserved in the limit $h \to 0$ if it could be properly defined. Thus it is tempting to think of white noise as the time derivative of a process with stationary independent increment; the Gaussian white noise would be the time derivative of the Wiener process and differentiating the Poisson process would yield in this spirit Poisson white noise. However, as discussed above, due to the fact that, see (2.104),

$$E\{\Delta W_t^2\} = E\{(W_{t+h} - W_t)^2\} \sim h \tag{3.27}$$

and

$$E\{\Delta \nu_t^2\} = E\{(\nu_{t+h} - \nu_t)^2\} \sim h \tag{3.28}$$

the variance of the difference quotients $\Delta W_t/h$ or $\Delta \nu_t/h$ diverges. This fact is often expressed by saying that the Wiener and Poisson processes are not differentiable in the mean square sense. We also know that both processes have a.s. sample paths which are not differentiable, i.e., both processes are also not differentiable realizationwise. In fact, it turns out that these processes are not differentiable in any ordinary sense. However, our considerations on the O-U process show a way out of this dilemma. Remember that in the same limit in which the O-U process X_t converges to Gaussian white noise $\bar{\sigma}\xi_t$,

$$X_t \to \bar{\sigma}\xi_t \quad \text{for} \quad \sigma \to \infty, \gamma \to \infty \quad \text{such that} \quad \sigma^2/\gamma^2 = \bar{\sigma}^2$$

the integrated O-U process $Y_t = \int_0^t X_s ds$ converges to the Wiener process: $Y_t \to \bar{\sigma} W_t$ (p. 53). This indicates that indeed the Gaussian white noise should be related to the Wiener process by differentiation. Since the correlation function of the Gaussian white noise is a generalized function, this suggests that though the processes with independent increments are not differentiable in any ordinary sense, they should be in some generalized sense. It can be shown that a rigorous definition of general white noise is possible, using the theory of generalized stochastic processes which are defined in a similar way to generalized functions [3.2b]. The important result is that there is a one to one relation between white-noise processes and ordinary processes with stationary independent increments, namely, "white noise = (d/dt) (process with stationary independent increments)". Since the latter class of processes is completely known [Ref. 3.3, p. 159], so is then the ensemble of possible white noises. The concept of generalized stochastic processes is explained in more detail in the Appendix A, where an explicit demonstration is also given that

$$\xi_t = \dot{W}_t \tag{3.29}$$

in the generalized functions sense.

Though Gaussian white noise is so very irregular, it is extremely useful to model rapidly fluctuating phenomena. Not surprisingly, in view of its properties – no continuous sample paths, infinite total power $S = 2\int_0^\infty (\bar{\sigma}^2/2\pi)\,d\nu = \infty$ – true white noise of course does not occur in nature. However, as can be seen by studying their spectra, thermal noise in electrical resistance, the force acting on a Brownian particle, or climate fluctuations, disregarding the periodicities of astronomical origin, are white to a very good approximation. These examples support the usefulness of the white-noise idealization in applications to natural systems. In any case, frequencies that are very much higher than the characteristic frequency of the system should be of no importance for the macroscopic properties of the system which, due to its inertia, acts so to speak as a low-pass filter. Noises with a very large effective bandwidth should be indistinguishable from white noise for practical purposes. (Experimental support for this will be presented in Chap. 7; a rigorous theoretical justification is furnished in Chap. 8.) Since $\nu_b = \pi/2\tau_{cor}$, this is of course only a rephrasing of the fact that if $\tau_{cor} \ll \tau_{macro}$, then it will be permissible to pass to the white-noise limit.

These considerations put the final touch to our phenomenological modeling of macroscopic systems, subjected to a rapidly fluctuating environment. In the idealization of δ-correlated external noise, the system is described by a SDE of the form:

$$\dot{X}_t = h(X_t) + \lambda g(X_t) + \sigma g(X_t)\xi_t \tag{3.30}$$

$$= f_\lambda(X_t) + \sigma g(X_t)\xi_t\,, \tag{3.31}$$

where we have suppressed the bar over σ to denote the intensity of the Gaussian white noise.

The next two chapters will deal mainly with the basic properties of (3.31). Let us first discuss the consequences of the white-noise idealization. What are the advantages of neglecting the small memory effects of the environment? After all, this leads to an object with some rather mind-boggling features and which does not belong to the realm of ordinary stochastic processes. For the sake of argument, suppose the following situation, not unusual in applications: the state of the system at time t has been determined accurately to be x. For the external parameter λ_t only its probability law is known, for instance that it is Gaussian distributed. Let us backtrack for a moment and consider the situation under a more realistic noise than white noise, e.g., let ξ_t be an O-U process. Then for a short time h into the future, the state of the system will be given by

$$X_{t+h} = x + f_\lambda(x)\, h + \sigma g(x)\, \xi_t h\,. \tag{3.32}$$

Of course, it would be convenient if the future stochastic evolution of the system could be predicted solely on the basis of the information we possess at the present time t on the state x of the system and on the environmental conditions as represented by the probability of ξ_t. In more mathematical terms, the probability that the system is in state y at some future time $t+h$ should depend only on the present state x and the stationary probability (density) $P_s(z)$, describing the environment, but not on the past history. Such a situation is the closest stochastic analog to the deterministic situation, where $X(t)$ of (3.1) is completely determined, once the initial condition $X(0)$ is given. This property is a verbal description of the defining feature of a Markov process. Before presenting a formal definition of this important class of stochastic processes, let us remark that the system can have the above property only if the environment is indeed already completely characterized by its one-dimensional probability density $p_s(z)$, and not, as is generally the case, by the infinite hierarchy of its n-dimensional probability densities. The only class of processes for which this is true are the processes with independent values at every instant of time, since for these completely random processes

$$p(z_1, t_1; \ldots; z_n, t_n) = \prod_{i=1}^{n} p_s(z_i)\,. \tag{3.33}$$

This is of course not an unexpected result: if the environment had a finite memory, then information on the past would indeed improve our prediction capabilities of the future stochastic evolution of the system. These heuristic considerations suggest that the system is Markovian if and only if the external fluctuations are white. At this stage we cannot yet formulate this result in the form of a mathematical theorem. Remember that white noise is an extremely irregular process, not possessing continuous sample paths. The Taylor expansion used in (3.32) to determine the state at time $t+h$ might well be meaningless. In other words, since the environment forgets immediately in what state it was the instant before, one might doubt that ξ_t has any bearings on the state of the system at some future time $t+h$. Thus, this problem is more delicate for Gaussian white noise and has to be analyzed with great circumspection. Furthermore, as also already emphasized in Sect. 1.5, one has to be careful as to the meaning of the formal SDE (3.30) and what one understands by "X_t is a solution of (3.30)." However, let us already mention for the reader who does not want to await the outcome of the rigorous mathematical analysis that indeed the following theorem holds: the process X_t, being a solution of (3.31) is Markovian, if and only if the external noise ξ_t is white. This result explains the importance and appeal of the white-noise idealization. If the system, coupled to a fluctuating environment, can be described by a Markov process, then we have the full arsenal of tools developed to deal with such stochastic processes at our disposition. There is definite ground for optimism as to our ability to carry out an analysis of the effects of external noise on nonlinear systems. If the system would have to be described, however, by a non-Markov process, the outlook would be rather bleak. The techniques to deal with the kind of non-Markov process that occurs in applications are rather underdeveloped. We shall come back to this point in Chaps. 8 and 9.

Note that for principal theoretical reasons a product of two generalized functions, or any other nonlinear operation on them cannot be defined. Therefore external white noise must necessarily appear in a linear way in the SDE (3.31). Otherwise such an equation would be meaningless. As mentioned, for a broad class of applications the phenomenological equation is indeed linear in the external parameter λ. If the dependence of $f(x)$ on λ is nonlinear, one can in most cases still pass to the white-noise limit. Again an SDE will be obtained with a random force term which is *linear* in λ_t. This will be discussed in detail in Chap. 8.

4. Markovian Diffusion Processes

In Chap. 3 we have heuristically argued that the state X_t of the system can be represented by a Markovian process, if and only if, the external noise is white. In this chapter, we shall first define Markovian processes mathematically. Then we shall turn our attention to the subclass of Markov processes having almost surely continuous sample paths, and which therefore is particularly important for our purposes. In many applications indeed it is physically desirable that the sample paths of both the fluctuating external parameters and of the system's state variables should possess this property; e.g., even if the temperature of a chemical reactor strongly fluctuates, one nevertheless expects that the concentrations of the reactants remain smooth functions presenting no discontinuous jumps in time. Henceforth, if nothing is said to the contrary, we shall thus always choose external noises having continuous sample paths. The question which remains to be clarified then is whether this also guarantees that the sample path of the state variables whose time evolution depends on these noises will be continuous. Since these variables obey differential equations the answer to this question is obviously yes if the external noise is nonwhite. It is considerably less evident for white external noise. However, if the passage to the white-noise idealization is to be useful, it should have the feature to conserve at least the continuity of the sample paths of X_t. The differentiability is lost, a point on which we shall comment later. It will be shown in the next chapter that a system subjected to Gaussian white noise can indeed be described by a Markov process with almost surely continuous sample paths. This class will therefore be treated in considerable detail in the present chapter.

4.1 Markovian Processes: Definition

Let us restate the loose verbal definition of a Markov process, employed in Sect. 3.2. A random process X_t is said to be Markovian if when the present state of the process X_t is known, then any additional information on its past history is totally irrelevant for a prediction of its future evolution. In different words: If the present is known, then past and future are (conditionally) independent.

To arrive at a precise mathematical definition of the Markov property, we have to cast the notion "past behavior of the process" in operational mathematical terms. The history of the process X_t consists obviously of the events that occurred up to the present time. In other words, the behavior of the system in the past can be characterized by the collection of events connected with the process X_t, i.e., the ensemble

$$\{X_\tau^{-1}(B),\ \tau\in[t_0,s],\ B\in\mathscr{B}\}=\{A\,|A=\{\omega\,|X_\tau(\omega)\in B\},\ \tau\in[t_0,s],\ B\in\mathscr{B}\}. \quad (4.1)$$

Let us denote the smallest σ field that contains this collection of events by $\mathscr{A}[t_0,s]$. This σ field $\mathscr{A}[t_0,s]$ is unique and contains the complete information on the past behavior of the stochastic process X_t. This follows directly from the definition (4.1) of this field of events. $\mathscr{A}[t_0,s]\subset\mathscr{A}$ is the smallest sub-σ-field with respect to which *all* the random variables X_t, $t\in[t_0,s]$, are measurable. Restating the above verbal definition of a Markov process in precise mathematical terms, we say:

A stochastic process $\{X_t,\ t\in\theta\}$, with $\theta=[t_0,T]\subset\mathbb{R}$ is called a Markov process if for $t_0\leqslant s\leqslant t\leqslant T$ and all $B\in\mathscr{B}$ the Markov property

$$P(X_t\in B\,|\mathscr{A}[t_0,s])=P(X_t\in B\,|X_s) \quad (4.2)$$

holds with probability 1. Note that the conditioning in lhs of (4.2) is with respect to the process's own past, in rhs with respect to the random variable X_s, i.e. the present state of the process. During the time span from t_0 to t the process evolves freely; no outside conditioning is imposed. A brief reflection shows that $\mathscr{A}[t_0,s]$ is generated by events of the form $\{\omega\,|X_{s_1}(\omega)\in B_1,\ldots,X_{s_n}(\omega)\in B_n\}$ with $t_0\leqslant s_1<\ldots<s_n\leqslant s$ and $B_i\in\mathscr{B}$. An equivalent condition to (4.2) is thus given by the hierarchy of equations:

$$P(X_t\in B\,|X_{t_1},\ldots,X_{t_n})=P(X_t\in B\,|X_{t_n}) \qquad \text{a.s.} \quad (4.3)$$

for $n\geqslant 1$, $t_0\leqslant t_1<\ldots<t_n<t\leqslant T$ and $B\in\mathscr{B}$.

Recalling our discussion on conditional probabilities (2.76), we know that a conditional distribution $Q_{X_t}(B\,|X_s)$ exists for the conditional probability $P(X_t\in B\,|X_s)$. The conditional distribution can be written in the form

$$Q_{X_t}(B\,|X_s)=P(B,t\,|X_s,s)\,, \quad (4.4)$$

where $P(B,t\,|x,s)$ is a function of four arguments, namely s, $t\in\theta$ with $t\geqslant s$ and $x\in\mathbb{R}$, $B\in\mathscr{B}$. It has the following properties:

i) $P(B,t\,|X_s,s)=P(X_t\in B\,|X_s)$

holds with probability 1. (Remember that a conditional probability is a random variable.)

ii) According to our general discussion of conditional probabilities and conditional distributions, $P(\cdot,t\,|x,s)$ is a probability on $\mathscr{B}$ for fixed s, t and x.

iii) $P(B,t\,|\cdot,s)$ is a $\mathscr{B}$-measurable function.

The above three properties are essentially those of a conditional distribution as is clear from Sect. 2.1.7 and are not particular to a Markov process. The following fourth one is, however, a characteristic feature of a Markov process:

iv) $P(B,t|x,s) = \int_{\mathbb{R}} P(B,t|z,t')P(dz,t'|x,s)$ (4.5)[1]

for $t_0 \leqslant s \leqslant t' \leqslant t \leqslant T$ and $B \in \mathscr{B}$.

This relation is known as the Chapman-Kolmogorov equation. Its proof follows straightforwardly from the Markov property (4.2):

$$
\begin{aligned}
P(B,t|X_s,s) &= P(X_t \in B|X_s) \\
&= P(X_t \in B|\mathscr{A}[t_0,s]) && \text{(Markov)} \\
&= E\{I_B(X_t)|\mathscr{A}[t_0,s]\} \\
&= E\{E\{I_B(X_t)|\mathscr{A}[t_0,s]\}|\mathscr{A}[t_0,t']\} && \text{((2.71), since for } t' \geqslant s,\ \mathscr{A}[t_0,t'] \supset \mathscr{A}[t_0,s]) \\
&= E\{E\{I_B(X_t)|\mathscr{A}[t_0,t']\}|\mathscr{A}[t_0,s]\} && \text{(2.71)} \\
&= E\{P(X_t \in B|\mathscr{A}[t_0,t']|\mathscr{A}[t_0,s]\} \\
&= E\{P(X_t \in B|X_{t'})|\mathscr{A}[t_0,s]\} && \text{(Markov)} \\
&= E\{P(B,t|X_{t'},t')|\mathscr{A}[t_0,s]\} \\
&= E\{P(B,t|X_{t'},t')|X_s\} && \text{(Markov)} \\
&= \int_{\mathbb{R}} P(B,t|z,t')P(dz,t'|X_s,s)\,. && \text{(2.73)}
\end{aligned}
$$

In the following, our main interest will lie with Markov processes that are smooth in the sense that probability densities exist. We shall consider from now on the case that

$$P(B,t|x,s) = \int_B p(y,t|x,s)\,dy \tag{4.6}$$

and call $p(y,t|x,s)$ the transition probability density. If $t = s$, then $p(y,s|x,s) = \delta(y-x)$. According to (2.79) the transition probability density is given by

$$p(y,t|x,s) = \frac{p(y,t;x,s)}{\int_{\mathbb{R}} p(y,t;x,s)\,dy} = \frac{p(y,t;x,s)}{p(x,s)} \tag{4.7}$$

if $p(x,s)$ is positive. Rewriting the Chapman-Kolmogorov equation for the density, we have:

$$p(y,t|x,s) = \int_{\mathbb{R}} p(y,t|z,t')p(z,t'|x,s)\,dz\,. \tag{4.8}$$

---

[1] The function $P(B,t|x,s)$, which also should fulfill $P(B,s|x,s) = I_B(x)$, is called the transition function of the Markov process X_t.

In words it says that for a Markov process the transition from (x, s) to (y, t) can be decomposed into two steps. First, there is a transition from the present state x to some arbitrary state z at some arbitrary intermediate time t'. Second, collect the transitions from all possible intermediary states into the state y at time t.

Together with the one-dimensional probability density, the transition probability density characterizes completely the process X_t. We have:

$$p(y_n, t_n; \ldots; y_1, t_1) = p(y_n, t_n | y_{n-1}, t_{n-1}) \cdot p(y_{n-1}, t_{n-1} | y_{n-2}, t_{n-2}) \cdot \\ \ldots \cdot p(y_2, t_2 | y_1, t_1) p(y_1, t_1) . \tag{4.9}$$

This shows that it is not necessary to determine the infinite hierarchy of probability densities in order to characterize a Markov process. The first two members $p(y, t)$ and $p(y, t | x, s)$, are sufficient. As is clear from their hierarchies, the examples of stochastic processes discussed in Chap. 2 are all Markov processes. The class of processes with independent values at each instant of time, i.e.,

$$p(y_n, t_n; \ldots; y_1, t_1) = \prod_{i=1}^{n} p(y_i, t_i) , \tag{4.10}$$

are trivially Markovian and will be called singular Markov processes in the following.

It should be realized that in general the Markov property allows for correlations at different instants of time and nonsingular Markov processes have a nonvanishing correlation time. In this sense, Markov processes form the simplest nontrivial class of random processes. The statement found from time to time that Markov processes are random processes without memory is an abuse of language. Only if the process is conditioned on the present state, given by the random variable X_s, then by definition there is no memory of the past. For a nonsingular stationary Markov process we have

$$\begin{aligned} p(y, t; x, s) &= p(y, t | x, s) p(x, s) \\ &= p(y, t-s | x, 0) p_s(x) \not\equiv p_s(y) p_s(x) \end{aligned} \tag{4.11}$$

and thus the correlation function $C(t-s)$ will in general not vanish for all $\tau = t-s$ larger than zero. It is usual to call a Markov process time homogeneous if its transition probability density fulfills

$$p(y, t | x, s) = p(y, t-s | x, 0) \equiv p(y, \tau | x) . \tag{4.12}$$

A time-homogeneous Markov process is stationary (in the strict sense) if $p(x, t) = p_s(x)$.

The motivation to discuss Markov processes was our heuristic derivation that the solution of the stochastic differential equation should possess the Markov property. Let us exploit the fact that in the sense of generalized functions Gaussian white noise is the derivative of the Wiener process, and formally multiply (3.31) by dt to obtain

$$dX_t = f_\lambda(X_t)\,dt + \sigma g(X_t)\,dW_t\,. \tag{4.13}$$

We shall postpone till the next chapter the detailed discussion of the rigorous mathematical meaning of such an equation and what we understand by X_t being a solution of (4.13). Let us only remark that in this form the SDE looks certainly more appealing than in the form (3.31), since only ordinary stochastic processes appear ("ordinary" is here understood as opposed to "generalized"). This raises the hope that it will be unnecessary to resort to the theory of generalized stochastic processes to handle (4.13).

4.2 Markovian Diffusion Processes: Definition

Suppose that at time t we have determined the process X_t to be in the state x. The form (4.13) of the SDE suggests then that in some sense, the increment ΔX_t of the process X_t is given, on a very small time interval at least, by

$$\Delta X_t = f_\lambda(x)\,\Delta t + \sigma g(x)\,\Delta W_t\,. \tag{4.14}$$

Defining $a = f_\lambda(x)$ and $\sqrt{D} = \sigma g(x)$, this relation can be written as

$$\Delta X_t = a\,\Delta t + \sqrt{D}\,\Delta W_t\,. \tag{4.15}$$

This looks like a Brownian motion with diffusion constant D in a flow with velocity a. However, a and $\sqrt{D}$ depend, so to speak, on the position of the Brownian particle. A more precise way is thus to say that locally in time, i.e., for a small time interval where $X_t \approx x = \text{const}$, the evolution of the state variable X_t resembles the motion of a Brownian particle with a diffusion constant $\sigma^2 g^2(x)$ in a fluid with a flow velocity $f_\lambda(x)$. Therefore, let us investigate if it is meaningful to define a class of Markov processes that look locally in time like a Wiener process with a systematic component. (More generally one can consider Markov processes that locally have the structure of a general process with stationary independent increments [4.1]).

For obvious reasons, this class of Markov processes goes by the name "diffusion processes". They are characterized by three properties. First, a diffusion process should have almost surely continuous sample paths. This is expressed by the following mathematical requirement: let X_t be a Markov process with transition probability density $p(y,t\,|\,x,s)$.

a) For every positive ε

$$\lim_{t\downarrow s} \frac{1}{t-s} \int_{|y-x|>\varepsilon} p(y,t|x,s)\,dy = 0\,. \tag{4.16}$$

This condition means that a large change in the state of the process X_t is improbable over a short time interval, i.e.,

$$P[|X_t - X_s| > \varepsilon | X_s = x] = o(t-s) . \tag{4.17}$$ [2]

Condition (4.16) ensures that a Markovian process has almost surely continuous sample paths (see, for instance, [Ref. 3.3, p. 188, Theorem III.5.3]). Before discussing the two other characteristic properties of a diffusion process, we shall dwell further on the implications of (4.16). While this condition guarantees the continuity of the sample paths, it says on the other hand that the realizations of a diffusion process are very wiggly. In fact, they are as wiggly as those of the Wiener process. This is of course to be expected in the light of the above discussion. Condition (4.16) implies that there is at least a nonzero probability that the "velocity" of X_t, i.e., the derivative of the sample paths with respect to time is infinite. If it were otherwise, then obviously $P[|X_t - X_s| > \varepsilon | X_s = x]$ would be identically zero for $t \neq s$, but $t-s$ sufficiently small, and not only $o(t-s)$. This wiggly character of the sample paths is necessary for a genuinely random process to be Markovian. Roughly speaking, random processes with smoother, i.e., differentiable, realizations cannot possess the Markov property, since the smoothness renders any (conditional) independence between future and past impossible. The only exceptions, i.e., the only Markov processes which have continuous *and* differentiable sample paths, are the deterministic ones given by

$$\dot{X}(t) = f(X(t)) , \tag{4.18}$$

with $X_{t_0} = C$, where the only random element is the initial condition C [1.65, 2.2].

The following two defining properties of a diffusion process express, roughly speaking, the fact that the mean value as well as the variance of the displacements of a Brownian particle in a flow are proportional to t. According to our above considerations this should be true locally in time for the increments of X_t conditioned on $X_s = x$. In precise mathematical terms these requirements read:

b) For every positive ε there exists a function $f(x,s)$, the so-called drift, such that

$$\lim_{t \downarrow s} \frac{1}{t-s} \int_{|y-x| \leq \varepsilon} (y-x) p(y,t|x,s) \, dy = f(x,s) . \tag{4.19}$$

c) For every positive ε there exists a function $g(x,s)$, called diffusion, such that

$$\lim_{t \downarrow s} \frac{1}{t-s} \int_{|y-x| \leq \varepsilon} (y-x)^2 p(y,t|x,s) \, dy = g^2(x,s) . \tag{4.20}$$

A Markov process fulfilling (a–c) is called a diffusion process.

In the two conditions (b, c) the truncated moments are used, since the moments themselves might not exist. However, in most practical applications the following condition is fulfilled:

---

[2] A function $u(t)$ is $o(t)$, if $\lim_{t \to 0} [u(t)/t] = 0$.

a′) There exists a positive δ such that

$$\lim_{t\downarrow s}\frac{1}{t-s}\int_{\mathbb{R}}|y-x|^{2+\delta}p(y,t|x,s)\,dy=0\,. \tag{4.21}$$

In other words, a higher than second, not necessarily integer, moment of the increment, conditioned on $X_s=x$, vanishes faster than $t-s$, i.e.,

$$E\{|X_t-X_s|^{2+\delta}|X_s=x\}=o(t-s)\,. \tag{4.22}$$

A fortiori we then have

$$\begin{aligned}o(t-s)&=\int_{|y-x|>\varepsilon}|y-x|^{2+\delta}p(y,t|x,s)\,dy\\&=\varepsilon^{2+\delta}\int_{|y-x|>\varepsilon}\frac{|y-x|^{2+\delta}}{\varepsilon^{2+\delta}}p(y,t|x,s)\,dy>\varepsilon^{2+\delta}\int_{|y-x|>\varepsilon}p(y,t|x,s)\,dy\,.\end{aligned} \tag{4.23}$$

This implies that condition (a) is fulfilled. Expression (4.21) is a sufficient condition for a Markov process X_t to have almost surely continuous sample paths. In a way similar to (4.23) we obtain from (4.21):

$$\int_{|y-x|>\varepsilon}|y-x|p(y,t|x,s)\,dy\leqslant\frac{1}{\varepsilon^{1+\delta}}\int_{\mathbb{R}}|y-x|^{2+\delta}p(y,t|x,s)\,dy=o(t-s) \tag{4.24}$$

and

$$\int_{|y-x|>\varepsilon}(y-x)^2p(y,t|x,s)\,dy\leqslant\frac{1}{\varepsilon^{\delta}}\int_{\mathbb{R}}|y-x|^{2+\delta}p(y,t|x,s)\,dy=o(t-s)\,. \tag{4.25}$$

This implies that the full moments can be used in (b, c) if (a′) holds, since (4.24, 25) prove that integration over $|y-x|>\varepsilon$ contributes only terms of an order less than $t-s$. We can thus replace (b, c) by

b′) There exists a function $f(x,s)$, such that

$$\lim_{t\downarrow s}\frac{1}{t-s}\int_{\mathbb{R}}(y-x)p(y,t|x,s)\,dy=f(x,s)\,. \tag{4.26}$$

c′) There exists a function $g(x,s)$, such that

$$\lim_{t\downarrow s}\frac{1}{t-s}\int_{\mathbb{R}}(y-x)^2p(y,t|x,s)\,dy=g^2(x,s)\,. \tag{4.27}$$

Conditions (a′ – c′) are sufficient for a Markov process X_t to be a diffusion process. If X_t is a time-homogeneous Markov process, i.e., $p(y,t|x,s)=p(y,t-s|x)$, then the drift f and the diffusion g^2 are time-independent functions.

Condition (b′) can be written as:

$$E\{(X_t - X_s)|X_s = x\} = f(x,s)(t-s) + o(t-s) . \tag{4.28}$$

This shows clearly that the drift represents the local systematic evolution of the system, *conditioned* on the fact that it is in state x at time s. For the variance of the increment conditioned on $X_s = x$, we obtain:

$$\begin{aligned} &E\{[(X_t - X_s) - E\{(X_t - X_s)\}]^2|X_s = x\} \\ &\quad = E\{(X_t - X_s)^2|X_s = x\} - E\{(X_t - X_s)|X_s = x\}^2 \\ &\quad = g^2(x,s)(t-s) + o(t-s) - f^2(x,s)(t-s)^2 + o(t-s) \\ &\quad = g^2(x,s)(t-s) + o(t-s) . \end{aligned} \tag{4.29}$$

This establishes that the definition (c′) describes the local variance around the systematic motion, *conditioned* on $X_s = x$. It seems at first glance surprising that the definition of a diffusion process, (4.16, 19, 20), involves only the first two (truncated) differential moments, i.e., $\lim_{t \downarrow s} (1/(t-s))\, E\{(X_t - X_s)^\nu|X_s = x\}$, $\nu = 1,2$, of the transition probability density; no appeal to any higher moments is made. This stems of course from the fact that this definition says X_t looks locally like the process $at + \sqrt{D}\, W_t$, i.e., a Wiener process with a systematic component, which is Gaussian distributed. As discussed in Sect. 2.1.5, it is a characteristic of the Gaussian distribution to be completely defined by its first two moments. This might lead one to conclude that diffusion processes, since they are already defined by their first two differential moments, are necessarily Gaussian processes. This is however *not* so. Loosely speaking, diffusion processes are Gaussian locally in time, but in general not globally. In a certain sense, almost all diffusion processes are non-Gaussian (Sect. 4.4). In fact, there is only *one* stationary Markov process that is Gaussian and belongs to the class of diffusion processes. This is the Ornstein-Uhlenbeck process which will be the topic of the next section.

4.3 The Ornstein-Uhlenbeck Process Revisited and Doob's Theorem

To prove the above assertion, we first show that the O-U process is a diffusion process, and then establish that it is the only stationary Gaussian one. It is easily verified that the O-U process fulfills conditions (a′ – c′). Its transition probability density is given by (2.121), i.e.,

$$\begin{aligned} p(y,t|x,s) &= p(y,t-s|x) \\ &= (2\pi\tilde{\sigma})^{-1/2} \exp\left(-\frac{1}{2}\frac{\{y - x\exp[-\gamma(t-s)]\}^2}{\tilde{\sigma}}\right) \end{aligned} \tag{4.30}$$

with

$$\tilde{\sigma} = (\sigma^2/2\gamma)\{1 - \exp[-2\gamma(t-s)]\} .$$

Hence,

$$\begin{aligned} E\{(X_t - X_s)\,|X_s = x\} &= E\{(X_t - x)\,|X_s = x\} \\ &= x\exp[-\gamma(t-s)] - x \\ &= -x\gamma(t-s) + o(t-s) \end{aligned} \tag{4.31}$$

and

$$\begin{aligned} E\{(X_t - X_s)^2|X_s - x\} &= E\{X_t^2|X_s = x\} - 2E\{X_t x\,|X_s = x\} + x^2 \\ &= \tilde{\sigma} + x^2\exp[-\gamma(t-s)]\{\exp[-\gamma(t-s)] - 2\} + x^2 \\ &= \sigma^2(t-s) + o(t-s)\,, \end{aligned} \tag{4.32}$$

and

$$\begin{aligned} &E\{(X_t - X_s)^4|X_s = x\} = E\{[X_t - x\mathrm{e}^{-\gamma(t-s)} + x(\mathrm{e}^{-\gamma(t-s)} - 1)]^4|X_s = x\} \\ &= E\{(X_t - x\mathrm{e}^{-\gamma(t-s)})^4|X_s = x\} + 6E\{(X_t - x\mathrm{e}^{-\gamma(t-s)})^2|X_s = x\} \\ &\quad \times x^2(\mathrm{e}^{-\gamma(t-s)} - 1)^2 + x^4(\mathrm{e}^{-\gamma(t-s)} - 1)^4 \\ &= 3\tilde{\sigma}^2 + 6\tilde{\sigma}x^2(\mathrm{e}^{-\gamma(t-s)} - 1)^2 + x^4(\mathrm{e}^{-\gamma(t-s)} - 1)^4 \\ &= o(t-s)\,, \end{aligned} \tag{4.33}$$

i.e., (a′) is fulfilled with $\delta = 2$. The second part of our assertion was that the O-U process is the only stationary Gaussian diffusion process. The class of Gaussian-Markov processes is completely known. It is characterized by *Doob's* famous theorem [4.2]: Let X_t be a stationary nonsingular Markov process, such that for arbitrary times s and t, $t \neq s$, the joint probability density of X_s and X_t is Gaussian. Then a constant $\gamma > 0$ exists such that for $t_1 < \ldots < t_n$ the $X_{t_1}, \ldots, X_{t_n}$ are jointly Gaussian distributed and the correlation function is given by

$$C(t-s) = (\sigma^2/2\gamma)\exp[-\gamma(t-s)] . \tag{4.34}$$

Remember that a Gaussian process is completely characterized by the mean value and the correlation function. This result establishes that, modulo an additive constant, the O-U process is the only stationary Gaussian diffusion process.

4.4 The Kolmogorov Backward Equation and the Fokker-Planck Equation

We shall now show that the class of (stationary) diffusion processes is much larger. Indeed conditions (a – c) allow for non-Gaussian behavior. The remarkable feature of diffusion processes is that their transition probability density, though non-Gaussian, is already *completely* determined by its *first two differential* moments. To prove this, we shall derive an evolution equation for $p(y,t\,|\,x,s)$, *solely* on the *basis* of *conditions (a – c)*. The general evolution equation for the transition probability density of a Markov process is of course the Chapman-Kolmogorov equation (4.8). It is, however, of little help to determine $p(y,t\,|\,x,s)$ since it is nonlinear in this quantity. We shall see in the following that in the case of a diffusion process it can be transformed into a linear partial differential equation for $p(y,t\,|\,x,s)$.

Let us begin by deriving an auxiliary result [Ref. 4.1, p. 32] which is, however, quite important in its own right. Suppose that the drift $f(x,s)$ and the diffusion $g(x,s)$ are continuous functions and let $v(x)$ be a bounded continuous function such that

$$\begin{aligned} u(x,s) &= E\{v(X_t)\,|X_s = x\} \\ &= \int_{\mathbb{R}} v(y)\,p(y,t|x,s)\,dy \end{aligned} \tag{4.35}$$

has continuous partial derivatives $\partial_x u(x,s)$ and $\partial_{xx} u(x,s)$. Then the partial derivative $\partial_s(u,s)$ exists and $u(x,s)$ satisfies the equation

$$-\partial_s u(x,s) = f(x,s)\,\partial_x u(x,s) + \tfrac{1}{2} g^2(x,s)\,\partial_{xx} u(x,s) \tag{4.36}$$

with the boundary condition

$$\lim_{s \uparrow t} u(x,s) = v(x)\,. \tag{4.37}$$

We shall present the proof of (4.36) in detail here in view of the importance of this equation in the theory of diffusion processes. Furthermore, the proof will shed light on the implications of conditions (a – c), the defining features of a diffusion process. Consider s_1 and s_2 with $s_1 \leqslant s \leqslant s_2 \leqslant t$. Then according to the definition of $u(x,s)$, (4.35), we have:

$$\begin{aligned} \Delta u \equiv u(x,s_1) - u(x,s_2) &= \int_{\mathbb{R}} v(y)\,p(y,t|x,s_1)\,dy - \int_{\mathbb{R}} v(y)\,p(y,t|x,s_2)\,dy \\ &= \int_{\mathbb{R}} v(y)\left[\int_{\mathbb{R}} p(y,t|z,s_2)\,p(z,s_2|x,s_1)\,dz\right]dy \\ &\quad - \int_{\mathbb{R}} v(y)\,p(y,t|x,s_2)\,dy \cdot \int_{\mathbb{R}} p(z,s_2|x,s_1)\,dz \end{aligned}$$

[in the first term we have used the Chapman-Kolmogorov equation (4.8) and in the second term we have multiplied by one in an unorthodox way]

$$\begin{aligned} &= \int_{\mathbb{R}} u(z,s_2)\,p(z,s_2|x,s_1)\,dz - \int_{\mathbb{R}} u(x,s_2)\,p(z,s_2|x,s_1)\,dz \\ &= \int_{\mathbb{R}} [u(z,s_2) - u(x,s_2)]\,p(z,s_2|x,s_1)\,dz\,. \end{aligned} \tag{4.38}$$

We decompose the state space into two parts and obtain:

$$\Delta u = \int_{|z-x|\leqslant\varepsilon} [u(z,s_2) - u(x,s_2)]\,p(z,s_2|x,s_1)\,dz$$
$$+ \int_{|z-x|>\varepsilon} [u(z,s_2) - u(x,s_2)]\,p(z,s_2|x,s_1)\,dz \equiv J_1 + J_2 \,.$$

Let us first deal with J_2. Since $v(x)$ was assumed to be bounded, $u(x,s)$ will inherit this property, i.e., $\sup_x |u(x,s_2)| < \infty$. Thus,

$$J_2 \leqslant 2\sup_z |u(z,s_2)|\cdot \int_{|z-x|>\varepsilon} p(z,s_2|x,s_1)\,dz$$

and by condition (a), Eq. (4.16), we have

$$J_2 = o(s_2 - s_1)\,. \tag{4.39}$$

We use Taylor's formula to write the first term as

$$J_1 = \int_{|z-x|\leqslant\varepsilon} [u'(x,s_2)(z-x) + \tfrac{1}{2}u''(x,s_2)(z-x)^2 + r_\varepsilon(x,z,s_2)]\,p(z,s_2|x,s_1)\,dz\,. \tag{4.40}$$

The error term $r_\varepsilon(x,z,s_2)$ has the following property

$$|r_\varepsilon(x,z,s_2)| \leqslant |z-x|^2 o_\varepsilon\,, \tag{4.41}$$

where

$$o_\varepsilon = \sup_{|z-x|\leqslant\varepsilon} |\partial_{xx}u(x+\vartheta(z-x),s_2) - \partial_{xx}u(x,s_2)|\,,$$
$$0<\vartheta<1 \quad \text{and} \quad o_\varepsilon \to 0 \quad \text{for} \quad \varepsilon\to 0\,.$$

We now use the two other conditions in the definition of a diffusion process, i.e. (4.19, 20), and have

$$\Delta u = [f(x,s)\,\partial_x u(x,s_2) + \tfrac{1}{2}g^2(x,s)\,\partial_{xx}u(x,s_2) + R_\varepsilon](s_2 - s_1) + o(s_2 - s_1)\,, \tag{4.42}$$

where $\lim_{\varepsilon\downarrow 0}\lim_{s_2\downarrow s_1} R_\varepsilon = 0\,.$ [cf. (4.41)].

Dividing both sides by $(s_2 - s_1)$ and taking the limit $s_2\downarrow s$, $s_1\uparrow s$ and $\varepsilon\downarrow 0$, we obtain (4.36).

Recall that

$$u(x,s) = \int_{\mathbb{R}} v(y)\,p(y,t|x,s)\,dy\,.$$

Hence (4.36) can be written in the form[3]

$$\int_{\mathbb{R}} v(y)\,[\partial_s p(y,t|x,s) + f(x,s)\,\partial_x p(y,t|x,s) \\ + \tfrac{1}{2}g^2(x,s)\,\partial_{xx} p(y,t|x,s)]\,dy = 0\,.$$

Since $v(y)$ is an arbitrary function, it follows that the transition probability density of a diffusion process obeys the following partial differential equation:

$$-\partial_s p(y,t|x,s) = f(x,s)\,\partial_x p(y,t|x,s) + \tfrac{1}{2}g^2(x,s)\,\partial_{xx} p(y,t|x,s)\,. \tag{4.43}$$

This equation as well as (4.36) is known as the Kolmogorov backward equation (KBE). The qualifier backward is used since the variation is, so to speak, with respect to the initial state x and the initial time s. The final state (y,t) is a parameter as far as the solution of the KBE is concerned. Roughly speaking, the KBE addresses the problem where the diffusion process X_t has to start at time s in order to arrive in the specified state y at time t. This problem plays in most practical applications only a minor role. A notable exception is genetics, where one is often interested in the fact that the system reaches a certain final state, for instance a mutant gene becomes fixed in the population [4.3]. However, in general the basic question is how does the system evolve in the future?, i.e., one would like to have an evolution equation in the "forward sense" where (x,s) is a parameter for the solution. Such a forward equation will be derived now. In the proof we shall have to invoke the following relation which is an immediate corollary of (4.36):

$$\lim_{s_2\downarrow s_1;\, s_1\uparrow s} \frac{1}{s_2-s_1}\int_{\mathbb{R}} [u(z)-u(x)]\,p(z,s_2|x_1,s_1)\,dz \\ = \lim_{s_2\downarrow s_1;\, s_1\uparrow s} \frac{1}{s_2-s_1}\int_{\mathbb{R}} [u(z)\,p(z,s_2|x_1,s_1)\,dz - u(x)] \\ = f(x,s)\,\partial_x u + \tfrac{1}{2}g(x,s)\,\partial_{xx} u\,. \tag{4.44}$$

This follows simply by observing that nothing changes in the proof of (4.36) if we start at the line (4.38) with a function u that does not depend on the argument time s. This result is needed in the derivation of the following fundamental result, namely the forward equation of the diffusion process.

If conditions (4.16, 19, 20) hold uniformly in x, then the transition probability density $p(y,t|x,s)$ of the diffusion process X_t satisfies

$$\partial_t p(y,t|x,s) = -\partial_y f(y,t)\,p(y,t|x,s) + \tfrac{1}{2}\partial_{yy} g^2(y,t)\,p(y,t|x,s)\,, \tag{4.45}$$

provided that the respective derivatives appearing in (4.45) exist and are continuous. The derivation of the forward equation goes as follows [4.1]: Let $v(x)$

[3] We have assumed that the derivatives of $p(y,t\mid x,s)$ exist and that the differentiation and integration can be interchanged.

be an arbitrary twice continuously differentiable function which vanishes identically outside a certain bounded interval. Then we have

$$\partial_t \int_{\mathbb{R}} p(y,t|x,s)\, v(y)\, dy = \lim_{t_1 \uparrow t;\, t_2 \downarrow t} \frac{1}{t_2 - t_1} \int_{\mathbb{R}} [p(y,t_2|x,s) - p(y,t_1|x,s)]\, v(y)\, dy$$

$$= \lim_{t_1 \uparrow t;\, t_2 \downarrow t} \int_{\mathbb{R}} p(y,t_1|x,s) \cdot \left[\frac{1}{t_2 - t_1} \int_{\mathbb{R}} p(z,t_2|y,t_1)\, v(z)\, dz - v(y) \right] dy\,.$$

[This equality is obtained by using the Chapman-Kolmogorov equation for the first term,

$$\int p(y,t_2|x,s)\, v(y)\, dy = \int\int p(y,t_2|z,t_1)\, p(z,t_1|x,s)\, v(y)\, dy\, dz\,,$$

and by renaming the variables: $y \leftrightarrow z$

$$= \int\int p(z,t_2|y,t_1)\, p(y,t_1|x,s)\, v(z)\, dz\, dy\,.]$$

Using (4.44), we obtain for the right-hand side

$$\text{rhs} = \int p(y,t|x,s)\, [f(y,t)\, \partial_y v(y) + \tfrac{1}{2} g^2(y,t)\, \partial_{yy} v(y)]\, dy\,.$$

Integration by parts yields

$$\int_{\mathbb{R}} \partial_t p(y,t|x,s)\, v(y)\, dy = \int_{\mathbb{R}} dy\, v(y) [-\partial_y f(y,t) + \tfrac{1}{2} \partial_{yy} g^2(y,t)]\, p(y,t|x,s)\,.$$

The boundary terms vanish since $v(y)$ is zero outside a bounded interval. Taking into account that $v(y)$ is an arbitrary function, the forward equation (4.45) follows. This evolution equation for $p(y,t|x,s)$ is linear in the transition probability density, as the Kolmogorov backward equation, in contrast to the Chapman-Kolmogorov equation from which both follow. It is known in the mathematical literature as the Kolmogorov forward equation since the variation is with respect to the "future" state y and time argument t. In the physical literature this equation is called the Fokker-Planck equation (FPE). The KBE and FPE show that a Markov diffusion process is indeed already completely defined by its first two differential moments: the KBE and FPE are partial differential equations for the transition probability densities, whose coefficients are given by the drift f and the diffusion g^2. Hence the transition probability density, as a solution of the KBE or the FPE, is completely and uniquely defined by its first two differential moments under certain regularity conditions on f and g^2. It is this surprising property that makes the concept of diffusion processes so powerful.

Multiplying the FPE (4.45) on both sides by the one-dimensional probability density $p(x,s)$ and then integrating over x, we obtain that the one-dimensional probability density $p(y,t)$ of a diffusion process also obeys an equation of type (4.45):

$$\partial_t p(y,t) = -\partial_y f(y,t) p(y,t) + \tfrac{1}{2}\partial_{yy} g^2(y,t) p(y,t) . \tag{4.46}$$

The inverse is of course *not* true. If the *one-dimensional* probability density $p(y,t)$ of some random process obeys an equation of the form (4.46), then the process is not necessarily Markovian [4.4]. If the diffusion process is time homogeneous, i.e., $p(y,t\,|\,x,s) = p(y,t-s\,|\,x,0) = p(y,\tau\,|\,x)$, then as remarked above, the drift and diffusion are time independent. Furthermore,

$$\partial_t p(y,t|x,s) = \partial_\tau p(y,\tau|x) \tag{4.47}$$

and

$$\partial_s p(y,t|x,s) = -\partial_\tau p(y,\tau|x) . \tag{4.48}$$

For time-homogeneous Markov processes, KBE and FPE read (writing t for τ):

$$\text{KBE:}\ \partial_t p(y,t|x) = f(x)\,\partial_x p(y,t|x) + \tfrac{1}{2} g^2(x)\,\partial_{xx} p(y,t|x) , \tag{4.49}$$

$$\text{FPE:}\ \partial_t p(y,t|x) = -\partial_y f(y) p(y,t|x) + \tfrac{1}{2}\partial_{yy} g^2(y) p(y,t|x) . \tag{4.50}$$

In the following, we shall consider mainly time-homogeneous diffusion processes, since one of our basic assumptions on the environment is that its variations contain no systematic component.

Let us now write down the FPE and KBE for the two examples of diffusion processes we have encountered so far. Brownian motion $\sqrt{D}\,W_t$ is a diffusion process and one might say trivially, since diffusion processes were defined as processes that look locally in time like a Wiener process. For this simple example FPE and KBE coincide and they are given by the classical diffusion equation:

$$\partial_t p(w,t|w_0) = \tfrac{1}{2} D\,\partial_{ww} p(w,t|w_0) . \tag{4.51}$$

The solution to this equation is of course given by (2.99). For the O-U process, the two equations read, using (4.31, 32):

$$\text{KBE:}\ \partial_t p(v,t|v_0) = -\gamma v\,\partial_v p(v,t|v_0) + \tfrac{1}{2}\sigma^2\partial_{vv} p(v,t|v_0) , \tag{4.52}$$

$$\text{FPE:}\ \partial_t p(v,t|v_0) = \partial_v \gamma v p(v,t|v_0) + \frac{\sigma^2}{2}\partial_{vv} p(v,t|v_0) , \tag{4.53}$$

and it is easily verified that (2.121) solves both equations.

4.5 Pawula's Theorem

Let us close this chapter on Markov processes by underscoring once more that diffusion processes are already completely defined by the first two differential moments of their transition probability, together with the property to have almost surely continuous sample paths. As was already emphasized when the definition of a diffusion process was completed, no appeal to higher order differential moments is made. Indeed, no such appeal is necessary as the explicit derivation of (4.36), of KBE (4.43) and of FPE (4.45) shows. This is of course completely in the spirit of the considerations leading to conditions (a – c), namely that diffusion processes are Markov processes that look locally in time like Gaussian process with independent increments: $ct + \sqrt{D}\,W_t$, i.e., a Brownian particle in a constant velocity flow. In such a process, all higher than second-order moments are $o(t)$. This might raise the question if it is possible to define further classes of Markov processes by, for instance, imposing the following conditions in analogy to the sufficient conditions (a′ – c′) for a diffusion process:

i) For $n \leqslant \nu$ there exists function $A_n(x,s)$ such that

$$\lim_{t\downarrow s} \frac{1}{t-s} E\{(X_t - X_s)^n | X_s = x\} = A_n(x,s) \not\equiv 0 . \tag{4.54}$$

ii) For $n > \nu$

$$\lim_{t\downarrow s} \frac{1}{t-s} E\{(X_t - X_s)^n | X_s = x\} = 0 . \tag{4.55}$$

The FPE for such a process would be

$$\partial_t p(y,t|x,s) = \sum_{n=1}^{\nu} \frac{(-1)^n}{n!} \partial_y^n A_n(y,t)\, p(y,t|x,s) , \tag{4.56}$$

according to the following lines of reasoning. Let $v(y)$ be an infinitely often differentiable function which vanishes outside a certain bounded interval. Then using the Chapman-Kolmogorov equation, we have

$$\begin{aligned} &\int_{\mathbb{R}} dy\, v(y) h^{-1}[p(y,t+h|x,s) - p(y,t|x,s)] \\ &\quad = h^{-1} \int_{\mathbb{R}} dy\, v(y) \int_{\mathbb{R}} dz\, p(y,t+h|z,t)\, p(z,t|x,s) \\ &\qquad - h^{-1} \int_{\mathbb{R}} dy\, v(y)\, p(y,t|x,s) . \end{aligned}$$

Using the Taylor expansion

$$v(y) = v(z) + \sum_{n=1}^{\infty} \frac{1}{n!} \partial_z^n v(z)(y-z)^n$$

and taking the limit $h \to 0$, we obtain

$$\int_{\mathbb{R}} dy\, v(y)\, \partial_t p(y,t|x,s) = \int_{\mathbb{R}} dy \sum_{n=1}^{\nu} \frac{1}{n!} A_n(y,t)\, p(y,t|x,s)\, \partial_y^n v(y)\,.$$

Integration by parts in the rhs yields (4.56).

However, any class of Markov processes defined like this with $\nu \geqslant 3$ is *empty*; this definition is logically inconsistent and (4.56) is not the forward equation of a Markov process. The flaw in the definition and in the derivation of (4.56) is that (4.55) implies of course that condition (a′) (4.21) is fulfilled with $\delta = \nu + 2$ (or $+1$ if ν is odd). Furthermore, (4.54) implies trivially that (b′, c′) are also fulfilled. It follows that the Markov process, defined by (4.54, 55), is a diffusion process and its transition probability density is governed by FPE (4.45). Thus if there is a $\nu \geqslant 3$, such that the differential moments of order $n > \nu$ vanish identically, then all differential moments $A_n(x,s)$ vanish identically for $n > 2$. Indeed, *Pawula* [4.5, 6] has given a beautiful direct demonstration of this fact based only on the Cauchy-Schwarz inequality for mathematical expectations.

Pawula's Theorem. If an even $\mu > 3$ exists, such that $A_\mu(x,s) \equiv 0$, then it follows that $A_n(x,s) = 0$ for all $n \geqslant 3$.

Since this theorem has not received the attention it deserves, we shall reproduce Pawula's proof here. Let $n \geqslant 3$ be odd. Then

$$\begin{aligned} |E\{(X_t - X_s)^n | X_s = x\}| &= |E\{(X_t - X_s)^{\frac{n-1}{2}} (X_t - X_s)^{\frac{n+1}{2}} | X_s = x\}| \\ &\leqslant (E\{(X_t - X_s)^{n-1} | X_s = x\})^{1/2} (E\{(X_t - x)^{n+1} | X_s = x\})^{1/2} \end{aligned} \tag{4.57}$$

implying

$$A_n^2(x,s) \leqslant A_{n-1}(x,s)\, A_{n+1}(x,s)\,. \tag{4.58}$$

For n even we obtain in a similar way

$$A_n^2(x,s) \leqslant A_{n-2}(x,s)\, A_{n+2}(x,s)\,. \tag{4.59}$$

Setting $n = r-1$, $r+1$ in (4.58) and $n = r-2$, $r+2$ in (4.59), where r is an even integer, we obtain:

$$\begin{aligned} &A_{r-2}^2(x,s) \leqslant A_{r-4}(x,s)\, A_r(x,s) && r \geqslant 6\,, \\ &A_{r-1}^2(x,s) \leqslant A_{r-2}(x,s)\, A_r(x,s) && r \geqslant 4\,, \\ &A_{r+1}^2(x,s) \leqslant A_r(x,s)\, A_{r+2}(x,s) && r \geqslant 2\,, \\ &A_{r+2}^2(x,s) \leqslant A_r(x,s)\, A_{r+4}(x,s) && r \geqslant 2\,. \end{aligned} \tag{4.60}$$

Using these four inequalities, we can move up the ladder of differential moments to arbitrary high n and down to $n = 3$ and conclude that if there is an even $\mu > 3$, with $A_\mu(x,s) \equiv 0$ then indeed $A_n(x,s) \equiv 0$ for $n \geqslant 3$. Thus, as far as the attempt to define classes of Markov processes by (4.54, 55) is concerned, only two such classes exist: first the class of diffusion processes where all differential moments of higher than second order vanish identically, and second, the class which is characterized by the fact that *all even* differential moments are not identically zero.

4.6 Non-Gaussian White Noise

Let us conclude with a brief remark on systems perturbed by non-Gaussian white noise. As mentioned in Sect. 3.2 and made intuitively clear, a process ξ_t is a white noise if and only if it is the derivative, in the generalized function sense, of a time-homogeneous process with independent increments. In this chapter we have considered only Gaussian white noise, the derivative of the Wiener process. This motivated the definition of a class of Markov processes, the diffusion processes, which look locally in time like the Wiener process with a systematic component: $at+\sqrt{D}\,W_t$. These diffusion processes can be characterized by the fact that $u(x,s)=\int_{\mathbb{R}} v(y)p(y,t\,|\,x,s)\,dy$ obeys the backward equation (4.36). In this spirit, one could envisage a class of Markov processes that look locally like a Poisson process. More generally, we could choose any time-homogeneous process with independent increments. Since the most general white noise is the derivative of such a time-homogeneous process with independent increment, any system disturbed by whatever kind of white noise would then be described by a Markov process of this class. Since, roughly speaking, the Wiener process is the only time-homogeneous process with independent increments which has a.s. continuous sample paths, these Markov processes would in general not have a.s. continuous sample paths. For more details see [4.1, 7]. Processes of this kind find applications in systems perturbed by shot noise and are, for instance, frequently employed in the description of neuron activity [4.8 – 10]. They can also be used to model internal fluctuations [4.11].

5. Stochastic Differential Equations

The presentation of Markov processes in the preceding chapter was motivated by our heuristic considerations that the temporal evolution of the state variable X_t of a system coupled to a random environment is Markovian if and only if the external noise is white. Furthermore, the form of the phenomenological equation with a Gaussian white noise, especially when rewritten as (4.13), suggested that we consider in particular a special class of Markov processes, the diffusion processes. To establish definitely that systems in a Gaussian white-noise environment can be described by diffusion processes, it remains to show that the concept of a solution to the stochastic differential equation can be mathematically well defined and that it is indeed a Markov process which fulfills conditions (4.16, 19, 20) as suggested by (4.13). The purpose of this chapter is to present in detail the theory of Ito and Stratonovich SDEs and their intimate connection with diffusion processes.

5.1 Stochastic Integrals: A First Encounter

When handling equations like (3.31), great circumspection has to be exerted. Indeed, taking (3.31) at face value, X_t inherits the extremely irregular features of the Gaussian white noise ξ_t and is certainly not defined in the ordinary sense. As already remarked, using the fact that ξ_t is the derivative in the generalized functions sense of the Wiener process W_t and rewriting (3.31) in the form (4.13) raises the hope that the theory of generalized stochastic processes can be avoided. Indeed, in (4.13) only ordinary processes appear. It is reasonable to expect that the concept "X_t is a solution of (4.13)" can be formulated in the framework of ordinary stochastic processes if the function $g(x)$ is sufficiently "smooth" as is the case in applications. The theory of stochastic differential equations, developed by *Doob* [5.1], *Ito* [5.2, 4] and *Stratonovich* [5.3], serves just this purpose. In the following, we shall present its basic ingredients and main results. Instead of looking at (4.13) in its differential form

$$dX_t = f(X_t)\,dt + \sigma g(X_t)\,dW_t, \tag{5.1}$$

it is convenient to switch to the equivalent integral form

$$X_t = X_0 + \int_0^t f(X_s)\,ds + \sigma \int_0^t g(X_s)\,dW_s. \tag{5.2}$$

(In this section we shall often set $\sigma = 1$, since for the more mathematical considerations the intensity of white noise is of no importance).

The task of giving a precise mathematical meaning to the SDE (5.1) is thus transformed into the task to define rigorously what we understand by the two integrals in (5.2). The first one presents no problem. Since the differential dX_t contains the differential dW_t of the Wiener process, in whatever way a solution to the equation is secured, X_t might be as irregular as the Wiener process but its behavior cannot be worse. Thus the sample paths of X_t might be almost surely nowhere differentiable but they will be continuous if f and g are smooth functions. This ensures that the first integral can be understood as an ordinary (Riemann) integral for each realization, i. e.,

$$\int_0^t f(X_s(\omega))\, ds = \lim_{n\to\infty} \sum_{i=1}^{n} f(X_{\tau_i}^{(n)}(\omega))(s_i^{(n)} - s_{i-1}^{(n)}) , \tag{5.3}$$

where the evaluation point $\tau_i^{(n)}$ is in the interval $[s_{i-1}^{(n)}, s_i^{(n)}]$ but otherwise arbitrary. More problematical is the second integral in (5.2), the one with respect to the Wiener process. As we have shown in Sect. 2.3, the sample paths of the Wiener process are so wiggly that almost surely they are of infinite length on an arbitrary finite time interval. In those cases we are interested in, the function $g(X_t)$ in the second integral will in general have sample paths of the same wiggly nature as the Wiener process. As discussed above, we expect X_t to inherit the features of W_t. Roughly speaking, the integrand of the second integral in (5.2) is an order of magnitude more "irregular" than the first integral. The irregularities of $g(X_t)$ and W_t will mutually reinforce themselves. This has rather annoying consequences for an interpretation of the second integral.

Consider the following simple standard example

$$\int_{t_0}^t W_s dW_s = ? \tag{5.4}$$

If this integral could be interpreted as an ordinary Riemann integral, then the answer is:

$$\int_{t_0}^t W_s dW_s = \tfrac{1}{2}(W_t^2 - W_{t_0}^2) . \tag{5.5}$$

Of course, this holds only if the conditions in the definition of a Riemann integral are fulfilled. Consider the approximating sums S_n,

$$S_n = \sum_{i=1}^{n} W_{\tau_i^{(n)}}(W_{t_i^{(n)}} - W_{t_{i-1}^{(n)}}) , \tag{5.6}$$

where $\{t_0 = t_1^{(n)} < \ldots < t_n^{(n)} = t\}$ is a series of partitions of the interval $[t_0, t]$ and where

$$\tau_i^{(n)} \in [t_{i-1}^{(n)}, t_i^{(n)}] \,, \tag{5.7}$$

but is *arbitrary* otherwise. If the S_n converge for finer and finer partitions, i.e., $\delta_n = \max_i(t_i^{(n)} - t_{i-1}^{(n)}) \to 0$, to one and the same random variable S, *independent* of the choice of the evaluations points $\tau_i^{(n)}$, then the Riemann integral exists and is given by the limit S. Let us therefore verify if

$$\underset{n\to\infty}{\text{qm-lim}}\, S_n = \tfrac{1}{2}(W_t^2 - W_{t_0}^2) \tag{5.8}$$

for all choices of $\tau_i^{(n)}$, which would establish (5.5). If qm-lim $X_n = X$ and qm-lim $Y_n = Y$, then it follows that qm-lim $(X_n + Y_n) = X + Y$. In light of this fact it is convenient to write S_n in the following form

$$\begin{aligned}
S_n &= \sum_{i=1}^{n} W_{t_{i-1}}(W_{t_i} - W_{t_{i-1}}) + \sum_{i=1}^{n} (W_{\tau_i} - W_{t_{i-1}})(W_{t_i} - W_{t_{i-1}}) \\
&= \sum_{i=1}^{n} \frac{1}{2}(W_{t_{i-1}} + W_{t_i} + W_{t_{i-1}} - W_{t_i})(W_{t_i} - W_{t_{i-1}}) + \sum_{i=1}^{n} (W_{\tau_i} - W_{t_{i-1}})(W_{t_i} - W_{\tau_i}) \\
&\quad + \sum_{i=1}^{n} (W_{\tau_i} - W_{t_{i-1}})^2 \\
&= \frac{1}{2}\sum_{i=1}^{n} (W_{t_i}^2 - W_{t_{i-1}}^2) - \frac{1}{2}\sum_{i=1}^{n} (W_{t_i} - W_{t_{i-1}})^2 \\
&\quad + \sum_{i=1}^{n} (W_{\tau_i} - W_{t_{i-1}})(W_{t_i} - W_{\tau_i}) + \sum_{i=1}^{n} (W_{\tau_i} - W_{t_{i-1}})^2 \\
&= S_1^{(n)} + S_2^{(n)} + S_3^{(n)} + S_4^{(n)} \,,
\end{aligned} \tag{5.9}$$

where for simplicity of notation the superscript n on t_i and τ_i is dropped. Obviously

$$\text{qm-lim}\, S_1^{(n)} = \tfrac{1}{2}(W_t^2 - W_{t_0}^2) \,, \tag{5.10}$$

and according to (2.106), we have for the second term

$$\text{qm-lim}\, S_2^{(n)} = -\tfrac{1}{2}(t - t_0) \,. \tag{5.11}$$

For the third contribution we have

$$\begin{aligned}
E\{S_3^{(n)2}\} &= E\left\{\sum_{i=1}^{n}\sum_{j=1}^{n} (W_{\tau_i} - W_{t_{i-1}})(W_{t_i} - W_{\tau_i}) \cdot (W_{\tau_j} - W_{t_{j-1}})(W_{t_j} - W_{\tau_j})\right\} \\
&= \sum_{i=1}^{n} E\{(W_{\tau_i} - W_{t_{i-1}})^2\} E\{(W_{t_i} - W_{\tau_i})^2\} \\
&\quad + \sum_{\substack{i=1 \\ i\neq j}}^{n}\sum_{j=1}^{n} E\{W_{\tau_i} - W_{t_{i-1}}\} E\{W_{t_i} - W_{\tau_i}\} E\{W_{\tau_j} - W_{t_{j-1}}\} E\{W_{t_j} - W_{\tau_j}\}
\end{aligned}$$

(since W_t has independent increments)

$$= \sum_{i=1}^{n} (\tau_i - t_{i-1})(t_i - \tau_i) \leqslant \delta_n (t - t_0) \underset{n\to\infty}{\longrightarrow} 0 \,. \tag{5.12}$$

Hence

$$\operatorname*{qm-lim}_{n\to\infty} S_3^{(n)} = 0$$

and the mean square limit of the first three terms exists and is independent of the choice $\tau_i^{(n)}$. However, the mean square limit of the last term *depends* on the choice of the evaluation points τ_i. We have

$$\operatorname*{qm-lim}_{n\to\infty} \left[S_4^{(n)} - \sum_{i=1}^{n} (\tau_i - t_{i-1}) \right] = 0 \,. \tag{5.13}$$

Indeed,

$$E\left\{\left[S_4^{(n)} - \sum_{i=1}^{n} (\tau_i - t_{i-1}) \right]^2 \right\} = \sum_{i=1}^{n} E\{(W_{\tau_i} - W_{t_{i-1}})^4\}$$

$$+ \sum_{\substack{i=1 \\ i\neq j}}^{n} \sum_{j=1}^{n} E\{(W_{\tau_j} - W_{t_{j-1}})^2 (W_{\tau_i} - W_{t_{i-1}})^2\} - 2 \sum_{i=1}^{n} E\{(W_{\tau_i} - W_{t_{i-1}})^2\} \sum_{i=1}^{n} (\tau_i - t_{i-1})$$

$$+ \left[\sum_{i=1}^{n} (\tau_i - t_{i-1}) \right]^2$$

$$= 2 \sum_{i=1}^{n} (\tau_i - t_{i-1})^2 + \sum_{i=1}^{n} \sum_{j=1}^{n} (\tau_i - t_{i-1})(\tau_j - t_{j-1})$$

$$- 2 \left[\sum_{i=1}^{n} (\tau_i - t_{i-1}) \right]^2 + \left[\sum_{i=1}^{n} (\tau_i - t_{i-1}) \right]^2$$

$$= 2 \sum_{i=1}^{n} (\tau_i - t_{i-1})^2 \leqslant 2 \delta_n (t - t_0) \underset{n\to\infty}{\longrightarrow} 0 \,.$$

Combining (5.10 – 13), it follows that

$$\operatorname*{qm-lim}_{n\to\infty} \left[S_n - \sum_{i=1}^{n} (\tau_i^{(n)} - t_{i-1}^{(n)}) \right] = \frac{1}{2} (W_t^2 - W_{t_0}^2) - \frac{1}{2} (t - t_0) \,. \tag{5.14}$$

This shows clearly that the limit of the approximating sums is dependent on the evaluation points $\tau_i^{(n)}$, since $\lim \sum_{i=1}^{n} (\tau_i^{(n)} - t_{i-1}^{(n)})$ can take *any value* in the interval $[0, t - t_0]$ depending on the explicit choice of $\tau_i^{(n)}$. Therefore (5.4) and a fortiori

$$\int G(W_s)\,dW_s \tag{5.15}$$

cannot be understood as an ordinary Riemann integral. However, at the same time that the simple example (5.4) reveals the difficulties with stochastic integrals of type (5.15) stemming from the extremely wiggly nature of W_t, it also furnishes a glimpse of the solution to the problem encountered. The preceding example shows that it is impossible to keep the requirement of the Riemann definition that the evaluation points $\tau_i^{(n)}$ can be arbitrarily chosen in the approximating sums. An unambiguous definition of the stochastic integral, at least for the above example (5.4), is achieved if the choice of $\tau_i^{(n)}$ is fixed once and for all. Naturally, this choice should be made in such a way that the integral resulting from such a definition has nice properties. For instance, with

$$\tau_i^{(n)} = (1-\alpha)\,t_{i-1}^{(n)} + \alpha t_i^{(n)} \qquad 0 \leqslant \alpha \leqslant 1\,, \qquad \alpha \text{ fixed} \tag{5.16}$$

we have

$$\sum_{i=1}^{n} (\tau_i^{(n)} - t_{i-1}^{(n)}) = \alpha \sum_{i=1}^{n} (t_i^{(n)} - t_{i-1}^{(n)}) = \alpha(t-t_0)\,. \tag{5.17}$$

If this choice is adopted, (5.14) yields:

$$\text{qm-lim}\, S_n = \tfrac{1}{2}(W_t^2 - W_{t_0}^2) + (\alpha - \tfrac{1}{2})(t-t_0)\,. \tag{5.18}$$

As far as applications are concerned, the infinity of possible integral definitions seems to be a rather dismal consequence of the white-noise idealization. The situation is however not as bad as it appears. Though of course any choice of α leads to a (mathematically) correct definition of the stochastic integral (5.4), most of them do not have any desirable features and would not be chosen as a basis for a stochastic calculus. In fact, there are only two reasonable choices: $\alpha = 0$ and $\alpha = 1/2$. The latter conserves the ordinary rules of calculus, namely the result (5.5) is recovered:

$$\alpha = 1/2\colon \int_{t_0}^{t} W_s\, dW_s = \tfrac{1}{2}(W_t^2 - W_{t_0}^2)\,. \tag{5.19}$$

If the external noise is described by a more realistic random process than Gaussian white noise, e.g., by an O-U process, the classical rules of calculus apply. As already mentioned, in this case the SDE can be considered as an ensemble of ordinary differential equations. (This point is taken up again in detail at the beginning of Chap. 8.) This suggests that if the SDE with white noise is the limit $\tau_{\text{cor}} \to 0$ of an SDE with a more realistic noise process, then an integral definition based on the choice $\alpha = 1/2$ would be reasonable. Thus the choice of the middle point $\tau_i = (t_{i+1} + t_i)/2$ as evaluation point is supported from an applied point of view. From a mathematical point of view, the choice $\alpha = 0$ appears to be more attractive. It makes maximal use of the important property of the Wiener process of having independent increments. With the left-hand point $\tau_i = t_{i-1}$ as evaluation point in the approximating sums, $G(W_{\tau_i})$ and $(W_{t_i} - W_{t_{i-1}})$ are independent. One feels intuitively that this should confer nice properties, from a viewpoint of probability theory, to a stochastic integral based on the choice $\alpha = 0$. Furthermore, if the external noise is *truly white*, then this is the only reasonable choice from a physical point of view. In a genuine white-noise environment, the instantaneous fluctuations of the parameter should not be correlated with the state of the system at the same instant of time. There is one further point in favor of the choice $\alpha = 0$. Recall that when we introduced the definition of a diffusion process, we were motivated to formulate the defining conditions (a – c) by looking at the following equations for the increments of X_t, conditioned on $X_s = x$,

$$X_{s+h} - X_s = f_\lambda(x)h + \sigma g(x)(W_{s+h} - W_s)\,. \tag{5.20}$$

This suggests that $\tau_i = t_{i-1}$ would be a particularly convenient choice to arrive at a *simple* connection between solutions of SDEs and diffusion processes. There seems to be only one disadvantage with $\alpha = 0$, namely we obtain the unfamiliar result:

$$\alpha = 0\colon \int_{t_0}^{t} W_s dW_s = \tfrac{1}{2}(W_t^2 - W_{t_0}^2) - \tfrac{1}{2}(t - t_0)\,. \tag{5.21}$$

However, this is no serious drawback. The classical rules of calculus have no special intrinsic value; it just so happens that they are the only ones we are familiar with. The above arguments indicate clearly that the choice of the left point as evaluation point in the approximating sums is rather advantageous. Indeed, the integral definition based on this convention and introduced by *Ito* in 1950 [5.4] plays a predominant role in mathematical literature. The other reasonable choice, $\tau_i = (t_{i+1} + t_i)/2$, was made by *Stratonovich* in his definition of a stochastic integral in 1966 [5.3]. It is mainly used, not surprisingly in the light of the above discussion, in the physical literature. The question which of the two mathematically correct definitions is more adequate in a concrete application is a problem of modeling. It will be addressed in detail once the two definitions and their consequences have been presented, furnishing the necessary elements for a rigorous discussion of this problem. Despite some clear and definite papers [1.89, 5.5] this issue is still shrouded by a fog of confusion in the literature. The solution is simple and easily arrived at if the precise meaning of the various definitions is respected. It will turn out that the Ito integral has the nicest mathematical properties and that any stochastic integral based on a different choice of α can be expressed as the sum of the corresponding Ito integral plus an ordinary Riemann integral. Hence the Ito integral is the ideal choice for the basis of a stochastic calculus. Therefore, we shall start by presenting the precise definition of the Ito integral for a large class of random processes G_t and expound its main features. After that we shall turn our attention to the Stratonovich integral and discuss the difference as well as the connection with the Ito integral.

5.2 The Ito Integral

Let us start by specifying the class of random processes G_t for which the Ito integral can be defined. Note that in the example of the preceding section, the essential differences between the choice $\alpha = 0$ and $\alpha = 1/2$ are that in the former case, (i) we need to know the Wiener process W_t only up to the instant s, $s \in [t_{i-1}, t_i]$ in order to calculate $W_{\tau_i^{(n)}}$. In fact, knowledge up to t_{i-1} suffices already; (ii) $W_{\tau_i^{(n)}}$ are independent of $[W_{t_i^{(n)}} - W_{t_{i-1}^{(n)}}]$.

Abstracting from this particular example and generalizing these two properties, one is led to consider the class of *nonanticipating* random processes G_t. Roughly speaking, nonanticipating processes are characterized by the property that G_t is known from the past history of the Wiener process up to time t and G_t does not depend on the future of the Wiener process, i.e., G_t is independent of all the increments $W_{t+s} - W_t$ of the Wiener process [$s \in (0, \infty)$].

This concept has been formalized in the following way. Consider the history of the Wiener process, i.e., the sub-σ-field $\mathcal{M}[t_0, t]$, given by

$$\mathcal{M}[t_0, t] = \sigma\{W_\tau, t_0 \leqslant \tau \leqslant t\}.$$

Let $\mathcal{F}_t$ denote the sub-σ-field:

$$\mathcal{F}_t = \sigma\{C, \mathcal{M}[t_0, t]\},$$

where C is a random variable which is independent of the Wiener process. It will serve later on as initial condition for our SDE. Note that the $\mathcal{F}_t$ form a family of increasing sub-σ-fields, i.e., $\mathcal{F}_s \subset \mathcal{F}_t$ for $s < t$. The future of the Wiener process can also be described by a sub-σ-field, namely $\mathcal{M}_t^+$, which is obviously given by,

$$\mathcal{M}_t^+ = \sigma\{W_{t+s} - W_t; 0 \leqslant s \leqslant \infty\}.$$

Since C does not depend on the Wiener process and since W_t has independent increments, the sub-σ-fields $\mathcal{F}_t$ and $\mathcal{M}_t^+$ are independent. The random process G_t is called nonanticipating (w.r.t. $\mathcal{F}_t$) if for all $t \in \theta$, G_t is measurable with respect to $\mathcal{F}_t$. In other words, G_t is not only a random variable w.r.t. the full σ field $\mathcal{A}$ of the underlying probability triple, but remains a random variable if only events associated with the history of the Wiener process are considered. This means that G_t can be a function or functional of the Wiener process up to the time t. For instance, $G_t = \max_{0 < s \leqslant t} W_s$ is a nonanticipating process, however $\tilde{G}_t = \max_{0 < s \leqslant 2t} W_s$ is obviously an anticipating stochastic process.

The Ito integral is defined for all nonanticipating processes whose sample paths are square integrable, i.e.,

$$\int_{t_0}^{t} G_s(\omega)^2 ds < \infty .$$

This integral is understood in the usual (Lebesgue or Riemann) sense. Following the customary procedure, one first defines the Ito integral for a simple case of nonanticipating functions, namely the piecewise constant step functions. Let G_t be a nonanticipating stochastic process, such that there exists a partition $t_0 < t_1 < \ldots < t_n = t$ (the t_i are nonrandom) with

$$G_t(\omega) = G_{t_{i-1}}(\omega) \quad \text{for} \quad t \in [t_{i-1}, t_i) \quad \text{and all} \quad \omega \in \Omega . \tag{5.22}$$

Then G_t is said to be a step function on $[t_0, t]$. We define

$$\int_{t_0}^{t} G_s dW_s = \sum_{i=1}^{n} G_{t_{i-1}}(W_{t_i} - W_{t_{i-1}}) . \tag{5.23}$$

This quantity has the usual linearity property

$$\int_{t_0}^{t} (aG_s^{(1)} + bG_s^{(2)}) dW_s = a \int_{t_0}^{t} G_s^{(1)} dW_s + b \int_{t_0}^{t} G_s^{(2)} dW_s \tag{5.24}$$

with $a, b \in \mathbb{R}$ and $G_s^{(1)}$, $G_s^{(2)}$ being nonanticipating step functions. Furthermore, it follows easily that if $E\{|G_t|\} < \infty$ for $t_0 \leqslant s \leqslant t$ then

$$E\left\{ \int_{t_0}^{t} G_s dW_s \right\} = 0 . \tag{5.25}$$

Indeed,

$$\begin{aligned} E\left\{ \int_{t_0}^{t} G_s dW_s \right\} &= E\left\{ \sum_{i=1}^{n} G_{t_{i-1}}(W_{t_i} - W_{t_{i-1}}) \right\} \\ &= \sum_{i=1}^{n} E\{G_{t_{i-1}}\} E\{(W_{t_i} - W_{t_{i-1}})\} , \end{aligned}$$

since $G_{t_{i-1}}$ is independent of $W_{t_i} - W_{t_{i-1}}$, G_t being a nonanticipating function. Using $E\{(W_{t_i} - W_{t_{i-1}})\} = 0$, (5.25) follows. Note that (5.25) does *not* necessitate that $E\{G_t\} = 0$! Similarly it can be shown that

$$E\left\{ \left(\int_{t_0}^{t} G_s dW_s \right)^2 \right\} = \int_{t_0}^{t} E\{G_s^2\} ds . \tag{5.26}$$

The definition of the Ito integral for an arbitrary nonanticipating random process G_t relies on the fact that such processes can be approximated arbitrarily well by step functions. The following theorem holds: Let G_t be a nonanticipating random process with

$$\int_{t_0}^{t} E\{G_s^2\} ds < \infty . \tag{5.27}$$

Then a series of nonanticipating step functions $G_t^{(n)}$ exists such that

$$\lim_{n\to\infty} \int_{t_0}^{t} E\{(G_s^{(n)} - G_s)^2\} ds = 0 , \tag{5.28}$$

i.e., the "distance" between G_t and $G_t^{(n)}$ tends to zero for $n \to \infty$. This approximation by step functions is used to define the Ito integral in general. It can be shown that

$$\operatorname{qm-lim}_{n\to\infty} \int_{t_0}^{t} G_s^{(n)} dW_s \tag{5.29}$$

exists and is *independent* of the particular series of step functions $G_t^{(n)}$ used to approximate G_t. This means that if $\tilde{G}_t^{(n)}$ is another series of nonanticipating step functions fulfilling (5.28), then

$$\operatorname{qm-lim}_{n\to\infty} \int_{t_0}^{t} G_s^{(n)} dW_s = \operatorname{qm-lim}_{n\to\infty} \int_{t_0}^{t} \tilde{G}_s^{(n)} dW_s . \tag{5.30}$$

It is therefore meaningful to define the Ito integral of G_t by

$$\int_{t_0}^{t} G_s dW_s = \operatorname{qm-lim}_{n\to\infty} \int_{t_0}^{t} G_s^{(n)} dW_s . \tag{5.31}$$

Obviously, the properties (5.24 – 26) carry over to this limit. The Ito integral can even be defined for nonanticipating random processes which do not satisfy (5.27). The convergence is however in a weaker sense, namely only in probability and not in the mean square:

$$\lim_{n\to\infty} P\left[\left| \int_{t_0}^{t} G_s^{(n)} dW_s - \int_{t_0}^{t} G_s dW_s \right| > \varepsilon \right] = 0 . \tag{5.32}$$

This is denoted by

$$\int_{t_0}^{t} G_s dW_s = \operatorname{st-lim} \int_{t_0}^{t} G_s^{(n)} dW_s . \tag{5.33}$$

If the nonanticipating random process G_t has *almost surely continuous realizations*, then the following particular choice can be made for the approximating step functions $G_t^{(n)}$:

$$G_t^{(n)} = \sum_{i=1}^{n} G_{t_{i-1}} I_{[t_{i-1}, t_i)}(t) , \tag{5.34}$$

that is

$$\int_{t_0}^{t} G_t dW_t = \text{st-lim} \sum_{i=1}^{n} G_{t_{i-1}}(W_{t_i} - W_{t_{i-1}}) \tag{5.35}$$

and if (5.27) holds,

$$\int_{t_0}^{t} G_t dW_t = \text{qm-lim} \sum_{i=1}^{n} G_{t_{i-1}}(W_{t_i} - W_{t_{i-1}}) . \tag{5.36}$$

Hence, if the stochastic nonanticipating process G_t has almost surely continuous realizations, then the Ito integral can indeed be defined as the limit (in probability or mean square) of the approximating sums, where the evaluation point τ_i is fixed to be the left-hand point of the partition: $\tau_i = t_{i-1}$.

Let us now define a random process Y_t by

$$Y_t(\omega) = \int_{t_0}^{t} G_s(\omega) dW_s(\omega) . \tag{5.37}$$

Then the nice properties of the Ito integral can be summarized as follows:

i) Y_t is a *nonanticipating process*;
ii) Y_t has *almost surely continuous sample paths*;
iii) If $\int_{t_0}^{t} E\{G_s^2\} ds < \infty$ for $t \in \mathbb{R}$ then

$$E\{Y_t\} = 0 \tag{5.38}$$

and

$$E\{Y_t Y_s\} = \int_{t_0}^{\min(t,s)} E\{G_u^2\} du ; \tag{5.39}$$

iv) $E\{Y_t | \mathscr{F}_s\} = Y_s$.

The last property is known as the martingale property. Besides Markov processes, martingales are the class of random processes that are extremely well studied from a mathematical viewpoint and which have gained considerable importance in recent years. For details on connections between martingales and diffusion processes, see [5.6].

The only point that might be slightly disturbing in a first encounter with the Ito integral is that it does not obey the rules of classical calculus as mentioned above. This is however only a question of habitude. To explain the fundamental rule of the Ito calculus, which replaces the chain rule of the classical calculus, it is convenient to introduce the notion of a stochastic differential. The relation (5.37) can be written in abbreviated form as

$$dY_t(\omega) = G_t(\omega) dW_t(\omega) . \tag{5.40}$$

This is a special form of a stochastic differential. In view of the fact that all these considerations were motivated by our intention to give a precise meaning to (5.1), consider the more general process X_t defined as:

$$X_t(\omega) = X_{t_0}(\omega) + \int_{t_0}^{t} F_s(\omega)\,ds + \int_{t_0}^{t} G_s(\omega)\,dW_s(\omega)\,, \tag{5.41}$$

where X_{t_0} is a random variable that is independent of $W_s - W_{t_0}, s \geqslant t_0$, and F_s is, like G_s, a given nonanticipating random process. Further, F_s is required to fulfill the condition $\int_{t_0}^{t} |F_s(\omega)|\,ds < \infty$, i.e., the first integral can be interpreted as an ordinary Riemann integral for the sample paths. The last one is an Ito integral. The process X_t is then nonanticipating. It has almost surely continuous sample paths and fulfills for $t_0 \leqslant s \leqslant t \leqslant T$ the relation:

$$X_t = X_s + \int_{s}^{t} F_{s'}\,ds' + \int_{s}^{t} G_{s'}\,dW_{s'}\,. \tag{5.42}$$

By definition, we shall say that such a process has the following *stochastic differential*:

$$dX_t = F_t\,dt + G_t\,dW_t\,. \tag{5.43}$$

Consider a stochastic process Z_t which is a function of the process X_t:

$$Z_t = u(t, X_t)\,, \tag{5.44}$$

where $u = u(t,x)$ is a real-valued function and has the continuous derivatives $\partial_t u$, $\partial_x u$ and $\partial_{xx} u$. Then the *Ito rule* states that Z_t has the following stochastic differential

$$dZ_t = [\partial_t u(t,X_t) + F_t \partial_x u(t,X_t) + \tfrac{1}{2} G_t^2 \partial_{xx} u(t,X_t)]\,dt + G_t \partial_x u(t,X_t)\,dW_t\,. \tag{5.45}$$

Roughly speaking, this means that in the Taylor expansion of $u(t,X_t)$ second-order terms have to be retained, since according to (2.110)

$$\text{“}(dW_t)^2 = dt\text{”}\,.$$

With the Ito rule (5.45), a consistent calculus of stochastic differentials can be constructed, referred to as the Ito calculus in the following.

5.3 Ito Stochastic Differential Equations and Diffusion Processes

This calculus constitutes the framework in which a precise mathematical meaning can be given to (5.1). Note that in (5.1) the functions f and g are *nonrandom* functions for fixed t and x. Randomness enters only indirectly via the random process X_t. In the following, we shall consider stochastic differentials of this form only. They are largely sufficient for our purposes to describe systems coupled to a Gaussian white-noise environment. We shall say that the stochastic differential of the form

$$dX_t = f(t, X_t)\,dt + g(t, X_t)\,dW_t \qquad X_{t_0} = C, \tag{5.46}$$

where f and g are real-valued smooth functions, is an Ito SDE with initial condition C(C is a random variable which is independent of $W_s - W_{t_0}$ for all $s \geqslant t_0$). The random process X_t is a solution of this SDE on the time interval $[t_0, T]$ if:

i) X_t is nonanticipating.

ii) The functions $F_t(\omega) = f(t, X_t(\omega))$ and $G_t(\omega) = g(t, X_t(\omega))$, which are nonanticipating because of (i), obey with probability one:

$$\int_{t_0}^{T} |F_t(\omega)|\,dt < \infty, \qquad \int_{t_0}^{T} G_t^2(\omega)\,dt < \infty.$$

This condition ensures that the right-hand side of (5.46) is well defined, as is clear in view of the definition of the Ito integral.

iii) $X_t = C + \int_{t_0}^{t} f(s, X_s)\,ds + \int_{t_0}^{t} g(s, X_s)\,dW_s$ (5.47)

holds with probability one for every $t \in [t_0, T]$.

It follows trivially from property (ii) of (5.37) that X_t has almost surely continuous sample paths. (To be precise, a version of X_t exists which does and it is the one we shall exclusively consider in the following.) Thus we have achieved our task to give a rigorous mathematical meaning to the phenomenological equations (3.31) and (4.13) respectively, modeling a system coupled to an extremely rapidly fluctuating environment. Let us now explore the main properties of a random process X_t defined by an Ito SDE. We expect it of course to be a diffusion process, but before ascertaining that, it is necessary to enquire as to the existence and uniqueness of a solution of (5.46).

5.3.1 Existence and Uniqueness of Solutions

In most applications we deal with autonomous SDE's, i.e., f and g do not depend explicitly on time. For this class of SDE's, the following rather general results concerning the existence and uniqueness hold [5.7]. If $f(x)$ and $g(x)$ are continuously differentiable functions [in short hand notation $f, g \in C^1(\mathbb{R})$], then

$$dX_t = f(X_t)\,dt + g(X_t)\,dW_t, \qquad X_{t_0} = C \tag{5.48}$$

has a unique solution which is defined up to some random explosion time η with $t_0 < \eta \leqslant \infty$. This condition on f and g is fulfilled in almost all applications. Indeed, f and g are commonly polynomials in x and hence $f, g \in C^1(\mathbb{R})$. The problem to determine the explosion time η, especially to decide if $\eta = \infty$ with probability one, i.e., no explosion, occurs, will be addressed later on in this chapter. Let us only remark that commonly in the mathematical literature the so-called growth condition is imposed:

$$f^2(t,x) + g^2(t,x) \leqslant K^2(1+x^2) , \tag{5.49}$$

with some positive constant K for all $x \in \mathbb{R}$. Since this implies that f and g do not grow faster than linearly with x, it is much too restrictive for applications. However, if (5.49) holds, then $\eta = \infty$ with probability one.

5.3.2 Markov Property of Solutions

Let f and g be such that a unique solution X_t of the Ito SDE (5.48) exists with initial condition $X_{t_0} = C$. According to (5.42), we have for the solution X_t of (5.48) the following relation:

$$X_t = X_s + \int_s^t f(X_u)\,du + \int_s^t g(X_u)\,dW_u \tag{5.50}$$

for $s \geqslant t_0$. Adopting the notation $X_t(C, t_0)$ to indicate the initial condition of the solution, (5.50) says that

$$X_t(C, t_0) = X_t(X_s, s) , \qquad t \geqslant s . \tag{5.51}$$

Let us emphasize again that C is in general a random variable. This is essential if, for instance, we want to obtain a stationary stochastic process as the solution of the SDE. Then it is necessary to start with an initial condition that is distributed according to the stationary probability density of X_t. If the solution of a SDE were Markovian, then it would obviously also be a sufficient condition to start with such a C, for X_t to be a stationary random process. This brings us to the question: do solutions of Ito SDE's possess the Markov property? Relation (5.50) persuasively suggests that they do indeed. Starting the random process $X_t(C, t_0)$, which is a solution of an Ito SDE, at time t_0 with the initial condition C and letting it evolve freely is equivalent to starting it at a later time s with the random variable X_s, to which it has evolved from $X_{t_0} = C$. In other words, if the random process is known at time s, i.e., the random variable X_s, then we need no information on the past to predict the future behavior of the solution of the Ito SDE. This leads us to conclude that if the solution of the Ito SDE (5.48) is unique, then the general property (5.42) of Ito SDE's should guarantee the Markov property of solutions. In fact, the following theorem holds: If the solution X_t of the Ito SDE exists and is unique, then X_t is a Markov process and its transition function $P(B, t \mid x, s)$ is given by:

$$P(B,t|x,s) = P(X_t \in B | X_s = x) = P(X_t(x,s) \in B) , \tag{5.52}$$

where $X_t(x,s)$ is the (unique) solution with initial condition $X_s = x$. The important feature of the above relation is that the transition function is given by an unconditional probability. The probability to find the process $X_t(C, t_0)$ at time t in the set B, conditioned by the fact that at some intermediate time s it took the value x, is equal to the probability that the process $X_t(x,s)$, starting at time s in x, is in B at time t. In the light of the above discussion it is intuitively obvious that the stated theorem is true. The formal proof of this fact is surprisingly technical. We prefer therefore to continue the exposition of Ito SDE's and to present an outline of the essential ideas in the rigorous proof of the Markov property in App. B.

5.3.3 Ito Equations and the Fokker-Planck Equation

Since the solution of an Ito SDE has almost surely continuous sample paths, we expect that X_t is not only Markovian but also a diffusion process. This could be established by verifying that X_t fulfills conditions ($a' - c'$), namely (4.21, 26 and 27). This is, however, rather tedious so we shall choose a different method to establish that X_t is a diffusion process. *McKean* [5.7] has given a most elegant and direct proof that the transition probability density of X_t exists and fulfills a FPE. This is of course sufficient to establish the diffusion character of the solution of (5.48). We shall now present McKean's beautiful proof (in a slightly restricted form, namely we assume $\eta = \infty$ with probability one, which is sufficient for our purposes).

Let X_t be a solution of the SDE (5.48), where now $f(x)$ and $g(x)$ are supposed to fulfill the stronger condition of being infinitely often differentiable[1] [denoted $C^\infty(\mathbb{R})$]. Then the transition probability density $p(y,t|x,s)$ exists and fulfills the FPE. The proof goes as follows.

Let $v(x,t)$ be an arbitrary infinitely often differentiable function with respect to x and t, which vanishes identically outside a bounded interval of the form $[x_1, x_2] \times [t_1, t_2]$, i.e.,

$$v(x,t) = 0 \quad \text{if} \quad x \notin [x_1, x_2] \quad \text{or} \quad t \notin [t_1, t_2] \quad (t_1 > 0) . \tag{5.53}$$

We shall say v has compact support. Since f and g do not depend on time, the process is obviously time homogeneous and we choose $t_0 = 0$ and $X_0 = x$.

Then consider the random process

$$Y_t = v(X_t, t) . \tag{5.54}$$

---

[1] This is still fulfilled in most applications. Let us mention, for completeness sake, that this is of course only a sufficient but not necessary condition for the solution of an Ito SDE to be a diffusion process. In particular, as is clear from the derivation of FPE, there are diffusion processes for which no FPE exists.

Using the Ito rule (5.45), we can calculate the stochastic differential of the process Y_t:

$$\begin{aligned} dY_t &= dv(X_t, t) \\ &= [\partial_t v(X_t,t) + f(X_t)\,\partial_x v(X_t,t) + \tfrac{1}{2} g^2(X_t)\,\partial_{xx} v(X_t,t)]\,dt \\ &\quad + g(X_t)\,\partial_x v(X_t,t)\,dW_t . \end{aligned} \tag{5.55}$$

The equivalent integral form reads

$$\begin{aligned} v(X_{t'}, t') - v(X_0, 0) = \int_0^{t'} [\partial_t v(X_t,t) + f(X_t)\,\partial_x v(X_t,t) \\ + \tfrac{1}{2} g^2(X_t)\,\partial_{xx} v(X_t,t)]\,dt + \int_0^{t'} g(X_t)\,\partial_x v(X_t,t)\,dW_t . \end{aligned} \tag{5.56}$$

Since $v(x,t)$ has compact support it follows that

$$\begin{aligned} &\text{i)}\ \ v(x,0) = 0 , \\ &\text{ii)}\ \ v(x,\infty) = 0 , \\ &\text{iii)}\ \ \int_0^{\infty} E\{(g(X_t)\,\partial_x v(X_t,t))^2\}\,dt < \infty . \end{aligned} \tag{5.57}$$

Thus according to (5.38)

$$E\left\{ \int_0^{\infty} g(X_t)\,\partial_x v(X_t,t)\,dW_t \right\} = 0 . \tag{5.58}$$

We let t' tend to infinity in (5.56) and then take the expectation to obtain:

$$0 = \int_0^{\infty} E\{[\partial_t v(X_t,t) + f(X_t)\,\partial_x v(X_t,t) + \tfrac{1}{2} g^2(X_t)\,\partial_{xx} v(X_t,t)]\}\,dt . \tag{5.59}$$

If the transition probability $P(B, t\,|\,X_s, s)$ of the process X_t, which is the solution of the SDE (5.48) with $f(x)$, $g(x) \in C^{\infty}(\mathbb{R})$, possesses a density then the expectation can be written using $p(y, t\,|\,x, 0)$:

$$0 = \int_0^{\infty} dt \int_{\mathbb{R}} dy\, p(y,t\,|\,x,0)\, [\partial_t v(y,t) + f(y)\,\partial_y v(y,t) + \tfrac{1}{2} g^2(y)\,\partial_{yy} v(y,t)] .$$

Integration by parts yields

$$\begin{aligned} 0 = \int_0^{\infty} dt \int_{\mathbb{R}} dy\, v(y,t) [-\partial_t p(y,t\,|\,x,0) - \partial_y f(y) p(y,t\,|\,x,0) \\ + \tfrac{1}{2} \partial_{yy} g^2(y) p(y,t\,|\,x,0)] . \end{aligned} \tag{5.60}$$

The boundary terms vanish since $v(y,t)$ has compact support. Here $v(y,t)$ is an arbitrary function and (5.60) implies therefore

$$\partial_t p(y,t|x,0) = -\partial_y f(y) p(y,t|x,0) + \tfrac{1}{2}\partial_{yy} g^2(y) p(y,t|x,0) , \tag{5.61}$$

i.e., the transition probability density of X_t satisfies a FPE. This result holds if the process X_t indeed possesses a smooth, i.e., twice differentiable, transition probability density $p(y,t\,|\,x,0)$, a point that remains to be verified. According to Weyl's lemma [5.7], (5.59) also implies the existence of an infinitely often differentiable transition probability density $p(y,t\,|\,x,0)$. The above derivation of the FPE rests therefore on mathematically solid ground and X_t is thus a diffusion process. McKean's proof reveals most strikingly the close connection between solutions of Ito SDE's and diffusion processes. In the light of the above proof, it is hardly exaggerated to say that the FPE for X_t is a direct consequence of the Ito rule. If we look at it the other way around, the unfamiliar Ito rule now becomes familiar and it is just the way the FPE transforms under a change of variables.

Let us remark without proof that if f and g depend explicitly on time then the SDE defines a diffusion process, if both functions are continuous in t, in addition to the requirements for existence and uniqueness of the solution. This establishes the fundamental result that under mild conditions, as far as applications are concerned, on f and g, the solutions of SDE's and diffusion processes are one and the same class of stochastic processes. This implies that any Markov process with almost surely continuous sample paths fulfilling conditions (4.16, 19, 20), which are no restrictions in applications, can be described by an Ito SDE of the form

$$dX_t = f(X_t, t)\,dt + \sigma g(X_t, t)\,dW_t . \tag{5.62}$$

The important point is that this stochastic equation is always *linear* in dW_t. This shows that there is a deep intrinsic reason for (3.31) to be linear in the Gaussian white noise ξ_t, apart from the purely technical reason that a product of two generalized functions cannot be defined. This holds also for the more general Markov processes briefly discussed at the end of Chap. 4. One can say that the linearity of the SDE in the white noise and the Markov property are inseparable.

5.4 Stratonovich Stochastic Integral

Let us now consider the second definition of a stochastic integral which plays a certain role in the literature, namely that derived by Stratonovich. The Ito integral is appealing because of its nice mathematical properties, especially the close connection between Ito SDE's and diffusion processes. The success of the Stratonovich integral stems from the fact that it is particularly well suited for certain needs in modeling physical systems. This point will be discussed in detail once a proper definition has been given. However, to sketch the situation briefly, note that the Ito integral implies that there is no dependence between the process

X_t and the random force $\dot{W}_t$ at the same instant of time t. Choosing the left points of the partition as evaluation points exploits, so to speak, maximally the fact that the Wiener process has independent increments. While this definition is certainly the only reasonable one, as already mentioned, if we deal with truly white noise one should not forget that white noise is an idealization for many applications to real systems.[2] A SDE with a white-noise term is obtained, as we discussed, via a limiting procedure. The starting point is a situation with real noise, having a short but nevertheless nonzero correlation time. This short but finite memory of the noise leads to a dependence between the process X_t and the random force ζ_t at the same instant of time t. As we discussed above, the limit from the real noise to the white-noise idealization has to be effected carefully, such that, roughly speaking, the impact of the external noise is conserved. We expect that a trace of the dependence between X_t and ζ_t should survive in the white-noise limit. We shall see that the Stratonovich integral takes into account just such a correlation between the system and the external noise. It is, so to speak, obtained as the white-noise limit of a real noise problem. Clearly, this is a problem of the modeling procedure: what SDE and thus what random process X_t describes most appropriately the system in the white-noise idealization of an extremely rapidly fluctuating environment.

5.4.1 Definition of the Stratonovich Integral and Its Relation with the Ito Integral

Before discussing these questions any further, it is necessary to give a precise definition of the Stratonovich integral (essentially following *Arnold* [2.2]). So far we have considered only scalar random processes. However, it will become apparent in the following that in order to be able to define a Stratonovich SDE for a scalar process, we need to consider the integral definition for a two-dimensional vector process. This is due to the fact that the approximating sums of the Stratonovich integral are slightly anticipating. In order to find the approximation to $G(X_s)$ for $s \in [t_{i-1}, t_i)$ in the approximating sum, it is not sufficient to know the Wiener process up to time s. Knowledge of the future behavior, up to t_i, is needed, since $X_{\tau_i^{(n)}} = (X_{t_i} + X_{t_{i-1}})/2$.

Let $\boldsymbol{Y}_t$ be a two-dimensional diffusion process[3], $\boldsymbol{Y}_t = \mathrm{col}(Y_t^{(1)}, Y_t^{(2)})$ with drift vector $\boldsymbol{f}(\boldsymbol{Y})$ and the diffusion matrix $\underset{\sim}{\boldsymbol{B}}(\boldsymbol{Y})$. Let $\boldsymbol{H}(\boldsymbol{x})$ be a two-dimensional real-valued vector function with continuous partial derivatives $\partial_{x_i}\boldsymbol{H}(\boldsymbol{x})$; $i = 1, 2$, and

---

[2] In biological applications, e.g., the ecology of insects with non overlapping generations, often a discrete time description is most appropriate. If the environmental fluctuations have a correlation time that is much shorter than the generation span, then the external parameter can be modeled by a random process with independent values for each generation. This is genuine discrete white noise.

[3] An n-dimensional diffusion process $Y_t = \mathrm{col}(Y_t^{(1)}, \ldots, Y_t^{(n)})$ is defined in a straightforward extension of the one-dimensional case as a Markov process which fulfills the conditions:

a) $\lim\limits_{t \downarrow s} \dfrac{1}{t-s} \displaystyle\int_{|y-s| \geqslant \varepsilon} p(y,t|x,s)\,dy = 0$;

$$\int_{t_0}^{t} E\{|H^{\mathrm{tr}}(Y_s)f(Y_s)|\}ds < \infty \tag{5.63}$$

and

$$\int_{t_0}^{t} E\{|H^{\mathrm{tr}}(Y_s)\underset{\sim}{B}(Y_s)H(Y_s)|\}ds < \infty\,. \tag{5.64}$$

Then the limit

$$\int_{t_0}^{t} H^{\mathrm{tr}}(Y_s)\circ dY_s = \underset{\delta_n\to 0}{\text{qm-lim}} \sum_{i=1}^{n} H^{\mathrm{tr}}\left(\frac{Y_{t_{i-1}}+Y_{t_i}}{2}\right)(Y_{t_i}-Y_{t_{i-1}})\,, \tag{5.65}$$

where $t_0 \leqslant t_1 < \ldots < t_n = t$ is a partition of $[t_0, t]$ with $\delta_n = \max(t_i - t_{i-1})$, is called a stochastic integral in the sense of Stratonovich. It is connected with the Ito integral, here defined as

$$\int_{t_0}^{t} H^{\mathrm{tr}}(Y_s)\,dY_s = \underset{\delta_n\to 0}{\text{qm-lim}} \sum_{i=1}^{n} H(Y_{t_{i-1}})(Y_{t_i}-Y_{t_{i-1}})\,, \tag{5.66}$$

via the formula

$$\int_{t_0}^{t} H^{\mathrm{tr}}(Y_s)\circ dY_s = \int_{t_0}^{t} H^{\mathrm{tr}}(Y_s)\,dY_s + \frac{1}{2}\sum_{j=1}^{2}\sum_{k=1}^{2}\int_{t_0}^{t} \partial_{x_k} H_j^{\mathrm{tr}}(\underset{\sim}{B})_{jk}\,ds\,. \tag{5.67}$$

This is easily verified by making a Taylor expansion of $H^{\mathrm{tr}}[(Y_{t_{i-1}}+Y_{t_i})/2]$ around $H^{\mathrm{tr}}(Y_{t_{i-1}})$ and using the definition of the diffusion matrix.

In order to define the Stratonovich stochastic differential equation, we set

$$\int_{t_0}^{t} G(X_s)\circ dW_s \equiv \int_{t_0}^{t} (0, G(X_s))\circ \begin{pmatrix} dX_s \\ dW_s \end{pmatrix}. \tag{5.68}$$

In the following, the Stratonovich integral will always be distinguished from an Ito integral by writing $\circ$ between integrand and dW_t. We say of a process X_t fulfilling

---

b) there exists an $\mathbb{R}^n$-valued function $f(s,x)$ such that

$$\lim_{t\downarrow s} \frac{1}{t-s} \int_{|y-x|\leqslant \varepsilon} (y-x)p(y,t|x,s)\,dy = f(s,x)\,;$$

c) there exists $n\times n$ matrix $\underset{\sim}{B}(s,x)$ such that

$$\lim_{t\downarrow s} \frac{1}{t-s} \int_{|y-x|\leqslant \varepsilon} (y-x)(y-x)^{\mathrm{tr}}p(y,t|x,s) = \underset{\sim}{B}(s,x)\,.$$

$|y|^2 = \sum y_i^2$ is the Euclidean norm of $\mathbb{R}^n$ and y^{tr} denotes the transpose of y.

$$X_t = X_{t_0} + \int_{t_0}^{t} f(X_s)\,ds + \sigma \int_{t_0}^{t} g(X_s) \circ dW_s \tag{5.69}$$

that it obeys the Stratonovich stochastic differential equation (S SDE)

$$dX_t = f(X_t)\,dt + \sigma g(X_t) \circ dW_t . \tag{5.70}$$

The Ito SDE for the composite diffusion process $\begin{pmatrix} X_t \\ W_t \end{pmatrix}$ reads of course:

$$\begin{aligned} dX_t &= f(X_t)\,dt + \sigma g(X_t)\,dW_t \\ dW_t &= dW_t , \end{aligned} \tag{5.71}$$

or in matrix notation

$$\begin{pmatrix} dX_t \\ dW_t \end{pmatrix} = \begin{pmatrix} f(X_t) \\ 0 \end{pmatrix} dt + \begin{pmatrix} \sigma g(X_t) & 0 \\ 1 & 0 \end{pmatrix} \begin{pmatrix} dW_t \\ dV_t \end{pmatrix} , \tag{5.72}$$

where V_t is a dummy Wiener process. For the diffusion matrix $\underset{\sim}{\boldsymbol{B}}$ we have

$$\underset{\sim}{\boldsymbol{B}} = \begin{pmatrix} \sigma g(x) & 0 \\ 1 & 0 \end{pmatrix} \begin{pmatrix} \sigma g(x) & 1 \\ 0 & 0 \end{pmatrix} = \begin{pmatrix} \sigma^2 g^2(x) & \sigma g(x) \\ \sigma g(x) & 0 \end{pmatrix} . \tag{5.73}$$

Hence the transformation formula (5.67) yields in this case:

$$\int_{t_0}^{t} \sigma g(X_s) \circ dW_s = \int_{t_0}^{t} \sigma g(X_s)\,dW_s + \frac{\sigma^2}{2} \int_{t_0}^{t} g'(X_s) g(X_s)\,ds . \tag{5.74}$$

This implies that the following Ito SDE

$$dX_t = \left[f(X_t) + \frac{\sigma^2}{2} g'(X_t) g(X_t) \right] dt + \sigma g(X_t)\,dW_t \tag{5.75}$$

is equivalent to the Stratonovich SDE (5.70) in the sense that they have one and the same solution. Let us discuss the main consequences of this transformation formula.

i) It is always possible to change from the Stratonovich interpretation of a SDE to the Ito interpretation by adding $\sigma^2 g' g\,dt/2$ or from the Ito version to the Stratonovich version by subtracting $\sigma^2 g' g\,dt/2$.

ii) The Stratonovich SDE (5.70) defines a diffusion process with drift $f + \sigma^2 g' g/2$ and diffusion g^2. This follows from (5.75).

iii) In the case of additive noise, i.e., $g(x) = \text{const}$, there is no difference between the Ito and the Stratonovich integral.

iv) In the case of multiplicative noise, i.e., $g(x) \neq \text{const}$, where the influence of the random force depends on the state of the process, the correlation between the process X_t and the random force W_t, implicit in the Stratonovich integral, leads to a systematic contribution in the evolution of X_t. It gives rise to the noise-induced drift $\sigma^2 g' g/2$. This term is also known as spurious drift since it does not appear in the phenomenological equation. There is, however, nothing spurious in this term, it has physical consequences.

v) The Stratonovich calculus obeys the classical chain rule. This point is verified, using the transformation formula (5.75), in App. C.

5.4.2 Ito or Stratonovich: A Guide for the Perplexed Modeler

As mentioned earlier, the importance of the Stratonovich integral stems from the fact that it models directly the correlations between the random environment and the system. These correlations in the white-noise idealization are a trace of the stochastic dependence between the state of the system and the environmental fluctuations when the latter have a nonvanishing correlation time. Indeed, X_t is a function of the history of the environment, $X_t = F[\lambda + \zeta_s, t_0 \leqslant s \leqslant t]$ and if $C(\tau) = E\{\zeta_s \zeta_{s+\tau}\}$ is not equal to zero for all $\tau > 0$ then X_t and ζ_t are dependent stochastic variables. However, all other choices of the evaluation point τ_i that differ from Ito's choice lead to correlations between the process and the external noise. That Stratonovich's choice models most directly the physical situation is already suggested by the fact that it is the only definition of a stochastic integral leading to a calculus with classical rules. This fact was first established by the following theorem of *Wong* and *Zakai* [5.8]. We shall not present its proof here since it is mainly quoted for historical reasons. Wong and Zakai's result will be obtained in a far easier way and under more general conditions in Chap. 8, where the neighborhood of white noise is studied via a perturbation method, inspired by techniques of *Blankenship* and *Papanicolaou* [5.9].

Wong-Zakai Theorem. Let $W_t^{(n)}$ be a sequence of random processes, which are continuous, of bounded variation, having piecewise continuous derivative and converging almost surely uniformly to the Wiener process W_t. Then under some mild conditions on f and g, the solutions of

$$dX_t^{(n)} = f(X_t^{(n)})\,dt + \sigma g(X_t^{(n)})\,dW_t^{(n)}\,, \tag{5.76}$$

where all integrals can be understood as ordinary Riemann integrals, converge almost surely uniformly to the solution of the following Ito SDE:

$$dX_t = \left[f(X_t) + \frac{\sigma^2}{2} g'(X_t) g(X_t) \right] dt + \sigma g(X_t)\,dW_t\,. \tag{5.77}$$

According to the transformation rule (5.75), the following *Stratonovich* SDE is equivalent to (5.77):

$$dX_t = f(X_t)\,dt + \sigma g(X_t) \circ dW_t\,. \tag{5.78}$$

This implies that if we start out with a phenomenological equation, containing real noise of the above form, i.e.

$$\dot{X}_t^{(n)} = f(X_t^{(n)}) + \sigma g(X_t^{(n)}) \dot{W}_t^{(n)} \tag{5.79}$$

and we pass to the white-noise limit so that an SDE of the form

$$\dot{X}_t = f(X_t) + \sigma g(X_t) \dot{W}_t \tag{5.80}$$

is obtained, the latter has then to be interpreted as a Stratonovich equation[4]. For practical calculations, it is convenient to switch over from the S SDE to the equivalent Ito form (5.77). The main advantage of the Ito version (5.77) of SDE (5.80) is the direct display of the characteristics of the diffusion process X_t by which a system coupled to an environment with extremely rapid fluctuations has to be modeled. Namely, the drift consists of the phenomenological part $f(x)$ plus the noise-induced drift $\sigma^2 g' g/2$; the latter indicates that in a fluctuating environment the systematic motion is modified also if the noise is only approximately white. The diffusion is given by that part of the phenomenological equation that multiplies the fluctuating parameter.

The unfamiliar situation that the calculus of differential equations with white-noise coefficients can be based on different definitions of the stochastic integrals, involved in solving such equations, has led to an astonishingly long-lived controversy in the physical literature as to which is the right definition. A lot of confusion has been created. We have the impression that quite a few authors were led astray by at least implicitly assuming some special intrinsic value for the classical rules of calculus. They have, however, none. Often a principle of invariance of the equations under "coordinate transformation" $y = u(x)$ is invoked to pick the Stratonovich integral as the "right" one and reject the Ito integral as the "wrong" one. It should be realized that this principle refers to an invariance of the form of the SDE under a nonlinear transformation of the system. Though it is a particular feature of the S SDE to be invariant, there is no physical virtue in this form invariance. It is only a different guise of the fact that the Stratonovich calculus obeys the familiar classical rules. The only quantities that have to be invariant under a transformation $y = u(x)$, where u is one to one, are the probabilities, as for instance

$$p(y,t)\,dy = p(x,t)\,dx\,. \tag{5.81}$$

This is of course guaranteed in *both* calculi.[5]

---

[4] In applications, the real noise will in general not be of the form $W_t^{(n)}$. However, as already mentioned, we shall see in Chap. 8 that the Wong and Zakai result holds for a large class of realistic nonwhite noises.

[5] Furthermore, for nonlinear systems there exist, so to speak, natural variables, namely the physical quantities that are measured (concentration, voltage conductance, etc.) and which display the nonlinear dynamics of the system. A nonlinear transformation to new "artificial" variables, which disguises the nonlinearities of the system, has physically only little value. As an illustration consider

The above presentation of the Ito and Stratonovich integrals shows clearly that there is no question which of the two integrals is the right one; both lead to a *consistent* calculus. The question is, how to interpret an SDE with Gaussian white noise that has been obtained via some limiting procedure, or to rephase the question, what are the coefficients of the diffusion process that models most adequately the system we want to describe. It is at this point that physical arguments have to be invoked to determine the appropriate drift and diffusion coefficient of the process. Formulated in this way, it becomes clear that the question Itò versus Stratonovich is void, as also stressed by *van Kampen* [5.10]; a diffusion process can be described either way. The Ito versus Stratonovich controversy has also been dealt with in a clear way by *Jazwinski* [5.11, p. 140]: "One might argue that the Stratonovich integral should be used because it is simpler, since it can be manipulated by the formal rules. This is intuitively appealing but beside the point. The Stratonovich integral does not offer any new mathematical insight or content. As a matter of fact, all the results concerning the Stratonovich integral are proved using Ito's theory. Of itself, the Stratonovich integral does not offer any additional physical insight. It does, under certain conditions, more directly model a physical process" and "Most important of all is the fact that the Ito integral is defined for a much broader class of functions than the Stratonovich integral."

Clearly the basic problem is *not* what is the right integral definition, but *how do we model real systems by diffusion processes.* If the starting point is a phenomenological equation in which some fluctuating parameters are approximated by Gaussian white noise, then the most appropriate diffusion process is the one that is defined by the Stratonovich interpretation of this equation.

On the other hand, for a broad class of biological and other systems, e.g., insect populations with nonoverlapping generations, the appropriate starting point is a discrete time description, such as:

$$X_{t_i} = X_{t_{i-1}} + f(X_{t_{i-1}})\Delta t + \sigma g(X_{t_{i-1}}) Q_{t_{i-1}}, \tag{5.82}$$

where $t_i = t_{i-1} + \Delta t$ for all i and the Q_{t_i} are Gaussian independent random variables with

$$E\{Q_{t_i}\} = 0 \tag{5.83}$$

and

$$E\{Q_{t_i}^2\} = \Delta t . \tag{5.84}$$

If times are considered that are large compared to the generation length Δt, then it is permissible to pass to the continuous time limit:

$$\dot{X}_t = f(X_t) + \sigma g(X_t)\dot{W}_t . \tag{5.85}$$

---

the simple case of a population undergoing Malthusian growth, $\dot{X} = \lambda X$ and $X(t) = X(0)\exp(\lambda t)$. It is hardly "comforting" for the population that its logarithm $\ln X(t)$ explodes only linearly with time.

It is obvious from the asymmetric form of (5.82) with respect to time that in this case the most appropriate diffusion process is the one that is defined by the Ito interpretation of (5.85). Thus we have two rough guidelines to direct our choice of the best diffusion model for a real system. If the stochastic differential equation has been obtained as the white-noise limit of a real noise equation, choose the Stratonovich interpretation; if it represents the continuous time limit of a discrete time problem, interpret it according to Ito. In any case, the ultimate test is the confrontation of the analytical results with the experimental facts. There are no universally valid theoretical a priori reasons why one or the other interpretation of an SDE should be preferred.

5.5 Classification of the Boundaries of a Diffusion Process

Our basic stochastic model describing the influence of extremely rapid environmental fluctuations is to represent the temporal evolution of the system by a diffusion process with a certain drift coefficient f and a certain diffusion coefficient $\sigma^2 g^2$. Both coefficients are uniquely determined by the phenomenological rate equation (1.13). The diffusion coefficient reflects the way in which the system is coupled to the environment and coincides indeed, as suggested by the notation, with the second term of the rhs of (1.13). The drift coefficient is either given by the first term on the rhs of (1.13), i.e., coincides with the deterministic evolution for a given value of the external parameter $\lambda + \zeta_t$ at each instant of time, or contains in addition the noise-induced drift. As explained above, a diffusion process with coefficients f and $\sigma^2 g^2$ can always be represented by the following Ito equation[6]:

$$dX_t = f(X_t)\,dt + \sigma g(X_t)\,dW_t . \tag{5.86}$$

In physical applications, the state space of X_t rarely coincides with the real line $\mathbb{R}$. The process is often restricted to some subset, in most cases the non-negative real numbers $\mathbb{R}_0^+$, if x is, e.g., a concentration-like variable. The question arises – what happens if the diffusion process reaches the boundaries of its state space? (If the boundary in question is plus or minus infinity, the process is said to explode.) Suppose for the following that the state space of the diffusion process X_t is the interval $[b_1, b_2]$, where possibly $|b_i| = \infty$. We assume furthermore that $f(x)$ and $g(x)$ satisfy a Lipshitz condition on every closed subinterval $[\bar{b}_1, \bar{b}_2]$ of the open interval (b_1, b_2), i.e., a constant K exists such that for $x, y \in [\bar{b}_1, \bar{b}_2]$ the inequality

$$|f(x) - f(y)| + |g(x) - g(y)| \leqslant K|x-y| \tag{5.87}$$

holds. A sufficient condition for (5.87) to be fulfilled is that f and g are once continuously differentiable, i.e., the same condition which guarantees existence and uniqueness of the solution of (5.48) up to same random "explosion" time η,

[6] If not specified otherwise, SDE will always be understood as an Ito equation henceforth.

cf. p. 94. Therefore we know that up to the random time η, when the diffusion process X_t reaches one of the boundaries for the first time, a unique solution exists, which is a diffusion process. The nicest diffusion processes are of course those which never reach the boundaries of their state space. Since such a process never "feels" the boundaries, the latter have no influence whatsoever on the solution of the SDE. Boundary conditions do not have to be imposed, indeed can*not* be imposed. Such boundaries are called inaccessible and if both b_1 and b_2 are of this type then $\eta = \infty$ with probability one and a unique solution of (5.48) exists on (b_1, b_2) for all times. Inaccessible boundaries come in two types:

i) so called natural boundaries, which are characterized by the fact that even if time goes to infinity they are attained only with probability zero. Let β be a point near the boundary, say b_1. If the diffusion process X_t is at $t = 0$ in the state $x_0 \in (b_1, \beta)$ then the boundary b_1 is by definition said to be natural if X_t reaches with probability one the point β earlier than the boundary b_1. This implies of course that the boundary is never reached: if the process X_t returns into the interval (b_1, β) after having attained β, then again it will once more reach with probability one β before b_1. The character of the boundaries is, as might be guessed, determined by the behavior of the drift and diffusion near the boundary. Indeed, the classification of boundaries is based on the integrability of certain functions made up of f and g. The analytical condition for the boundary b_i to be natural is that

$$\phi(x) = \exp\left(-\int_{\beta}^{x} \frac{2f(z)}{\sigma^2 g^2(z)}\, dz\right) \tag{5.88}$$

is not integrable in the vicinity of the boundary. For instance b_1 is natural, if

$$L_1(b_1) = \int_{b_1}^{\beta} \phi(x)\, dx \tag{5.89}$$

is equal to infinity: $L_1(b_1) = \infty \Leftrightarrow b_1$ natural.

ii) The other behavior of a diffusion process near an inaccessible boundary is that if X_t starts at $t = 0$ at $x_0 \in (b_1, \beta)$, then it either leaves this interval in a finite time τ_{x_0} and in this case always via the right point, or it never leaves this interval and then $X_t \to b_1$ for $t \to \infty$. Such a boundary is called attracting. It is characterized by the fact that

$$L_1(b_1) = \int_{b_1}^{\beta} \phi(x)\, dx < \infty \tag{5.90}$$

and

$$L_2(b_1) = \int_{b_1}^{\beta} \frac{1}{\sigma^2 g^2(y)} \int_{b_1}^{y} \exp\left(-\int_{\beta}^{x} \frac{2f(z)}{\sigma^2 g^2(z)}\, dz\right) dx \exp\left(\int_{\beta}^{y} \frac{2f(z)}{\sigma^2 g^2(z)}\, dz\right) dy \tag{5.91}$$

equal to infinity, i.e.,

$$L_1(b_i) < \infty\,, \quad L_2(b_i) = \infty \Leftrightarrow b_i \,\text{attracting}\,.$$

If neither $L_1(b_i) = \infty$ nor $L_2(b_i) = \infty$ then with nonzero probability certain sample paths of the diffusion process X_t attain the boundary b_i of the state space in finite time. In such a case, the process feels the presence of a boundary and the behavior of the diffusion process is not solely defined by the drift and diffusion, but also by the boundary conditions that have in general to be imposed. The basic boundary conditions that may be prescribed are the following:

i) *absorption*: once the diffusion process reaches the boundary b_i, it stays there;

ii) *reflection*: when the diffusion process attains the boundary b_i, it returns in a continuous way either instantaneously into the interval (b_1, b_2), so-called instantaneous reflection, or it stays some positive length of time at the boundary, so-called delayed reflection;

iii) *jump reflection*: once the diffusion process reaches the boundary b_i, it jumps back to a random point in the interior of (b_1, b_2).

More complicated boundary conditions can be obtained by combining these basic ones.

The accessible boundaries, for which $L_1(b_i) < \infty$ and $L_2(b_i) < \infty$, also divide into two subclasses. Processes in the first subclass are characterized by such a drift and diffusion that it becomes increasingly more difficult for the process X_t to reach a point β in the interior of (b_1, b_2) the closer it comes to the boundary. In mathematical terms: if X_t starts at $t = 0$ at $x_0 \in (b_1, \beta)$, then the probability that X_t reaches β in finite time goes to zero as x_0 tends to b_1. So the boundary acts naturally as an absorbing boundary. Only the boundary conditions (i) and (iii) are compatible with it and can be imposed on the process. The analytical condition for b_1 to be an absorbing boundary is that

$$L_3(b_1) = \int_{b_1}^{\beta} \frac{1}{g^2(y)} \frac{1}{\phi(y)} dy \tag{5.92}$$

is equal to infinity, i.e.,

$$b_1 \ \text{absorbing} \ \Leftrightarrow L_1 < \infty\,, \quad L_2 < \infty\,, \quad L_3 = \infty\,.$$

If all three quantities L_1, L_2 and L_3 are finite for the boundary under consideration, then the boundary is called a regular boundary and any boundary condition can be prescribed. The classification is for convenience summarized in the following table:

natural	$L_1 = \infty$			no b.c.
attracting	$L_1 < \infty$,	$L_2 = \infty$		no b.c.
absorbing	$L_1 < \infty$,	$L_2 < \infty$,	$L_3 = \infty$	b.c. (i) or (iii)
regular	$L_1 < \infty$,	$L_2 < \infty$,	$L_3 < \infty$	any b.c.

We call b an intrinsic boundary of the diffusion process X_t, if either $g(b) = 0$ or $|f(b)| = \infty$. It is obvious from the definition of the quantities L_1, L_2 and L_3

that any boundary b, which is not intrinsic, is a regular boundary. Indeed, for a nonintrinsic boundary b there is a neighborhood in which $|f(x)| < \infty$ and $g(x) > 0$ due to the continuity of the drift and diffusion coefficients, implying that $\phi(x)$ is a bounded, strictly positive function and that none of the quantities L_1, L_2 or L_3 can diverge.

The classification scheme which we have presented above is the one adopted in the Russian literature [5.12]. We shall designate it in the following as the Gihman-Skorohod (G-S) classification. There is a second scheme that has found wide spread use in the literature and which was introduced by *Feller* [1.87]. (For reviews, see [5.13, 14].) The terminology and the grouping of boundary behavior in this scheme is slightly different from the G-S scheme. To discuss the differences and similarities between both classifications we shall use the prefixes G-S and F for the two schemes:

i) a G-S natural boundary can be an F natural or an F entrance boundary;
ii) a G-S attracting boundary is an F natural boundary;
iii) a G-S absorbing boundary is an F exit boundary;
iv) a G-S regular boundary is an F regular boundary.

The only difference, apart from naming the same behavior by a different word, occurs for inaccessible boundaries. A boundary b_i is F natural if

$$h_2(x) = g^{-2}(x) \left(\exp \frac{2}{\sigma^2} \int^x \frac{f(u)}{g^2(u)} du \right) \int_{x_0}^{x} \phi(z) dz \qquad x_0 \in (b_1, b_2) \tag{5.93}$$

and

$$h_1(x) = \phi(x) \int_{x_0}^{x} g^{-2}(z) \left(\exp \frac{2}{\sigma^2} \int^z \frac{f(u)}{g^2(u)} du \right) dz \qquad x_0 \in (b_1, b_2) \tag{5.94}$$

are not integrable near b_i.
A boundary b_i is F entrance if

$$L_1(b_i) = \infty$$

and if $h_2(x)$ is integrable near b_i.
The characteristic of a F entrance boundary is that it cannot be reached from the interior of the state space and that any probability, initially assigned to this boundary, flows into the open interval (b_1, b_2). In other words, if the diffusion process is started on an F entrance boundary, it quickly moves into the interior of the state space and never returns to the boundary. An F natural boundary is also inaccessible from the inside of the state space, but in contrast to an F entrance boundary any probability that is initially assigned to an F natural boundary is trapped there forever. Clearly, a diffusion process cannot be started from an F natural boundary.

In the following we shall adhere mainly to the G-S scheme, since it fits most of our purposes better than the Feller scheme. We shall often drop the prefix G-S; it will always be understood, unless explicitly specified otherwise by the prefix F, that the boundary behavior is classified in Gihman-Skorohod terms.

6. Noise-Induced Nonequilibrium Phase Transitions

All the essential mathematical tools needed to answer the question of how nonlinear macroscopic systems behave in a fluctuating environment are now assembled. At this point let us briefly recall and summarize the essential features of our approach for the convenience of the reader.

We consider systems that can be modeled by a phenomenological equation of the type

$$\dot{X} = f_\lambda(X) , \tag{6.1}$$

where X is a variable characterizing the state of the system and λ denotes an external parameter which depends on the state of the environment. The assumptions under which (6.1) constitutes a valid description have been discussed in Chap. 1. To take into account the fluctuations of the environment, we replace the parameter λ in the phenomenological equation by a stationary random process $\lambda_t = \lambda + \sigma\zeta_t$, where the external noise ζ_t has zero mean value and intensity σ^2. For a detailed discussion of the properties of the surroundings see Chap. 1. To describe rapid environmental fluctuations we have adopted the white-noise idealization for ζ_t, i.e., $\lambda_t = \lambda + \sigma\xi_t$. This idealization has been a constant source of confusion and bewilderment in the literature. In our treatment, to avoid the ambiguities and misconceptions that have plagued earlier theories of multiplicative external white-noise more often than not, we have discussed at great length in mathematically precise terms the white-noise limit and its implications in the description of nonlinear systems subjected to fluctuating surroundings. It was found that systems coupled to a rapidly fluctuating environment can be modeled by Markov processes that are solutions of stochastic differential equations. The two concepts necessary for a satisfactory formalization of this fact, namely diffusion processes and Ito SDE or Stratonovich SDE, were discussed in detail in the two preceding chapters. The main upshot of that discussion is that systems of the type (6.1), where $f_\lambda(X)$ is linear in the external parameter – a case that covers a broad class of applications –

$$f_\lambda(X) = h(X) + \lambda g(X) ,$$

can be modeled under the influence of rapid external noise by a diffusion process whose differential moments are $f(X)$ and $\sigma^2 g^2(X)$ or $f(X) + \sigma^2 g'(X) g(X)/2$ and $\sigma^2 g(X)$, respectively. In other words, the description can be based on the Ito SDE

$$dX_t = [h(X_t) + \lambda g(X_t)]\, dt + \sigma g(X_t)\, dW_t , \tag{6.2}$$

or the Stratonovich SDE

$$dX_t = [h(X_t) + \lambda g(X_t)]\,dt + \sigma g(X) \circ dW_t\,. \tag{6.3}$$

The Fokker-Planck equation (FPE), which governs the evolution of the transition probability density $p(y,t|x)$[1], reads for those two cases:
Ito:

$$\partial_t p(y,t|x) = -\partial_y f_\lambda(y)\,p(y,t|x) + \frac{\sigma^2}{2}\partial_{yy} g^2(y)\,p(y,t|x)\,; \tag{6.4}$$

Stratonovich:

$$\partial_t p(y,t|x) = -\partial_y \left[f_\lambda(y) + \frac{\sigma^2}{2} g'(y)g(y) \right] p(y,t|x) + \frac{\sigma^2}{2}\partial_{yy} g^2(y)\,p(y,t|x)\,. \tag{6.5}$$

Using this modeling procedure we shall now investigate if and how external noise modifies the macroscopic behavior of nonlinear systems.

6.1 Stationary Solution of the Fokker-Planck Equation

Since environmental fluctuations can be modeled by a stationary random process, as discussed in Sect. 1.5, we expect that in general a system subjected to external noise for sufficiently long time will also settle down to stationary behavior. Stationary behavior here means that as time goes to infinity the system will attain a probability density $p_s(x)$, whose shape does not change any more with time: the probability to find the system in the neighborhood of a particular state x becomes time independent. However, the sample paths $X_t(\omega)$ will, of course, in general not approach a steady-state value $x(\omega)$: the state of the system *does* still fluctuate, it changes from one instant of time to the other. These fluctuations are however such that X_t and $X_{t+\tau}$ have the same probability density, namely $p_s(x)$.

In the following, we shall determine the stationary probability density $p_s(x)$ which characterizes the steady-state behavior of the system under external white noise, and discuss its main properties. In particular, we shall be interested in any modification in the macroscopic behavior of the system caused by environmental fluctuation. $p_s(x)$ is the stationary solution of FPE which can be written in the following form:

$$\partial_t p(x,t|x_0,0) + \partial_x J(x,t|x_0,0) = 0\,, \tag{6.6}$$

---

[1] The processes are time homogeneous. The conditional probability density $p(y,t+s|x,s)$ depends thus only on the time difference.

where

$$J(x,t|x_0,0) = f(x)p(x,t|x_0,0) - \frac{\sigma^2}{2}\,\partial_x g^2(x)p(x,t|x_0,0)\,.$$

Here J can be interpreted as a probability current and (6.6) is then just a continuity equation, expressing the conservation of the probability. The stationary FPE then reads

$$\partial_x J_s(x) = 0 \tag{6.7}$$

and implies that the stationary probability current is constant on the state space $[b_1, b_2]$:

$$J_s(x) = \text{const} \quad \text{for} \quad x \in [b_1, b_2]\,.$$

In the stationary case, there are no sources or sinks in the state space for the probability. We have that the probability current in the interior of the state space is equal to the current across the boundaries, which we call J:

$$J \equiv J_s(x) = J_s(b_1) = J_s(b_2)\,.$$

Hence we have for the stationary probability density $p_s(x)$:

$$-f(x)p_s(x) + \frac{\sigma^2}{2}\,\partial_x g^2(x)p_s(x) = -J\,. \tag{6.8}$$

To solve (6.8), we define the auxiliary function $q(x) = g^2(x)p_s(x)$. The stationary FPE then reads

$$\partial_x q(x) = \frac{2}{\sigma^2}\,\frac{f(x)}{g^2(x)}\,q(x) - \frac{2}{\sigma^2}\,J\,,$$

and its solution is easily calculated to be

$$q(x) = N\exp\left(\frac{2}{\sigma^2}\int^x \frac{f(u)}{g^2(u)}\,du\right) - \frac{2J}{\sigma^2}\int^x \exp\left(\frac{2}{\sigma^2}\int_z^x \frac{f(u)}{g^2(u)}\,du\right)dz\,.$$

Thus we obtain for the stationary probability density

$$p_s(x) = \frac{N}{g^2(x)}\exp\left(\frac{2}{\sigma^2}\int^x \frac{f(u)}{g^2(u)}\,du\right) - \frac{2}{\sigma^2 g^2(x)}\,J\int^x \exp\left(\frac{2}{\sigma^2}\int_z^x \frac{f(u)}{g^2(u)}\,du\right)dz\,. \tag{6.9}$$

The integration constant N is determined by the normalization condition, whereas J, being the probability current at the boundaries of the state space, depends on the nature of those boundaries. When the boundaries are natural (or regular with instantaneous reflection imposed as a boundary condition), there is no flow of probability out of the state space and hence $J=0$. In the following, we shall always consider situations for which this condition is fulfilled. In that case

$$p_s(x) = \frac{N}{g^2(x)} \exp\left(\frac{2}{\sigma^2} \int^x \frac{f(u)}{g^2(u)} du\right). \tag{6.10}$$

To be a stationary probability density, (6.10) has to be normalizable, i.e.,

$$N^{-1} = \int_{b_1}^{b_2} \frac{1}{g^2(x)} \exp\left(\frac{2}{\sigma^2} \int^x \frac{f(u)}{g^2(u)} du\right) < \infty. \tag{6.11}$$

If one of the boundaries is an attracting boundary, an absorbing boundary, or a regular boundary, with absorption imposed as a boundary condition, then $p_s(x) \equiv 0$ for $x \in (b_1, b_2)$, i.e., the trivial solution of the stationary FPE (6.7). We shall also say in this case that no (regular) stationary probability density exists.

For a diffusion process with drift $f + \sigma^2 g' g/2$ and diffusion $\sigma^2 g^2$, i.e., corresponding to the Stratonovich SDE,

$$dX_t = f(X_t)dt + \sigma g(X_t) \circ dW_t, \tag{6.12}$$

it is easily obtained that the stationary probability density (boundaries such that $J=0$) is:

$$p_s(x) = \frac{N}{g(x)} \exp\left(\frac{2}{\sigma^2} \int^x \frac{f(u)}{g^2(u)} du\right). \tag{6.13}$$

In summary, the stationary behavior of a system described by the SDE

$$dX_t = f(X_t)dt + \sigma g(X_t) dW_t \tag{6.14}$$

is given by

$$p_s(x) = N g^{-\nu}(x) \exp\left(\frac{2}{\sigma^2} \int^x \frac{f(u)}{g^2(u)} du\right) \tag{6.15}$$

if $J=0$ and $p_s(x)$ can be normalized. Here $\nu = 1$ refers to the Stratonovich interpretation of (6.14) and $\nu = 2$ to the Ito version. It may happen that a SDE interpreted according to Ito admits a stationary solution while interpreted according to Stratonovich it does not, or vice versa, since the formulae for $p_s(x)$ differ by a factor $g^{-1}(x)$. If such a discrepancy between the Ito and Stratonovich versions occurs, it should be interpreted as a "red warning light". In general, it signals that the model used to describe the system has some dangerous or pathological features, for instance, g is not continuously differentiable as in the case $g(x) \sim \sqrt{x}$, $x \in [0, \infty)$. In any case, if there is a qualitative difference in the stationary behavior between the Ito or Stratonovich interpretation of SDE, one has to be doubly careful in justifying the modeling procedure. In all applications discussed later on, there is no qualitative difference between the two versions. Indeed, we consider it a highly desirable feature of a model to yield a SDE that is robust against a change in the interpretation of the equation as Ito or Stratonovich.

Suppose now that the diffusion process given by the SDE (6.14) interpreted in the Ito sense admits a stationary solution that is strictly positive in the interior of the state space, i.e., $p_s(x) > 0$ for all $x \in (b_1, b_2)$. This means, in particular, that the state space (b_1, b_2) of the process X_t does not divide into two or more intervals which do not communicate with each other. It implies that no matter at what point x_0 the process X_t is started in the state space, it will visit the neighborhood of any point $x \in (b_1, b_2)$ infinitely often. This requires in particular that the fluctuations vanish nowhere in the interior of the state space, i.e., that $g(x) > 0$ for $x \in (b_1, b_2)$. Then a solution $p(x, t)$ of the FPE with an arbitrary initial condition $p(x, 0)$ converges to $p_s(x)$ for t tending to infinity. To see this, consider the following functional of $p(x, t)$:

$$\begin{aligned} \phi(t) &= \int_{b_1}^{b_2} dx\, p(x,t) \ln [p(x,t)/p_s(x)] \\ &= \int_{b_1}^{b_2} dx\, p(x,t) \{\ln [p(x,t)/p_s(x)] + p_s(x)/p(x,t) - 1\} \geqslant 0 \,, \end{aligned} \tag{6.16}$$

which is nonnegative for all t since $\ln(1/y) \geqslant 1 - y$ for $y > 0$[2].

The equality holds only for $y = 1$, i.e., $p(x, t) = p_s(x)$. We shall now show that

$$\dot{\phi} \equiv d_t \phi(t) \leqslant 0 \,. \tag{6.17}$$

A functional which exhibits the properties (6.16, 17) with $\phi = 0$ and $\dot{\phi} = 0$ if and only if $p(x, t) = p_s(x)$ is known as a Liapounov functional. It is a well-known theorem [1.73] that the existence of such a functional implies the global asymptotic stability of the stationary state. In more physical language, $\phi(t)$ is an $\mathscr{H}$ function for the FPE and (6.17) is the corresponding $\mathscr{H}$ theorem. Let us now prove that $\phi(t)$ monotonely decreases and thus establish that any solution of FPE approaches $p_s(x)$ as $t \to \infty$ if $p_s(x) > 0$ for $x \in (b_1, b_2)$.

We differentiate (6.16) with respect to time and use the fact that the one-dimensional probability density obeys the FPE:

---

[2] Two write (6.16), simply remember that $\int_{b_1}^{b_2} [p_s(x) - p(x,t)]\, dx = 0$, $\forall t$ since $\int_{b_1}^{b_2} p(x,t)\, dx = 1$.

$$\dot{\phi} = \int dx \left[-\partial_x f(x) p(x,t) + \frac{\sigma^2}{2} \partial_{xx} g^2(x) p(x,t) \right] \ln [p(x,t)/p_s(x)] \, .$$

Since we consider only processes with no flux across the boundaries the boundary terms vanish, as mentioned above, and so through integration by parts we obtain

$$\begin{aligned}\dot{\phi} &= \int dx\, p(x,t) \left[f(x)\, \partial_x + \frac{\sigma^2}{2} g^2(x)\, \partial_{xx} \right] \cdot \ln [p(x,t)/p_s(x)] \\ &= \int dx\, p(x,t) \left[f(x) \frac{p_s(x)}{p(x,t)} \cdot \partial_x \frac{p(x,t)}{p_s(x)} + \frac{\sigma^2}{2} g^2(x) \cdot \partial_x \frac{p_s(x)}{p(x,t)} \cdot \partial_x \frac{p(x,t)}{p_s(x)} \right. \\ &\quad \left. + \frac{\sigma^2}{2} g^2(x) \frac{p_s(x)}{p(x,t)} \cdot \partial_{xx} \frac{p(x,t)}{p_s(x)} \right] \\ &= \int dx \left[f(x) p_s(x) \cdot \partial_x \frac{p(x,t)}{p_s(x)} + \frac{\sigma^2}{2} g^2(x) p_s(x) \cdot \partial_{xx} \frac{p(x,t)}{p_s(x)} \right] \\ &\quad - \int dx \frac{\sigma^2}{2} g^2(x) \frac{p_s^2(x)}{p(x,t)} \left(\partial_x \frac{p(x,t)}{p_s(x)} \right)^2 . \end{aligned} \tag{6.18}$$

The first integral is zero, as can be seen after integration by parts:

$$\begin{aligned}&\int dx \frac{p(x,t)}{p_s(x)} \left[-\partial_x f(x) p_s(x) + \frac{\sigma^2}{2} \partial_{xx} g^2(x) p_s(x) \right] \\ &\quad = \int dx \frac{p(x,t)}{p_s(x)} \partial_t p_s(x) = 0 \, . \end{aligned}$$

Hence $\dot{\phi}$ reduces to the second integral in (6.18) whose integrand is obviously nonnegative. This completes the proof of (6.17). It is also clear from (6.18) that $\dot{\phi} = 0$ if and only if

$$\partial_x [p(x,t)/p_s(x)] = 0 \, , \quad \text{i.e.,} \quad p(x,t) = \text{const} \cdot p_s(x) \, . \tag{6.19}$$

Since both sides of (6.19) are probability densities and hence are normalized to one, we conclude that the constant is equal to one. Relations (6.16, 17) have the following consequence, already announced above: if the diffusion process X_t is started with a probability density that differs from the stationary one, it will approach the stationary density as time tends to infinity. Indeed, $\phi(t)$ is strictly positive and monotonely decreases with time, since $\dot{\phi} < 0$. This implies that $\phi(t) \to 0$ for $t \to \infty$ and since $\phi = 0$ if and only if $p(x,t) = p_s(x)$, we have $\lim_{t \to \infty} p(x,t) = p_s(x)$.

Furthermore, it can be shown that if the stationary probability density exists and the diffusion process X_t is started with it, then it is an ergodic process. To be precise the following theorem holds: If the integral (6.11) is finite, then for integrable functions $\varphi(x)$ the following equality holds with probability one:

$$\lim_{T\to\infty} \frac{1}{T} \int_0^T \varphi(X_t(\omega))dt = \int_{b_1}^{b_2} \varphi(x) p_s(x)\, dx$$
$$= E\{\varphi(X)\}. \tag{6.20}$$

This means that the mathematical expectation of a stationary diffusion process X_t can be determined by observing just one arbitrary sample path of the process and taking the time average as defined by the lhs of (6.20). Furthermore, the stationary probability density itself can be obtained in this way. To see this, consider the indicator function I_x^ε of the interval $[x-\varepsilon, x+\varepsilon]$, i.e.,

$$I_x^\varepsilon(z) = \begin{cases} 1 & z \in [x-\varepsilon, x+\varepsilon] \\ 0 & z \notin [x-\varepsilon, x+\varepsilon] \end{cases}. \tag{6.21}$$

Then the ergodic theorem implies that

$$\lim_{T\to\infty} \frac{1}{T} \int_0^T I_x^\varepsilon(X_t(\omega))dt = \int_{x-\varepsilon}^{x+\varepsilon} p_s(z)\, dz \qquad \text{a.s.} \tag{6.22}$$

This relation establishes that $p_s(x)\,dx$ equals the fraction of time an arbitrary sample path of the diffusion process spends in an infinitesimal neighborhood of x.

6.2 The Neighborhood of Deterministic Behavior: Additive and Small Multiplicative Noise

Let us now consider a nonlinear macroscopic system which has been coupled to its environment for a sufficiently long time to have settled down to a stationary state. If the surroundings are unvarying, then the steady states of the system are the zeros of the right-hand side of the deterministic phenomenological equation

$$\dot{X} = h(X) + \lambda g(X). \tag{6.23}$$

In the following, we shall also suppose that the deterministic system is stable in the sense that the solution $X(t)$ does not blow up to infinity. To be precise, for every $x_0 \in (b_1, b_2)$ there exists a constant $C < \infty$, in general dependent on λ, such that

$$|X(t)| \leqslant C \;\forall t \tag{6.24}$$

if $X(0) = x_0$. This is fulfilled if a $K > 0$ exists such that

$$h(x) + \lambda g(x) < 0 \quad \text{for all} \quad x > K \tag{6.25}$$

and $h(x) + \lambda g(x) > 0$ for all $x < -K$, respectively.

If X is a concentration–like variable and therefore has to be nonnegative, then the rhs of (6.23) has to obeys the following condition:

$$h(0) + \lambda g(0) \geqslant 0 \quad \text{for all} \quad \lambda \,. \tag{6.26}$$

If both b_1 and b_2 are finite, we require $h(b_1) + \lambda g(b_1) \geqslant 0$ and $h(b_2) + \lambda g(b_2) \leqslant 0$ for all λ. The solution of a first-order, one-variable differential equation is a monotone function with respect to time, since $\dot{X}$ takes one and only one well-defined value for every X. Thus (6.24) implies that (6.23) admits at least one stable steady state. If the deterministic phenomenological equation admits more than one steady state, then stable and unstable ones alternate. If there are two or more stable stationary states, then the state space divides in nonoverlapping regions, the "basins of attraction" of the various stable states. This is very easily seen if we write the phenomenological equation in the form

$$\dot{X} = -\partial_X V_\lambda(X) \,, \quad \text{where} \quad V_\lambda(x) = -\int^x [h(z) + \lambda g(z)]\, dz \tag{6.27}$$

is called the potential of (6.23). Obviously the steady states are the extrema of the potential $V_\lambda(x)$ and the normal modes $\omega(\bar{X})$ of the linear stability analysis are given by

$$\omega(\bar{X}) = -\partial_{XX} V_\lambda(\bar{X}) \,. \tag{6.28}$$

Hence, the stable steady states correspond to the minima of $V_\lambda(x)$ and the unstable steady states to the maxima. Imagine x to be the coordinate of a ball which moves with an extremely large friction in the landscape given by the potential $V_\lambda(x)$. Depending on its initial position, the ball comes to rest in one of the valleys, i. e., the minima of $V_\lambda(x)$. Once on the valley bottom, it will not leave the valley if it is subjected to infinitesimally small perturbations. Indeed, it will always return to its rest state in the local potential minimum as long as the perturbations do not make it cross the potential barrier separating two valleys. On the other hand, a ball at rest on a mountain top will never return to this rest state if subjected to perturbations, even if they are infinitesimally small.

Let us now return to our main problem and investigate how the stationary behavior of a system is modified in a fluctuating environment. In this case, the "state" of the system is given by a random variable. To unify our language for the description of deterministic and stochastic steady-state behavior, we shall say that in the former case the system is described by a degenerate random variable of the form

$$X(\omega) = \bar{X}_i \quad \text{if} \quad X(0)(\omega) \in A(\bar{X}_i) \,,$$

where $A(\bar{X}_i)$ denotes the basin of attraction of the i-th steady state. Adopting the convention (2.15), we characterize this (degenerate) random variable by its probability law. In the deterministic case, the stationary "probability density" consists of delta peaks centered on the steady state $\bar{X}_i$. The weight of the delta peaks is given by the initial preparation of the system. External noise obviously has a disorganizing influence; the sharp delta peaks will be broadened. If we think in terms of the above picture of a ball in a mountainous landscape, then the external fluctuations will jiggle the ball around and tend to spread out the probability of where the ball is found. In a featureless, flat landscape, the ball would undergo a sort of Brownian motion. Actually we have however a counteracting systematic force pushing the ball back to the valley bottom. Hence equilibrium will be struck between the two forces and we expect the following stationary behavior: the probability density has a maximum at the coordinate that corresponds to the minimum of the potential and has a certain spread around it, depending on the strength of the external noise. If there is more than one minimum and if there is no effective upper bound on the external fluctuations, then we expect a multimodal probability density with peaks corresponding to the various minima of the potential. Indeed, if the external noise is Gaussian distributed as we suppose here, fluctuations become rapidly more unlikely the larger they are, but even for an arbitrarily large fluctuation the probability is not strictly zero, only extremely small. If we wait long enough a fluctuation will occur that takes the ball over the potential barrier into the other valley. If the barrier is very high and the variance of the external noise small, it might take an astronomically large time for such a fluctuation to occur. However, if we take the limit time tending to infinity seriously, and we have to in order to determine the form of the stationary probability density, the fluctuation will have occurred and the ball will have visited, in general even infinitely often, all the potential valleys. As is clear from this discussion, such a picture is expected to hold only if either the influence of the external noise is independent of the state of the system, or if the variance is extremely small, such that the system spends most of the time in the minima and makes the crossing from one minimum to the other relatively rapidly. Apart from this restriction, however, a general picture emerges. The state of the system, i.e., the random variable, either as the functional form of the mapping from the sample space Ω into the state space $[b_1, b_2]$, or with (2.15) as the functional form of its probability law, is given by an interplay between the dynamics of the system and the external fluctuations.

Let us first consider the case that the intensity of the white noise is extremely small, i.e., $\sigma^2 \ll 1$. Defining

$$U(x) = \int^{x} \frac{f(u)}{g^2(u)}\, du$$

we can write (6.15) as

$$p_s(x) = N \exp\left[\frac{2}{\sigma^2} U(\bar{x}_m)\right] \exp\left\{\frac{2}{\sigma^2}\left[U(x) - U(\bar{x}_m) - \frac{\nu\sigma^2}{2} \ln g(x)\right]\right\}. \quad (6.29)$$

Here $\bar{x}_m$ is the location of the highest maximum of $U(x)$, which we suppose to lie in the interior of the state space (b_1, b_2):

$$U(x) < U(\bar{x}_m) \quad \text{for} \quad x \neq \bar{x}_m .$$

If σ^2 tends to zero and $x \neq \bar{x}_m$, the second factor becomes exponentially small, so that the dominant contribution to the stationary probability density comes from a neighborhood of the order of σ^2 around the highest maximum of $U(x)$. Such a situation can be analyzed by the method of steepest descent [6.1] and one obtains after some tedious calculation

$$E\{X\} = \bar{x}_m + \frac{\sigma^2}{2U''(\bar{x}_m)} \left(\frac{1}{2} \frac{U'''(\bar{x}_m)}{U''(\bar{x}_m)} - \frac{g'(\bar{x}_m)}{g(x_m)} \right) + O(\sigma^4) \tag{6.30}$$

and

$$E\{\delta X^2\} = -\frac{2\sigma^2}{U''(\bar{x}_m)} + O(\sigma^4) . \tag{6.31}$$

If the external noise is additive, i.e., $g(x) = \text{const} = c$, then

$$U(x) = -\frac{1}{c^2} V_\lambda(x) \tag{6.32}$$

and the highest maximum of U and of the probability density $p_s(x)$ coincides with the position of the deepest potential well for all σ^2, $\bar{x}_m = \bar{x}$; no shift occurs. In this situation, on the basis of the deterministic description alone, it can already be decided which one of the locally stable steady states is globally speaking the most stable one, namely $V(\bar{x}_m) < V(x)$ for $x \neq \bar{x}_m$. The other locally stable steady states are called metastable. If, however, the influence of the external noise depends on the state of the system, i.e., if we deal with multiplicative noise, then

$$U(x) \neq -\frac{1}{c^2} V_\lambda(x) \tag{6.33}$$

and the deterministic steady state, giving rise to the highest maximum of $U(x)$, is not necessarily the one with the deepest potential well in the deterministic description. This is in close analogy with findings for internal fluctuations in macroscopically large, but finite systems [6.2]. It indicates that the criterion of absolute stability for the deterministic steady states depends explicitly on the nature of the random perturbations the system is subjected to. Absolute stability and the phase coexistence point, i.e., the point in phase space where two peaks of $p_s(x)$ have equal height, cannot be determined solely on the basis of the deterministic description, a point already stressed by *Landauer* [6.3].

6.3 Transition Phenomena in a Fluctuating Environment

After analyzing the limiting case of extremely small noise intensities σ^2, we now study the stationary behavior of macroscopic systems for arbitrary noise intensities. In particular, we shall be interested in transition phenomena that occur under the influence of external noise. This confronts us immediately with the questions, what do we mean by a transition in a macroscopic system coupled to a random environment and how do we detect such a transition? The phenomenon of nonequilibrium phase transitions in a system with deterministic external constraints is by now a familiar one and was briefly discussed in Chap. 1. The behavior of a nonlinear system as a function of an external parameter is best described by the corresponding bifurcation diagram. Over a certain range of values of the external parameter the steady state changes only quantitatively, if at all. However, at certain values a qualitative change occurs in the form of a second- or first-order nonequilibrium phase transition (Chap. 1). If the external constraints fluctuate, then the steady state of the system is described by a genuine random variable as discussed above. In analogy to the deterministic situation, it is quite natural to say that a transition has occurred whenever the steady state of the system changes qualitatively, i.e., loosely speaking, the random variable changes qualitatively. To formulate this in a more precise way, remember that a random variable is a function from the sample space into the state space. A *transition occurs* precisely at that point in parameter space, consisting of the mean value of the external noise, its variance, its correlation time, etc., *where the functional form of the mapping from the sample space Ω into the state space* $[b_1, b_2]$ *changes qualitatively*. Adopting convention (2.15) this corresponds to a qualitative change in the probability law characterizing the random variable. In our case this probability law is given by (6.15), the exact expression for the stationary probability density of a system subjected to Gaussian white noise. How can we detect such a qualitative change? Intuitively one would look at the deterministic situation for guidance and try to extend the criteria used there in a natural way to the stochastic case. This is indeed the best way to proceed. In the deterministic case, a nonequilibrium phase transition occurs when the potential $V_\lambda(x)$, cf. (6.27), changes qualitatively. For instance, the number of local extrema changes. This fact has found its precise formulation in catastrophe theory [6.4, 5]. In our unifying language, any such change leads to a qualitative change in the (degenerate) random variable describing steady-state behavior. It is thus most natural to consider the extrema of the stationary probability density $p_s(x)$ as indicators for a transition in the stochastic case. [Transitions involving changes in the nature of a boundary are not detected by this indicator (Sect. 6.4).] This choice is not only the most direct extension of the deterministic concepts, but also the most appropriate compared with other possibilities that come to mind in the stochastic case as, e.g., the moments of the distribution. The latter are not reliable indicators to decide whether a qualitative change in the random variable has occurred. First, the moments do not always uniquely determine the probability distribution [Ref. 2.3, p. 166]. Second, in more physical terms the averaging procedure washes out a lot of information. To illustrate this with a typical example, consider the time-dependent Landau equation, often used to describe equilibrium critical phenomena [6.6]:

$$dX_t = (\lambda X_t - X_t^3)\,dt + \sigma dW_t\,.$$

In the deterministic case, $\sigma = 0$, a critical point occurs at $\lambda = 0$. For λ negative the system has only one steady state, $\bar{x} = 0$, i.e., the potential $V_\lambda(x)$ has only one minimum. For λ positive, $\bar{x} = 0$ becomes a maximum of $V_\lambda(x)$ and two minima develop at $\bar{x} = \pm(\lambda)^{1/2}$, i.e., the system now has two stable and one unstable steady states. In the stochastic case, which corresponds to additive noise here, the steady-state behavior of the system is described by a random variable whose probability law is given by

$$p_s(x) = N \exp\left[-V_\lambda(x)/\sigma^2\right]\,.$$

It is obvious that also in the stochastic case a qualitative change in the steady state occurs at $\lambda = 0$. This transition is accurately reflected by the behavior of the extrema of $p_s(x)$. If however the moments are used, no transition phenomenon is detected. In particular, the first moment remains *zero* even for positive values of λ. Clearly, it is *not* the mean value that corresponds to the macroscopic states or phases of the system, *but* the maxima of $p_s(x)$.

This example confirms that the most direct extension of the deterministic concepts as presented above is also the most appropriate. A qualitative change in steady-state behavior is unambiguously reflected in the extrema of the probability density. The one exception is the transition from a degenerate to a genuine random variable. Here, the variance is the best indicator. To avoid any possible misunderstanding, let us emphasize at this point that we do not concentrate exclusively on the extrema of $p_s(x)$, on the most probable values. In particular, we do not imply that the maxima dominate the stationary probability distribution. External noise is of macroscopic nature and thus is not small in the sense of internal fluctuations. Naturally this makes the transition zone larger and the peaks broader, but it does not render impossible experimental observation of the transition phenomena. To make the point as clear as possible, let us reiterate that in a natural and direct extension of the deterministic notions a *transition occurs* if the *steady state* of the system *as given by the random variable changes qualitatively*. (The following chapters will give evidence that this is a physical and operational definition, i.e., transitions defined in this way can be observed experimentally.) The extrema of the stationary probability density are from this point of view merely a practical way to monitor such a qualitative change. The *number* and *position* of the extrema of $p_s(x)$ in the stochastic case and of $V_\lambda(x)$ in the deterministic case are the most distinguishing features of the steady-state behavior of the system.

The extrema of $p_s(x)$ also have, however, a particular significance from a physical point of view. The considerations made above for small noise intensities $\sigma^2 \ll 1$ indicate that they are the continuation of the deterministic steady states. This interpretation is buttressed by the fact that the diffusion process X_t is ergodic, if $p_s(x)$ is normalizable. As we have seen above, this implies that $p_s(x)\,dx$ equals the fraction of time an arbitrary sample path of the diffusion process spends in an infinitesimal neighborhood of x. The maxima of $p_s(x)$ are therefore the states in whose neighborhood the system spends relatively much time. These are the states that are preferentially observed if an experiment or field study is carried out. In analogy with the picture of a particle moving in a landscape, defined by a potential, the maxima of $p_s(x)$ correspond to the valleys of a potential. The minima of $p_s(x)$ are the states that the system leaves rather quickly and they can be said to correspond to the mountain tops of a potential. We can introduce a "probabilistic" potential by writing the stationary probability density in the form

$$p_s(x) = N \exp[-2\mathscr{V}(x)/\sigma^2], \tag{6.34}$$

where of course

$$\mathscr{V}(x) = -\left[\int^x \frac{f(z)}{g^2(z)} dz - \nu \frac{\sigma^2}{2} \ln g(x)\right]. \tag{6.35}$$

In the case of additive external noise, the probabilistic potential $\mathscr{V}(x)$ and the deterministic potential $V_\lambda(x)$ coincide modulo an inessential constant. In this case, the extrema of $p_s(x)$ correspond exactly to the valleys and mountain tops of the deterministic potential landscape. These considerations justify the identification of the extrema of the stationary probability density with the macroscopic steady states of the system. The maxima represent the stable steady states, the minima the unstable steady states. Note that this identification of the extrema with steady states is legitimate only if the probability flux J_s vanishes in the steady state. Otherwise probability could pile up in certain regions of state space, due to the fact that the flow is slower there than in other regions, and thus create maxima of $p_s(x)$. However, as mentioned earlier, the systems we consider here and which are in general the only ones of importance in applications, have boundaries across which no flux of probability occurs. Hence J_s vanishes indeed identically. Our identification of the extrema of $p_s(x)$ with the macroscopic steady states thus rests on firm ground. As in the case of equilibrium transitions and nonequilibrium transitions with internal fluctuations, the extrema of the probability density correspond to the phases of the system. Indeed, if the stationary probability density has only one maximum, the system fluctuates around one macroscopic state, i.e., occurs only in one phase. However, if the stationary probability density displays two or more maxima, the system has the possibility to be in two phases for identical external conditions. Due to the fact that external noise is in general not macroscopically small as internal fluctuations are, there are a few quantitative differences. In the case of external noise the picture is somewhat fuzzier, though qualitatively the same. The peaks corresponding to the phases of the system are broader than the nearly delta peaks of the internal fluctuations. Though broadened, the phases are observable experimentally. Also the system switches back and forth between the different phases more rapidly than under the influence of internal fluctuations.

Before we proceed any further, let us summarize the salient features of our discussion so far:

i) A transition occurs when the functional form of the random variable describing the steady state of the system changes qualitatively.

ii) This qualitative change is most directly reflected by the extrema of the stationary probability law, except if the transition is due to a change in the nature of a boundary (Sect. 6.4).

iii) The physical significance of the extrema, apart from being the most appropriate indicator of a transition, is their correspondence to the macroscopic phases of the system. The extrema are the order parameter of the transition.

The extrema of $p_s(x)$ are easily found from the following relation:

$$[h(x_m) + \lambda g(x_m)] - \nu \frac{\sigma^2}{2} g(x_m) g'(x_m) = 0 . \quad (6.36)$$

This is the basic equation for an analysis of the influence of rapid external noise on the steady-state behavior of macroscopic nonlinear systems. Before discussing it in detail, let us remark on the differences between the Ito and Stratonovich interpretation of a given SDE describing the system. If, as we suppose in general and as was discussed above, the stationary probability density exists for the two versions, there is no qualitative difference as to the extrema of $p_s(x)$. The difference is only quantitative, namely a factor two in the intensity of the white noise. The basic equation (6.36) contains two terms. The one in brackets, set equal to zero, corresponds to the equation for the deterministic steady states (6.23). The second term describes the influence of the external noise. We again have to distinguish between the two cases of additive $g(x) = 1$, and multiplicative $g(x) \neq 1$, noise respectively. As mentioned before, in the first case the influence of environmental fluctuations does not depend on the state of system. Consequently, the extrema of $p_s(x)$ *always* coincide with the deterministic steady states, independent of the intensity of the external white noise. Indeed, we even have the stronger result that the probabilistic and deterministic potential are identical modulo a constant. Additive external white noise does not modify qualitatively the stationary behavior of one-variable systems. It only jiggles the particle around in the potential landscape, but does not influence the potential itself. Therefore it has only a disorganizing effect, which we expect in any case from an external noise, smearing $p_s(x)$ out around the deterministic steady states.

In the second case, the multiplicative noise case, the effect of the environmental fluctuations *does* depend on the state of the system. This means that not only the particle is jiggled around in the potential landscape, but also the ground is heaving up and down randomly. If σ^2 is sufficiently small, then the roots of (6.36) do not differ in number and position from the deterministic steady states as was shown above by a steepest-descent analysis. The external noise is not sufficiently strong to change the potential qualitatively; valleys stay valleys and mountain tops stay mountain tops. However, their relative heights and depths may be altered and the most stable state under multiplicative external noise is not necessarily the one with the deepest potential well, even for $\sigma^2 \ll 1$, as discussed above. If, however, the intensity σ^2 of the noise increases, then we come to a point where the second term in (6.36) can no longer be neglected. In fact, if σ^2 is sufficiently large, the extrema of $p_s(x)$ can be essentially different in *number* and *position* from the deterministic steady state, provided $g(x)$ is nonlinear in a suitable way; for instance $f(x) = h(x) + \lambda g(x)$ is a polynomial of degree n and $g'(x)g(x)$ a polynomial of degree m greater than n. Formulated differently, when the intensity σ^2 crosses a certain threshold value the shape of $p_s(x)$, i.e., the random variable describing the stationary behavior of the system, can change drastically; a transition occurs. Thinking in terms of the potential landscape we can say that multiplicative external noise can create new potential wells. This means that in addition to the disorganizing effect, which it shares with the additive noise, multiplicative noise can create new states. It can induce new nonequilibrium phase transitions which are not expected from the usual phenomenological descriptions. These transitions correspond to situations in which the system no longer adjusts its macroscopic behavior to the average properties of the environment, but responds in a definite, more active way. The random variable, giving the steady state of the system, is qualitatively quite different from the (degenerate) random variable corresponding to the deterministic steady-state situation. Since this new class of nonequilibrium phase transitions is solely due to the external noise, we shall call these transitions noise-induced nonequilibrium phase transitions or for short, *noise-induced transitions*. This and the following chapters will be devoted to elucidate the special features of these transitions, to study their consequences in physico-chemical and biological systems, to go beyond the white-noise idealization and to analyse the dynamical properties of such transitions.

6.4 The Verhulst System in a White-Noise Environment

The concept of noise-induced transitions was introduced on the Verhulst model [6.7] and as a start to the study of these phenomena on concrete examples it is natural to consider this most simple, though nontrivial, system first.

The Verhulst model was originally proposed to describe the growth of a biological population, but the phenomenological equation which corresponds to it, namely

$$\dot{X} = \lambda X - X^2 , \tag{6.37}$$

has in the course of time found applications in many different fields. For the sake of concreteness and simplicity however, we shall limit ourselves here to the standard interpretation of (6.37). Accordingly, the variable X is a measure of the size of a biological population and its physically acceptable state space corresponds to the interval $[0, \infty)$. The Malthusian growth parameter is λ which corresponds to the difference between birth and death rates of the population. The second term is a self-restriction term which expresses the fact that the resources of the population are limited; for instance the available food supply of the territory restricts the otherwise unlimited growth of the population under deterministic constant environmental conditions. The solution of (6.37) is

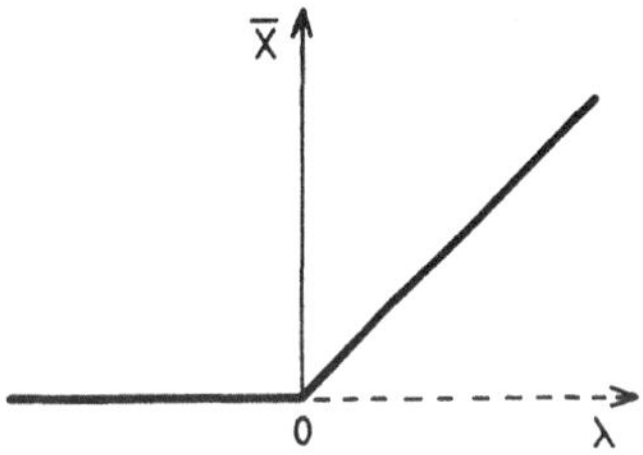

Fig. 6.1. Bifurcation diagram of the physically acceptable steady states $\bar{x}$ of the Verhulst model. Full line: stable states; broken line: unstable states

$$X(t) = X(0)\,e^{\lambda t}\{1 + X(0)\,[(e^{\lambda t} - 1)/\lambda]\}^{-1}.$$

For $\lambda < 0$ (6.37) admits only the stationary state solution $\bar{x} = 0$, which is stable. At $\lambda = 0$, this solution becomes unstable and a new branch of stable steady states, $\bar{x} = \lambda$, bifurcates. This branch emerges in a continuous but nondifferentiable way and hence we say that at $\lambda = 0$, the system undergoes a second-order phase transition (Fig. 6.1).

In the following, we shall consider the situation where environmental fluctuations are rapid compared with $\tau_{\text{macro}} = \lambda^{-1}$ which defines the macroscopic time scale of evolution. Changes in the environmental state act on the system through the external parameter λ. According to our discussion in Chaps. 1 and 3, we assume that the parameter λ can be written as $\lambda_t = \lambda + \sigma\xi_t$ in which λ is its average value, ξ_t is Gaussian white noise and σ measures the intensity of the external noise. Thus (6.37) is replaced by the stochastic differential equation:

$$\begin{aligned} dX_t &= (\lambda X_t - X_t^2)\,dt + \sigma X_t\,dW_t \\ &= f(X_t)\,dt + \sigma g(X_t)\,dW_t\,. \end{aligned} \tag{6.38}$$

As we have seen in Chap. 5, the choice of an appropriate diffusion process for this system, i.e., the one corresponding to the Ito or the one corresponding to the Stratonovich interpretation of (6.38), has to be decided according to the particular situation which one wants to model. Here we shall develop both interpretations on an equal footing. In the case of the Ito interpretation of (6.38) the FPE for the probability density $p(x,t)$ of the diffusion process reads:

$$\partial_t p(x,t) = -\partial_x[(\lambda x - x^2)\,p(x,t)] + \frac{\sigma^2}{2}\,\partial_{xx}x^2 p(x,t)\,. \tag{6.39}$$

For the Stratonovich interpretation of (6.38), we have

$$\partial_t p(x,t) = -\partial_x\left[\left(\lambda x - x^2 + \frac{\sigma^2}{2}x\right)p(x,t)\right] + \frac{\sigma^2}{2}\,\partial_{xx}x^2 p(x,t)\,. \tag{6.40}$$

As remarked above, the physical state space to which the diffusion process needs to be restricted is the nonnegative real half line. It can be seen that zero and ∞ are also intrinsic boundaries of the process since $g(0) = 0$ and $f(\infty) = -\infty$. According to (5.89) these intrinsic boundaries are natural boundaries if, with $\beta > 0$,

$$L_1(0) = C\int_0^{\beta} x^{-(2\lambda/\sigma^2)-2+\nu} \exp(2x/\sigma^2)\,dx = \infty \tag{6.41}$$

and

$$L_1(\infty) = C\int_{\beta}^{\infty} x^{-(2\lambda/\sigma^2)-2+\nu} \exp(2x/\sigma^2)\,dx = \infty\,. \tag{6.42}$$

Condition (6.42) is always fulfilled, i.e., infinity is a natural boundary for all values of λ and σ. The probability of explosion, even if time goes to infinity, is thus always zero. The situation for the lower boundary zero is more complicated. Equation (6.41) shows that zero is a natural boundary if $\lambda > \sigma^2/2$ in the Ito interpretation and if $\lambda > 0$ in the Stratonovich interpretation. In that case neither of the two intrinsic boundaries is accessible and conveniently no boundary conditions have to be imposed on the solution of the FPE. For $\lambda < 0$ (Stratonovich) and $\lambda < \sigma^2/2$ (Ito) we find that

$$L_2(0) = \infty\,,$$

which implies that 0 is an attracting boundary.[3]

In this section we shall discuss the stationary-state solution $p_s(x)$ of (6.39, 40). One has:

$$p_s(x) = N x^{(2\lambda/\sigma^2)-\nu} \exp\left(-\frac{2x}{\sigma^2}\right) \quad \text{with} \quad \begin{cases} \nu = 1 & \text{Stratonovich} \\ \nu = 2 & \text{Ito} \end{cases}\,. \tag{6.44}$$

One may notice that $p_s(x)$ is integrable over $[0, \infty)$, that is the stationary state solution exists, if and only if

$$2\frac{\lambda}{\sigma^2} - \nu + 1 > 0\,,$$

that is $\lambda > 0$ in the Stratonovich interpretation and $\lambda > \sigma^2/2$ in the Ito interpretation. As was to be expected, this coincides with the condition that $b_1 = 0$ is a natural boundary. The norm N is then given by

---

[3] The fact that the population never attains the boundary zero in a finite time is a consequence of the use of a continuous variable to describe the population. Obviously, if the population drops to a vanishingly small size the phenomenological equation based on this idealization breaks down. A way to overcome this difficulty is to define the probability of extinction as

$$\lim_{t\to\infty} \int_0^{\varepsilon} p(x,t)\,dx \quad \text{with} \quad 1 \gg \varepsilon > 0\,. \tag{6.43}$$

$$N^{-1} = \left[\left(\frac{2}{\sigma^2}\right)^{2(\lambda/\sigma^2)-\nu+1}\right]^{-1} \Gamma\left(\frac{2\lambda}{\sigma^2} - \nu + 1\right). \tag{6.45}$$

In the case that the stationary probability density (6.44) does not exist, one should notice that zero is not only an intrinsic boundary but also a stationary point: drift and diffusion vanish simultaneously for $x = 0$. Since it is attracting, the stationary probability mass will be entirely concentrated on zero. Speaking in terms of probability densities one can say that

$$p_s(x) = \delta(x) \quad \text{for} \quad \lambda < 0 \text{ (Stratonovich)} \quad \text{and} \quad \lambda < \frac{\sigma^2}{2} \text{ (Ito)} .$$

As explained in the preceding section, the extrema of $p_s(x)$ are the most appropriate indicator for a transition in the steady-state behavior of the system. They may be identified with the macroscopic steady states of the systems and are the order parameter for nonequilibrium phase transitions. In the case of the Verhulst model they are the zeros of

$$\lambda x_m - x_m^2 - \frac{\nu\sigma^2}{2} x_m = 0 , \tag{6.46}$$

namely:

$$x_{m1} = 0$$

$$x_{m2} = \lambda - \nu\sigma^2/2 \quad \text{which exists only if} \quad \lambda > \nu\sigma^2/2 . \tag{6.47}$$

Here x_{m2} is always a maximum and x_{m1} is a maximum for $0 < \lambda < \nu\sigma^2/2$. This discussion clearly shows that in a fluctuating environment the Verhulst model has, in contrast to the deterministic case, *two* transition points: one at $\lambda_S^{(1)} = 0$ (Stratonovich) or $\lambda_I^{(1)} = \sigma^2/2$ (Ito) corresponding to the transition in the nature of the boundary zero, and another transition taking place at $\lambda_S^{(2)} = \sigma^2/2$ (Stratonovich) or $\lambda_I^{(2)} = \sigma^2$ (Ito). This latter transition corresponds to an abrupt change in the shape of the probability density, the maximum occurs at a nonzero value of the population. Summarizing the behavior of the probability density which is sketched in Fig. 6.2, we have:

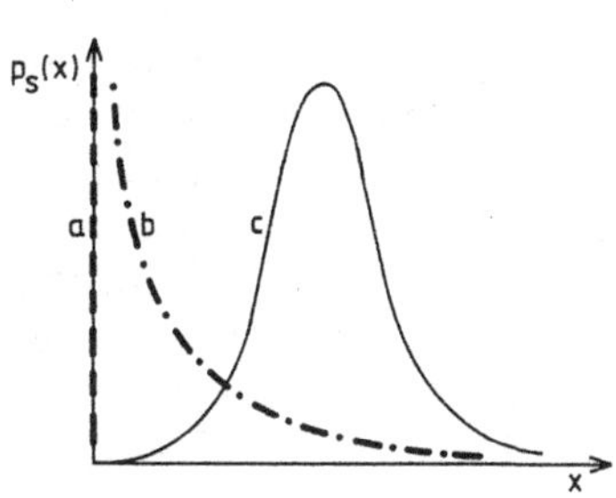

Fig. 6.2. Sketch of the stationary probability density of the Verhulst model (for different values of the external noise intensity σ^2). (*Curve a*) $\lambda_S<0$ or $\lambda_I<\sigma^2/2$; (*Curve b*) $0<\lambda_S<\sigma^2/2$ or $\sigma^2/2<\lambda_I<\sigma^2$; (*Curve c*) $\lambda_S>\sigma^2/2$ or $\lambda_I>\sigma^2$

i) If λ_S is negative ($\lambda_I < \sigma^2/2$) the stationary point zero is stable.

ii) The point $\lambda_S = 0$ ($\lambda_I = \sigma^2/2$) is a transition point, since the stationary point $x_0 = 0$ becomes unstable for $\lambda_S > 0$ ($\lambda_I > \sigma^2/2$) and a new genuine stationary probability density appears.

iii) The stationary density has a divergence at zero if $0 < \lambda_S < \sigma^2/2$ ($\sigma^2/2 < \lambda_I < \sigma^2$), i.e., it keeps part of the property of the delta function. Though zero is no longer a stable stationary point, it remains the most probable value. In a certain sense, the delta function begins to leak away to the right if λ crosses this transition point.

iv) If λ_S becomces greater than $\sigma^2/2$ ($\lambda_I > \sigma^2$), the character of the density changes again. The point $\lambda_S = \sigma^2/2$ ($\lambda_I = \sigma^2$) is therefore a second transition point. This shows that remarkably the system can be made to undergo a transition by keeping the average state of the environment constant but increasing or decreasing the intensity of its fluctuations. Such transitions belong to the general class of phenomena which we call noise-induced transitions.

It is also interesting to examine the behavior of the mean and variance of the population. We find in the Stratonovich interpretation:

$$E\{X\} = \int_0^\infty x p_s(x)\, dx = \lambda \,, \tag{6.48}$$

$$E\{X^2\} = \int_0^\infty x^2 p_s(x)\, dx = \lambda^2 + \lambda\sigma^2/2 \,, \tag{6.49}$$

$$E\{\delta X^2\} = E\{(X - E\{X\})^2\} = \lambda\sigma^2/2 \,. \tag{6.50}$$

Although the character of the stationary probability density changes at $\lambda_S = \sigma^2/2$ ($\lambda_I = \sigma^2$), this is not reflected by the behavior of the mean value and the variance, an expected result in the light of Sect. 6.3. The curve $E\{X\}$ vs λ is just the same as for the stable stationary solution of the deterministic model. In contrast, the extrema display only the second transition point, since the first is due to the change in the nature of $b_1 = 0$ from attracting to natural; their dependence with respect to λ can be viewed as a modification of the deterministic bifurcation diagram amounting to a shift of the transition point from $\lambda = 0$ to $\lambda = \sigma^2/2$. One notes that the second transition coincides with the point at which $[E\{(\delta X)^2\}]^{1/2}$ equals $E\{X\}$. This can be interpreted by saying that for $0 < \lambda < \sigma^2/2$ the fluctuations dominate over the autocatalytic growth of the population and extinction, though no longer certain, remains the most probable outcome. Note, however, that the population of course never actually reaches zero since this boundary is natural. Nevertheless, an appreciable amount of probability mass is accumulated in a vanishingly small neighborhood of zero such that the distribution function $F(x)$ emerges with a vertical tangent at $x = 0$. Hence the probability of extinction as defined in (6.43) is nonzero. For $\lambda > \sigma^2/2$ the autocatalytic growth wins over the influence of the fluctuations. In the neighborhood of zero this manifests itself by the fact that the probability of extinction drops to zero and the distribution function now emerges with horizontal tangent.

In a white-noise environment the Verhulst model has, as we have seen, two transition points, characterized by different order parameters. First, one has the point $\lambda_S = 0$ ($\lambda_I = \sigma^2/2$) where genuine growth becomes possible. This transition corresponds to the change from a degenerate random variable for steady-state behavior to a genuine stochastic variable. This type of qualitative change, due to the fact that the nature of the boundary $b_1 = 0$ switches from attracting to natural, is most naturally monitored via the moments, in particular the variance characterizing the width of the probability distribution. Second, one has the noise-induced point $\lambda_S = \sigma^2/2$ ($\lambda_I = \sigma^2$) which corresponds to a qualitative change in a genuine stochastic variable and involves no change in the nature of a boundary; the probability of extinction abruptly drops to zero. This second transition can be understood as a *noise-induced shift* of the deterministic transition, the transition from extinction to survival at $\lambda = 0$. Note that a linearization of the Verhulst equation can only describe the first transition, the change in the nature of the boundary $b_1 = 0$. This is due to the fact that $b_1 = 0$ is at the same time a stationary point of the SDE. Since the diffusion vanishes at a stationary point, the latter is always an intrinsic boundary. The loss of stability of the stationary point coincides with the change in the nature of the boundary to a natural boundary [5.12]. Either one can be determined by a linear analysis. The linearization obviously fails for the second transition point. Noise-induced transitions that do not involve a change in the nature of a boundary are intrinsically a nonlinear phenomenon. Since the external fluctuations are not macroscopically small, i.e., are not of order $O(V^{-\alpha})$, the state of the system is determined by the interplay of the full nonlinear dynamics and the external noise. A qualitative change in the genuine random variable, which describes the steady-state behavior of the system, involves therefore nonlinear effects and necessitates a fully nonlinear treatment.

The above type of noise-induced transition, namely a shift of transition phenomena which are already present in the usual deterministic bifurcation diagrams, is characterized by the fact that it can occur for arbitrarily small values of noise intensity, if the system is sufficiently close to the deterministic instability point. The fact that not only a shift of the deterministic transition occurs, but also that the deterministic transition point is split into two is caused by the particularity of the Verhulst model, that a boundary, namely zero, coincides for all values of λ and σ^2, with a stationary point of the SDE. The shift type of noise-induced transitions is expected to occur as a rather common phenomenon in the neighborhood of instability points in systems subjected to multiplicative white noise. While this kind of noise-induced transition is interesting, namely emphasizing the fact that the knowledge of the average environmental state is insufficient to predict the macroscopic behavior of the system, we shall now show that external noise can lead to even more profound modifications in the macroscopic behavior of nonlinear systems.

6.5 Pure Noise-Induced Transition Phenomena: A Noise-Induced Critical Point in a Model of Genic Selection

6.5.1 The Model

The most drastic modification of macroscopic properties which may result from the influence of external noise is of course the induction of a transition in a system that is incapable of any transition phenomenon whatsoever under deterministic conditions. In order to present this phenomenon and to discuss its properties, we shall use in this section one of the simplest models which displays pure noise-induced transitions. It corresponds to the deterministic phenomenological equation

$$\dot{X} = \alpha - X + \lambda X(1-X), \quad X \in [0,1] \tag{6.51}$$

in which X is a state variable and λ a parameter coupled to the environment. Though this model was initially introduced in a purely theoretical framework by *Arnold* et al. [6.8], it is not artificial and particularly in the field of population genetics, it finds a quite realistic interpretation. We shall postpone this interpretation and its justification which is rather involved till Sect. 6.5.4. From now on, to be brief, we shall already call this model the *genetic model*. Applications in other fields are however also possible. We shall start which a chemical realization of (6.51), which from a physico-chemical point of view permits us to exemplify some of the remarkable features of pure noise-induced transitions. Consider the following reaction scheme:

$$A + X + Y \underset{k_2}{\overset{k_1}{\rightleftarrows}} 2Y + A^* \tag{6.52}$$

$$B + X + Y \underset{k_4}{\overset{k_3}{\rightleftarrows}} 2X + B^* . \tag{6.53}$$

Obviously, the reactions conserve the total number of X and Y particles

$$\tilde{X}(t) + \tilde{Y}(t) = N = \text{const} . \tag{6.54}$$

Using this relation and defining the dimensionless parameters

$$\alpha = \frac{k_2 A^*}{k_2 A^* + k_4 B^*}, \quad \lambda = \frac{k_3 B + k_4 B^* - k_1 A - k_2 A^*}{k_2 A^* + k_4 B^*}, \tag{6.55}$$

one finds that the time evolution of the fraction $X = \tilde{X}/N$ obeys the kinetic equation (6.51). The form of the dependence of the multiplicative parameter λ on the externally controlled variables A, B, A^*, B^* allows for a range of variation which in principle extends from $-\infty$ to $+\infty$.

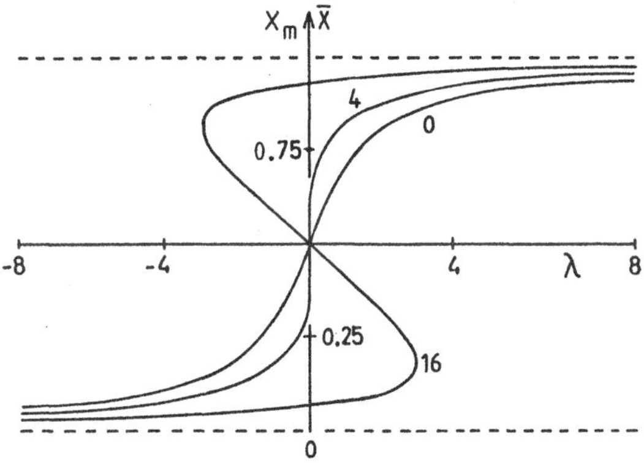

Fig. 6.3. Extrema of the stationary probability density $p_s(x)$ of the genetic model as a function of λ for increasing values of σ^2. The curve labeled zero (switching curve) corresponds to the deterministic case; it is single valued for all values of λ. For $\sigma^2 = \sigma_c^2 = 4$ (Stratonovich interpretation) $x_m = 1/2$ $\lambda_c = 0$ is a noise-induced critical point

For the sake of simplicity we take $\alpha = 1/2$; the physically meaningful steady-state value $\bar{X}$ is then:

$$\bar{X} = [\lambda - 1 + (\lambda^2 + 1)^{1/2}]/(2\lambda) .$$

This is a one to one mapping from the interval $(-\infty, +\infty)$ onto the interval $[0,1]$ as can be seen in Fig. 6.3 (the curve labeled 0). It is easily found by linear stability analysis that in the neighborhood of $\bar{X}$, the inverse of the relaxation time $\tau_{\text{macro}}^{-1} \equiv \omega = -(1+\lambda^2)^{1/2}$ is negative for all λ's. Furthermore, since $\dot{X}(X) < 0$ for $X > \bar{X}$ and $\dot{X}(X) > 0$ for $X < \bar{X}$, these stationary states are asymptotically globally stable. From a thermodynamic point of view, it is worth remarking that the asymptotic global stability of the systems' stationary states holds whatever the magnitude of the nonequilibrium constraints imposed by the environment on the system. In other words, the thermodynamic branch is the unique stable branch of stationary states possible under deterministic environmental conditions. Whatever the difference between the actual value of the products to substrates ratio $(A^*B^*)/(AB)$, even if this difference is quite large compared to the equilibrium mass action value

$$\left(\frac{A^*B^*}{AB}\right)_{\text{eq}} = \frac{k_1 k_3}{k_2 k_4} , \tag{6.56}$$

in the course of time the system always evolves towards states belonging to the thermodynamic branch. This system has the desired feature that in a constant environment no instability occurs. Any transition phenomena which may possibly be observed in a fluctuating environment are thus a pure noise effect corresponding to a qualitative change of the macroscopic properties.

6.5.2 A Noise-Induced Critical Point

Suppose now that the system is coupled to noisy surroundings in such a way that the external parameter λ becomes a fluctuating parameter. A convenient interpretation of the origin of these fluctuations in the chemical interpretation is that

A and B are fluctuating quantities while A^* and B^* are in large excess so that their fluctuations can be neglected to a good degree of approximation. We assume again that the external fluctuations are extremely rapid and write $\lambda_t = \lambda + \sigma\xi_t$. We obtain the SDE (in which without loss of generality we have put $\alpha = 1/2$)

$$dX_t = [\tfrac{1}{2} - X_t + \lambda X_t(1-X_t)]\,dt + \sigma X_t(1-X_t)\circ dW_t\,, \tag{6.57}$$

which we shall interpret in the sense of Stratonovich. The reader can easily verify that the Ito interpretation of (6.57) leads qualitatively to the same results. As already remarked above, the difference is a factor 2 in the noise intensity.

According to (5.75) the following Ito equation is equivalent to the Stratonovich SDE (6.57):

$$dX_t = \left[\frac{1}{2} - X_t + \lambda X_t(1-X_t) + \frac{\sigma^2}{2}X_t(1-X_t)(1-2X_t)\right]dt \\ + \sigma X_t(1-X_t)\,dW_t\,. \tag{6.58}$$

Conveniently the boundaries 0 and 1 of the state space are as in the Verhulst model intrinsic boundaries for the diffusion process X_t, since $g(0) = g(1) = 0$. Furthermore, it is easily verified, using the analytic condition (5.89), that both boundaries are natural (in G-S sense) for the whole range of the parameters λ and σ. Therefore the stationary probability density of the diffusion process, defined by (6.58), reads

$$p_s(x) = Nx^{-1}(1-x)^{-1}\exp\left\{\frac{2}{\sigma^2}\left[-\frac{1}{2x(1-x)} - \lambda\ln\left(\frac{1-x}{x}\right)\right]\right\} \tag{6.59}$$

and is always normalizable on $[0, 1]$. Further,

$$N = \tfrac{1}{2}\exp(2/\sigma^2)\cdot K_0^{-1}(2/\sigma^2)\,,$$

for the case $\lambda = 0$, which we shall mainly consider here. K_0 is the modified Bessel function. To investigate the occurrence of transition phenomena in the system coupled to external noise, in agreement with our general discussion in Sect. 6.3 we study the behavior of the extrema of the stationary probability density to detect any qualitative changes in the steady-state behavior of the system. No transitions involving changes in the nature of the boundaries occur in this system. The extrema x_m of $p_s(x)$ can be calculated from (6.36), which yields:

$$\frac{1}{2} - x_m + \lambda x_m(1-x_m) - \frac{\sigma^2}{2}x_m(1-x_m)(1-2x_m) = 0\,. \tag{6.60}$$

Note that the degree of this polynomial is increased by one compared to that for the deterministic steady states. For simplicity let us first discuss the case $\lambda = 0$.

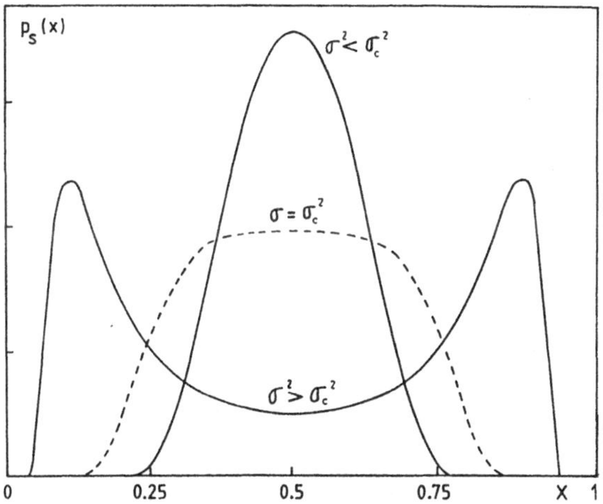

Fig. 6.4. Sketch of the stationary probability density $p_s(x)$ of the genetic model below, at (broken curve) and beyond the critical intensity σ_c^2

The stationary solution of the deterministic phenomenological equation (6.51) is then

$$\bar{X} = 1/2 \, . \tag{6.61}$$

In a rapidly fluctuating environment we have instead from (6.60)

$$x_{m1} = 1/2 \quad \text{and} \quad x_{m\pm} = 1/2[1 \pm (1 - 4/\sigma^2)^{1/2}] \, . \tag{6.62}$$

Thus for $\sigma^2 > 4$ the stationary probability density possesses three extrema, of which $x_{m\pm}$ are maxima. The deterministic state $x_{m1} = 1/2$ which is the most probable state for $\sigma^2 < 4$ has turned into a minimum (Fig. 6.4). Going back to our original reaction scheme, the phenomenon can be visualized as follows (Fig. 6.4): for $\sigma^2 \to 0$ one has an equal amount of X and Y particles in the reactor. This means that if X and Y are particles of different colors, say yellow and blue, the reactor would appear green. If σ^2 is finite but below the critical value, neither X nor Y will dominate in the reactor and accordingly it will show a flickering green shade. Above the critical variance $\sigma_c^2 = 4$, it will either be mostly blue or mostly yellow and spend an equal amount of time in both states. Below the threshold, exchanging X and Y particles has no effect on the macroscopic state of the reactor. On the contrary, above the threshold the symmetry of this exchange has been broken. Exchanging X and Y particles corresponds to the observation of two well-distinct macroscopic states.

The situation is qualitatively the same for the asymmetric case $\lambda \neq 0$. Even if the deterministic steady-state solution lies close to either one of the two boundaries of the state space, nevertheless the probability density will always become bimodal once the intensity σ^2 of the external noise crosses a certain threshold value. The latter increases with $|\lambda|$. The genetic model thus always exhibits a transition which is a pure noise effect. For $\lambda = 0$, this transition is a soft one. At $\sigma^2 = 4$, $x_{m1} = 1/2$ is a double maximum and the distance between x_{m+} and x_{m-}

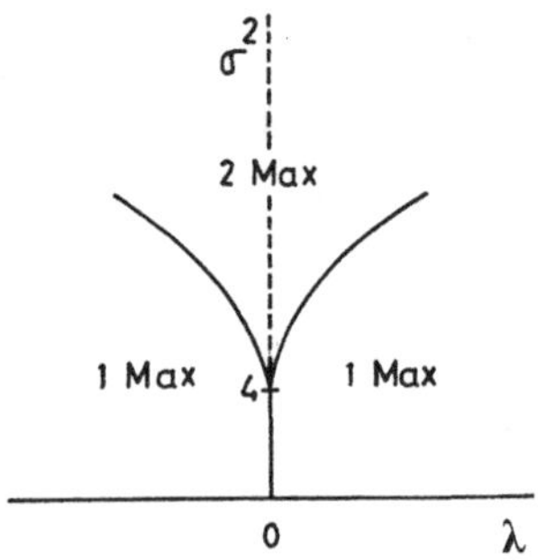

Fig. 6.5. Cusp catastrophe

tends to zero like $(\sigma^2-\sigma_c^2)^{1/2}$ for $\sigma^2 \downarrow \sigma_c^2$. This indicates that external Gaussian white noise induces critical behavior in the genetic model with a critical point at $\lambda=0$, $x=1/2$, $\sigma^2=4$. This is confirmed by the behavior of $p_s(x)$ for the asymmetric cases with $\lambda \neq 0$. For $\lambda>0$ ($\lambda<0$), the peak corresponding to the steady state of the deterministic equation moves towards 1 (towards 0) with growing σ^2 and if σ^2 exceeds $\sigma_{th}^2\,(|\lambda|)>4=\sigma_c^2$, a second peak appears at a finite distance from the original one, near the other boundary of the state space. If we keep σ^2 fixed and bigger than 4 and vary λ along the real line, the situation resembles a first-order transition as is clear from the sigmoidal form of the curve for the extrema of $p_s(x)$, e.g., $\sigma^2=16$ in Fig. 6.3. This shows that indeed $\sigma_c^2=4$ is a critical variance beyond which a hysteresis phenomenon for the extrema occurs. The above facts can be summarized in the statement that as to the extrema of $p_s(x)$ we have a cusp catastrophe in the (λ,σ^2) half-plane with critical point at $(0,4)$ (Fig. 6.5).

This qualitative change in the steady-state behavior of the system can of course be traced back to the above-mentioned fact that the degree of the polynomial (6.60), giving the extrema of the probability density, is increased by one compared with the equation for the deterministic steady states. A nice way to visualize this qualitative change is to consider the stochastic potential. Below the transition this potential has only one valley and resembles therefore the deterministic potential. The only effect of the noise is a disorganizing effect leading to a broadening of the stationary probability density. As sketched in Fig. 6.6 at the transition point $\sigma_c^2=4$, the bottom then heaves up and two new potential wells are created, since the boundaries 0 and 1 have to remain natural boundaries. This shows that in contradistinction to additive noise, multiplicative noise has not only a disorganizing effect but may stabilize new macroscopic states in the system. The existence of such a potential makes it possible to treat noise-induced phase transitions in the language of catastrophe theory [6.9].

To conclude, let us go back to the chemical interpretation (6.52, 53) of model (6.51) which permits us to situate the originality of this pure noise-induced transition from a thermodynamic point of view. Under deterministic environmental conditions, a general thermodynamic result ([1.14] and Sect. 1.2) rules out completely the possibility of any bistability phenomenon at equilibrium in an ideal chemical system, i.e., in the case of the present model when the products to substrates ratio (A^*B^*/AB) is equal to its mass action value (6.56). It is worthwhile to point out here that obviously this result does not hold in terms

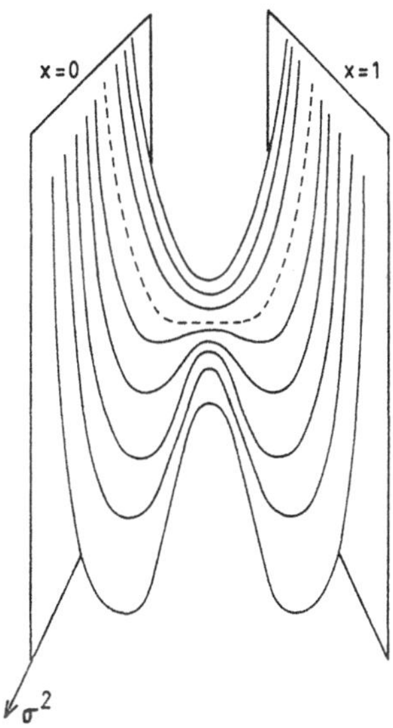

Fig. 6.6. Sketch of the stochastic potential $\mathscr{V}(x)$ for increasing values of σ^2 and $\lambda = 0$. At σ_c^2 (broken curve), the minimum at $x = 1/2$ becomes flat; beyond this point it transforms into a maximum while the minima at $x_{m\pm}$ (6.62) become more pronounced as σ^2 continues to increase. Near the planes $x = 0$ and $x = 1$, $\mathscr{V}(x) \to \infty$, reflecting the fact that these values are natural boundaries for X_t

of the average value of the environment. If condition (6.56) is replaced by the condition

$$\frac{E\{A^*\}E\{B^*\}}{E\{A\}E\{B\}} = \frac{k_1 k_3}{k_2 k_4}$$

on the averages of the substrates and products, no restriction results that is strong enough to bar the multiplicative parameter λ, defined by (6.55), from fluctuating over its state space $(-\infty, +\infty)$. Thus even for an environment obeying this *pseudo-equilibrium* condition, a noise-induced transition leading to bistability occurs. This illustrates marvellously that one must be prepared to renounce certain convenient ideas when dealing with environments which fluctuate rapidly and strongly: common belief has it that due to their rapidity these fluctuations are averaged out and that for this reason physico-chemical systems submitted to them adapt essentially to the constant average state of the environment. If this belief reflected the true state of affairs, the above condition would imply that the system is in equilibrium on the average and no transitions occur. As the preceding example most clearly demonstrates, this is far from the real physical situation.

6.5.3 Critical Exponents for Noise-Induced Critical Behavior

In order to establish the close ties of the noise-induced transitions just described with the more classical phase-transition phenomena, let us now calculate their critical exponents. Since the system is spatially homogeneous we expect mean-field theory results. For those readers who are not familiar with the classical theory of equilibrium phase transitions, we briefly recall the essential results of the so-called mean-field theory in App. B.

As already explained in Sect. 6.3, the order parameter for noise-induced transitions corresponds to the extrema of the stationary probability density, i.e., the values for the macroscopic "phases" of the system. To be precise, for the genetic model we choose $m = |x_m - \bar{x}|$ as the order parameter. This amounts only to a translation of the origin by $-1/2$. The role of the temperature T is played here by the intensity σ^2 of the noise. The analog of the applied external field h is the parameter describing the average state of the environment. In the genetic model, it is obviously the selection coefficient λ. To determine the critical exponent β, we have to find the behavior of m as a function of σ^2 near $\sigma_c^2 = 4$. Then (6.62) yields

$$m = \tfrac{1}{2}[(\sigma^2 - 4)/\sigma^2]^{1/2} \sim (\sigma^2 - \sigma_c^2)^{1/2} \quad \text{for} \quad \sigma^2 \gtrsim \sigma_c^2 . \tag{6.63}$$

Hence β takes the classical value 1/2. To determine the critical exponent δ, we have to find the behavior of m for small λ at $\sigma^2 = \sigma_c^2$. From (6.60) we obtain for $m = x_m - 1/2$:

$$-m + \lambda(1/4 - m^2) + \sigma^2 m(1/4 - m^2) = 0 \tag{6.64}$$

and rearranging terms, (6.64) reads

$$m^3 + \lambda m^2/\sigma^2 + (4 - \sigma^2)m/(4\sigma^2) - \lambda/(4\sigma^2) = 0 . \tag{6.65}$$

Specifying $\sigma^2 = \sigma_c^2 = 4$, (6.65) yields

$$\begin{aligned} P &\equiv m^3 + \frac{\lambda}{4} m^2 - \frac{\lambda}{16} \\ &= m^3 + a_2 m^2 + a_0 = 0 . \end{aligned}$$

The solution of this cubic equation is given by

$$\begin{aligned} m_1 &= (s_1 + s_2) - a_2/3 \\ m_{2/3} &= -1/2(s_1 + s_2) - a_2/3 \pm \frac{\mathrm{i}\sqrt{3}}{2}(s_1 - s_2) , \end{aligned}$$

where

$$s_{1,2} = (2^{-5}\lambda - 3^{-3}\, 2^{-6}\lambda^3 \pm u)^{1/3}$$

and

$$u = 2^{-5}\lambda(1 - 3^{-3}\lambda^2)^{1/2} \quad \text{which for small } \lambda , \quad |\lambda| \ll 1$$

can be written as

$$u = 2^{-5}\lambda(1 - 2^{-1}\, 3^{-3}\lambda^2 + \dots) .$$

In the same approximation, i.e., up to order λ^3:

$$s_1 = [2^{-4}\lambda - 2^{-5}3^{-3}\lambda^3 + O(\lambda^4)]^{1/3}$$

$$s_2 = [O(\lambda^4)]^{1/3} .$$

Therefore we finally obtain

$$m_1 = [2^{-4}\lambda - 2^{-5}3^{-3}\lambda^3 + O(\lambda^4)]^{1/3} - 2^{-2}3^{-1}\lambda \tag{6.66}$$

and the critical exponent δ also takes the classical value: $\delta = 3$.

To calculate γ, we have to determine the behavior of the "susceptibility" $(\partial m/\partial\lambda)_{\lambda=0}(\sigma^2)$ as a function of σ^2 near $\sigma_c^2 = 4$.
Denoting $\partial/\partial\lambda$ by $\wedge$, we obtain from (6.65)

$$3m^2\hat{m} + m^2/\sigma^2 + 2\lambda m\hat{m}/\sigma^2 + (4-\sigma^2)\hat{m}/(4\sigma^2) - 1/(4\sigma^2) = 0 .$$

We set $\lambda = 0$, and take into account that $m^2 = 1/4 - 1/\sigma^2$ for $\lambda = 0$ and $\sigma^2 > \sigma_c^2$ to obtain

$$2\left(\frac{1}{4} - \frac{1}{\sigma^2}\right)\hat{m} + \frac{1}{\sigma^2}\left(\frac{1}{4} - \frac{1}{\sigma^2}\right) - \frac{1}{4\sigma^2} = 0$$

or

$$\left(\frac{\partial m}{\partial\lambda}\right)_{\lambda=0}(\sigma^2) = \hat{m} = \frac{2}{\sigma^2(\sigma^2-4)} ,$$

i.e.,

$$\left(\frac{\partial m}{\partial\lambda}\right)_{\lambda=0}(\sigma^2) \sim (\sigma^2 - \sigma_c^2)^{-1} .$$

For $\sigma^2 < \sigma_c^2$ we have

$$\frac{4-\sigma^2}{4\sigma^2}\hat{m} - \frac{1}{4\sigma^2} = 0$$

and thus

$$\left(\frac{\partial m}{\partial\lambda}\right)_{\lambda=0}(\sigma^2) \sim (\sigma_c^2 - \sigma^2)^{-1} .$$

This establishes that $\gamma = \gamma' = 1$, i.e., all critical exponents of the pure noise-induced critical point are given by the classical values. This shows that equilibrium phase transitions, nonequilibrium phase transitions and noise-induced phase transitions are indeed close kin. There is a deep unity in the fundamental phenomenon, namely to be a phase transition, and except for the qualifiers equilibrium, nonequilibrium and noise induced no further distinction is warranted.

6.5.4 Genic Selection in a Fluctuating Environment

To complete the presentation of our basic model (6.51) for pure noise-induced transitions, we shall show that it can be used to describe a well-defined mechanism of genetic selection in population dynamics. We consider a single haploid population and focus on a particular genetic locus for which two alleles A and a are possible. The number of individuals in the population having genotype A and a are respectively N_A and N_a. We assume that the total number of individuals $N = N_A + N_a$ in the population is constant due to population regulating mechanisms such as food supply, predators, etc. We assume further that the total population N is large so that we may neglect the internal statistical fluctuations which become important when the population size is small.

We shall restrict ourselves to populations with nonoverlapping generations. Let the length of one generation be Δt. We are interested in the variations of the frequencies $X = N_A/N$ and $(1-X) = N_a/N$ of the two alleles in the population from generation to generation. These frequencies change under the influence of two factors: natural selection which favors the allele best adapted to the environment and mutations which transform one allelic form into the other. We shall consider the case that these two processes operate slowly and cause only small changes per generation. Then their effects are additive [Ref. 6.10, p. 150]. If u_A and u_a are the mutation rates per generation from A to a and from a to A, respectively, we have for the change in frequency of A:

$$\begin{aligned}\Delta_{\mathrm{mut}} X(t) &= -u_A X(t) + u_a[1 - X(t)] \\ &= -(u_A + u_a)X(t) + u_a\,. \end{aligned} \tag{6.67}$$

The fact that one allele is better adapted to the environment than the other means that it has a higher reproductive success, in other words a large growth rate per generation w:

$$N_A(t+\Delta t) = w_A N_A(t)\,, \qquad N_a(t+\Delta t) = w_a N_a(t) \tag{6.68}$$

with

$$w_A = 1 + s_t/2 \quad \text{and} \quad w_a = 1 - s_t/2\,,$$

where s_t is the selection coefficient per generation. We obtain for the change in frequency due to natural selection:

$$\Delta_{\mathrm{n.s.}}X(t) = \frac{N_{\mathrm{A}}(t+\Delta t)}{N_{\mathrm{A}}(t+\Delta t)+N_{\mathrm{a}}(t+\Delta t)} - X(t)$$
$$= \frac{(1+s_t/2)X(t)}{(1+s_t/2)X(t)+(1-s_t/2)[1-X(t)]} - X(t)$$
$$= \frac{s_tX(t)-s_tX^2(t)}{1-s_t/2+s_tX(t)} . \tag{6.69}$$

For the total change of frequency of A from one generation to the next we have

$$\Delta X(t) = \Delta_{\mathrm{mut}}X(t) + \Delta_{\mathrm{n.s.}}X(t)$$
$$= -(u_{\mathrm{A}}+u_{\mathrm{a}})X(t) + u_{\mathrm{a}} + s_t\frac{X(t)[1-X(t)]}{1-s_t/2+s_tX(t)} \tag{6.70}$$

and if the selection coefficient per generation is small

$$\Delta X(t) = -(u_{\mathrm{A}}+u_{\mathrm{a}})X(t) + u_{\mathrm{a}} + s_tX(t)[1-X(t)][1+s_t/2-s_tX(t)+\ldots] . \tag{6.71}$$

If the environment exhibits a certain random variability, then the selection coefficient per generation s_t will fluctuate from generation to generation. We consider time spans which are long compared to Δt, the length of one generation, but short compared to the time scale of any systematic evolution of the environment. Thus can s_t be modeled by a stationary random process. We shall discuss here the simplest case where the environmental fluctuations are independent from one generation to the next. In moderate zones, this is true to a very good degree of approximation for climate fluctuations. Then s_t is discrete white noise with

$$E\{s_t\} = s = \lambda\Delta t \tag{6.72}$$

and

$$E\{(s_t-s)(s_{t+\mathrm{i}\Delta t}-s)\} = C(\mathrm{i}\Delta t) = \sigma^2\Delta t\,\delta_{i,0} . \tag{6.73}$$

Here we have introduced the selection rate λ. The fitness of an allele depends in general on a multitude of environmental factors. Thus s_t is the cumulative effect of a large number of small additive contributions. Invoking the central limit theorem we can therefore assume that s_t is Gaussian. Using the mutation rates v_{A} and v_{a}, i.e.,

$$u_{\mathrm{A}} = v_{\mathrm{A}}\Delta t \quad \text{and} \quad u_{\mathrm{a}} = v_{\mathrm{a}}\Delta t , \tag{6.74}$$

we can write (6.71) as

$$\Delta X_t = [-(v_A + v_a)X_t + v_a]\Delta t + s_t X_t(1-X_t)[1 + s_t/2 - s_t X_t + \ldots] . \tag{6.75}$$

In the continuous time limit $\Delta t \to 0$, the Markov process X_t converges towards a diffusion process. To characterize the latter we have to find the first two differential moments. The drift is given by

$$\lim_{\Delta t \to 0} \frac{1}{\Delta t} E\{\Delta X_t | X_t = x\}$$

and the diffusion by

$$\lim_{\Delta t \to 0} \frac{1}{\Delta t} E\{\Delta X_t^2 | X_t = x\} .$$

From (6.75) we obtain

$$E\{\Delta X_t | X_t = x\} = [-(v_A + v_a)x + v_a]\Delta t + \lambda \Delta t x(1-x) + \sigma^2 \Delta t x(1-x)(\tfrac{1}{2} - x) + o(\Delta t) \tag{6.76}$$

and

$$E\{\Delta X_t^2 | X_t = x\} = \sigma^2 \Delta t x^2 (1-x)^2 + o(\Delta t) . \tag{6.77}$$

Hence the changes in the frequency of allele A in a haploid population due to the processes of mutation and natural selection in a stationary random environment are described by the following Ito SDE

$$dX_t = [v_a - (v_A + v_a)X_t + \lambda X_t(1-X_t) + \frac{\sigma^2}{2} X_t(1-X_t)(1-2X_t)]dt + \sigma X_t(1-X_t) dW_t . \tag{6.78}$$

Equation (6.78) also describes the changes in the frequency of allele A in a diploid population if no dominance occurs, i.e., the properties of the heterozygote Aa are the average of the properties of the homozygotes AA and aa [Ref. 6.10, pp. 148, 150]. If the mutation rates v_A and v_a are equal, (6.78) goes over into (6.57) via a simple rescaling of time. Reinterpreting results from Sect. 6.5.2 now from a genetic point of view, we are led to some startling conclusions. Even if on the average both alleles are equally fit, $\lambda = 0$, i.e., in a deterministic environment, no selection would occur, one must expect in a random environment to find predominantly *only one* of the alleles if $\sigma^2 > 4$. Indeed, the population will be found to correspond to either one of the most probable states, x_{m+} or $x_{m-} = 1 - x_{m+}$, of the stationary state probability density associated with (6.78). In other words, though there is no systematic selection pressure, in an ensemble of populations relatively pure populations will dominate, if the intensity of the environmental fluctuations is sufficiently large.

This throws some new light on the influence of environmental fluctuations on the preservation of protein polymorphism. It has been perceived [6.11 – 16] (for a review see [6.17]) that random temporal variations in selection intensities may constitute an important factor in the mechanism at the origin of protein polymorphism which *Kimura* and his proponents [6.18] attribute essentially to random sampling. For the above model, its properties are such that qualitatively the outcome very much depends on the intensity of environmental variability: as long as $p_s(x)$ admits only one extremum, i.e., $\sigma^2 < 4$, the population evolves in the course of time essentially in the neighborhood of the state $x = 1/2$ where indeed polymorphism dominates. To the contrary, for large values of σ^2 the transition from one maximum to the other, i.e., from one macroscopic stationary state to the other, becomes more and more improbable; the bottleneck between the maxima is indeed narrower the bigger σ^2:

$$p_s(1/2) = 2\exp(-2/\sigma^2)K_0^{-1}(2/\sigma^2) \approx -2\exp(-2/\sigma^2)/\ln(2/\sigma^2)$$

for large σ^2, i.e., $p_s(1/2) \to 0$ for $\sigma^2 \to \infty$. For large enough σ^2 the transition from one peak to the other is so rare an event that it would be extremely unlikely to happen in a time interval of ten or even a hundred generations. Thus one is led to the conclusion that increasing environmental variability favors, at least in haploid populations and diploid populations with no dominance, the stabilization of one of the genotypes with respect to the other. Quite strikingly when the average value of the environment is not neutral, i.e., $\lambda \neq 0$, this effect may lead to the stabilization of the normally considered as "unfit" genotype.

6.6 Time-Dependent Behavior of Fokker-Planck Equations: Systems Reducible to a Linear Problem

The preceding sections dealt in detail with the stationary behavior of nonlinear systems coupled to a fluctuating environment. In the remainder of this chapter, we shall address the transient behavior of such systems. In other words, we shall investigate how they approach the stationary state starting from an arbitrary initial condition. This problem is considerably harder than the analysis of stationary behavior. In general no explicit formula for the time-dependent solution of the FPE exists, even for one-variable systems, in contrast to the stationary solution.

The exact time-dependent solution of a SDE is, however, easily obtained if the drift and diffusion coefficient are linear functions. It is thus worthwhile for a study of transient behavior to determine those nonlinear SDE's which can be transformed into a linear SDE by a (bijective) change of variable. For some systems belonging to this class, it is then possible to derive the exact time-dependent solution of the corresponding FPE in an explicit manner.

6.6.1 Transformation to Linear SDE

According to *Gihman, Skorohod* [5.12] the nonlinear SDE (6.14) can be transformed into the linear SDE

$$dY_t = (\alpha + \beta Y_t)dt + (\gamma + \vartheta Y_t)dW_t, \tag{6.79}$$

where α, β, γ and ϑ are real constants, if

$$\frac{d}{dX}\left(\frac{\frac{d}{dX}[g(X)Z'(X)]}{Z'(X)}\right) = 0 \tag{6.80}$$

with

$$Z(X) = f(X)/g(X) - \tfrac{1}{2}g'(X). \tag{6.81}$$

The appropriate transformed variable is given by

$$Y = C\exp[\vartheta B(X)] \tag{6.82}$$

with

$$\vartheta = -\left\{\frac{d}{dX}[g(X)Z'(X)]\right\}/Z'(X) \tag{6.83}$$

and

$$B(X) = \int^X \frac{dZ}{g(Z)}. \tag{6.84}$$

If $\vartheta = 0$,

$$Y = \gamma B(X) + C, \tag{6.85}$$

where C is an arbitrary constant. The solution of (6.79) is

$$Y_t = \exp\left[\left(\beta - \frac{\vartheta^2}{2}\right)t + \vartheta W_t\right] \cdot \left\{Y_0 + \int_0^t \exp\left[-\left(\beta - \frac{\vartheta^2}{2}\right)s - \vartheta W_s\right](\alpha - \vartheta\gamma)ds \right.$$
$$\left. + \int_0^t \exp\left[-\left(\beta - \frac{\vartheta^2}{2}\right)s - \vartheta W_s\right]\gamma dW_s\right\}. \tag{6.86}$$

Inspecting (6.86), we conclude that the case $\vartheta = 0$ is the most convenient for our purposes. For the subclass of models fulfilling this particular condition, W_t does not appear in the exponential function. In other words, the Gaussian Wiener process is subjected only to linear transformations for $\vartheta = 0$ (6.86) and thus Y_t is a Gaussian process, provided Y_0 is Gaussian or a nonrandom constant.[4] Since a Gaussian process is completely determined by $E\{Y_t\}$ and $E\{(Y_t - \delta Y_t)(Y_s - \delta Y_s)\}$,

[4] Remember Y_0 has to be independent of the Wiener process (p. 93).

the transition probability density of Y_t and of X_t using the inverse transformation can easily be found in an exact and explicit way.

6.6.2 Examples: The Verhulst Model and Hongler's Model

It is easily verified that the Verhulst model fulfills condition (6.80). The transformation $y = 1/x$ leads to the linear SDE

$$dY_t = \left[\left(\frac{\nu}{2}\sigma^2 - \lambda\right) Y_t + 1\right] dt + \sigma Y_t dW_t \,. \tag{6.87}$$ [5]

According to (6.86) the solution of this SDE is

$$Y_t = \exp\left[\left(\frac{\nu-1}{2}\sigma^2 - \lambda\right) t + \sigma W_t\right] \times \left\{ Y_0 + \int_0^t \exp\left[-\left(\frac{\nu-1}{2}\sigma^2 - \lambda\right) s - \sigma W_s\right] ds \right\} . \tag{6.88}$$

The Verhulst model does not belong to the subclass $\vartheta = 0$. Thus unfortunately the Wiener process appears in the exponential function. This constitutes obviously a nonlinear transformation of a Gaussian process and implies, as remarked above, that solution (6.88) is not a Gaussian process. This makes it rather difficult to evaluate the transition probability density of Y_t and consequently that of X_t. In other words, (6.88) is not a suitable starting point to find the time-dependent solution of the FPE in the Verhulst model. Different techniques have to be employed and we shall come back to this problem in the next section.

As to the prototype system for pure noise-induced transitions, we find by applying condition (6.80) to the genetic model that it cannot be transformed into a linear SDE. However, some indications on the time-dependent properties of this model can be obtained by analyzing the following system introduced by *Hongler* [6.19]:

$$dU_t = -\frac{1}{2\sqrt{2}} \tanh(2\sqrt{2}\, U_t)\, dt + \frac{\sigma}{4} \operatorname{sech}(2\sqrt{2}\, U_t) \circ dW_t \,. \tag{6.89}$$

Though there is no physico-chemical process which can immediately be associated with (6.89), this model shows features similar to the genetic model, most importantly a noise-induced critical point. This is not surprising since in a certain sense Hongler's model is close to the genetic model. Expanding the hyperbolic functions up to second order around zero, we obtain:

---

[5] $\nu = 1$ corresponds to the Stratonovich interpretation of the original Verhulst equation, $\nu = 2$ to the Ito version.

$$dU_t = -U_t dt + \sigma(1/4 - U_t^2) \circ dW_t, \tag{6.90}$$

which under the simple shift or coordinate $x = 1/2 + u$ yields (6.57) with $\lambda = 0$.

The analysis of the pure noise-induced critical point found in the genetic model revealed that this transition involves only a local change in the shape of the stationary probability density, namely near $x = 1/2$. Outside this neighborhood, in particular near the natural boundaries of the system, the density is not affected by the transition. In view of this fact and of the property that the genetic model and Hongler's model coincide in the neighborhood of $x = 1/2$, it is not astonishing that Hongler's model exhibits the same kind of noise-induced transition. This makes it plausible that the dynamics of the noise-induced transition are qualitatively similar in both models. This conjecture will be confirmed later on. The nice feature of (6.89) is that it fulfills conditions (6.80). The transformation

$$z = \sinh(2\sqrt{2}x) \tag{6.91}$$

transforms it into the O-U equation

$$dZ_t = -Z_t dt + \frac{\sqrt{2}}{2}\sigma dW_t. \tag{6.92}$$

Hongler's model belongs thus to the subclass $\vartheta = 0$. We have seen that the transition probability density of (6.92) is given by

$$p(z,t|z_0) = \left[2\pi\frac{\sigma^2}{4}(1-e^{-2t})\right]^{-1/2} \exp\left(-\frac{1}{2}\frac{(z-z_0 e^{-t})^2}{\frac{\sigma^2}{4}(1-e^{-2t})}\right) \tag{6.93}$$

with the initial condition

$$p(z,0|z_0) = \delta(z-z_0).$$

Transforming back to the original variable, one obtains

$$p(x,t|x_0) = p(z(x),t|z(x_0))\left|\frac{dz}{dx}\right|$$

$$= \frac{2\sqrt{2}\cosh(2\sqrt{2}x)}{\left[2\pi\frac{\sigma^2}{4}(1-e^{-2t})\right]^{1/2}} \exp\left(-\frac{1}{2}\frac{[\sinh(2\sqrt{2}x) - \sinh(2\sqrt{2}x_0)e^{-t}]^2}{\frac{\sigma^2}{4}(1-e^{-2t})}\right). \tag{6.94}$$

We shall come back to (6.94) in Sect. 6.8 where we discuss in more detail the dynamical properties of noise-induced transitions.

6.7 Eigenfunction Expansion of the Transition Probability Density

6.7.1 Spectral Theory of the Fokker-Planck Operator and the Sturm-Liouville Problem

In the preceding section we made a preliminary attempt to obtain information on the transient behavior of nonlinear systems in the presence of external noise. We determined the particular class of systems where the time-dependent solution of the SDE can easily be obtained. While the Verhulst model belongs to this class, this is not the case for the more interesting genetic model. However, even if the time-dependent solution of the SDE is explicitly known, the evaluation of the time-dependent solution of FPE, i.e., the transition probability density, is tractable only for a subclass, of which the Verhulst equation is not a member. It thus becomes necessary to attack this problem in a more general way. For our purposes, it suffices to consider only situations in which a genuine stationary probability density exists and is unique. In particular, this is fulfilled when the boundaries are either natural or regular boundaries with instantaneous reflection, which covers most situations in applications. This implies that there is no probability flux at the boundaries, i.e.,

$$\left[-\frac{\sigma^2}{2}\partial_x g^2(x)p(x,t)+f(x)p(x,t)\right]_{x=b_i}=0\,. \tag{6.95}$$

The systems we are interested in are described by a time-homogeneous diffusion process, i.e., the drift and diffusion do not depend on time. The FPE can therefore be solved by a separation of variables:

$$p(x,t|x_0)=\psi(x|x_0)\exp(-\mu t)\,. \tag{6.96}$$

For the sake of shortness of notations, we shall drop the conditioning on x_0 in the following. Injecting (6.96) in the FPE we obtain

$$-\mu\psi(x)=-\partial_x f(x)\,\psi(x)+\frac{\sigma^2}{2}\partial_{xx}g^2(x)\,\psi(x) \tag{6.97}$$

with the boundary condition

$$\left[-\frac{\sigma^2}{2}\partial_x g^2(x)\,\psi(x)+f(x)\,\psi(x)\right]_{x=b_i}=0\,. \tag{6.98}$$

Obviously $\psi(x)\equiv 0$ is a solution of (6.97) for all values of μ. This is the trivial solution. In general (6.97) will not admit a nontrivial solution for all values of μ. Those specific values of μ, for which a function $\psi_\mu(x)$ exists which does not vanish identically in the interval (b_1, b_2) and which fulfills (6.97, 98) are called the eigenvalues. The corresponding solutions are called the eigenfunctions. The set of all eigenvalues is called the spectrum of the operator. The problem of finding the time-dependent solution of the FPE thus reduces to an eigenvalue problem. In order for an eigenvalue problem to be well posed, the space of admissible functions for the operator has to be specified. The solutions of the FPE should of course be probability densities. This implies in particular that they have to be normalized to one for all times:

$$\int_{b_1}^{b_2} p(x,t)\,dx = 1\,.$$

The functions on which the FP operator acts should thus be integrable on the state space (b_1, b_2), i.e.,

$$\int_{b_1}^{b_2} \psi(x)\,dx < \infty\,. \tag{6.99}$$

The space of functions which fulfill (6.99) is denoted by $L_1(b_1, b_2)$. This is the appropriate function space for our eigenvalue problem [6.20].

If the eigenvalue problem had been formulated for the Kolmogorov backward equation, which is an equally valid approach to study transient behavior, the appropriate function space is the space of functions continuous on $[b_1, b_2]$, denoted by $C[b_1, b_2]$ [6.20]. The eigenvalue problem for the FPE is closely related to the so-called Sturm-Liouville problem, which has been abundantly studied in the literature [6.21 – 24]. The second-order equation

$$[P(x)\,Y'(x)]' + Q(x,\mu)\,Y(x) = 0 \tag{6.100}$$

with the boundary conditions

$$\alpha Y(b_1) + \beta Y'(b_1) = 0\,, \qquad \gamma Y(b_2) + \delta Y'(b_2) = 0 \tag{6.101}$$

is called a Sturm-Liouville equation. Here $P(x)$ and $Q(x,\mu)$ are continuous functions on $[b_1, b_2]$, $P(x)$ is positive and Q depends continuously also on the parameter μ. Further, α, β, γ and δ are real constants. The application of the Sturm-Liouville theory to diffusion processes is presented in [6.25]. To cast the eigenvalue problem for the FPE into a Sturm-Liouville problem, set

$$\psi(x) = \varphi(x)\,p_s(x)\,. \tag{6.102}$$

Injecting this expression into (6.97 and 98), we obtain

$$\begin{aligned} -\mu\varphi p_s &= -f'\varphi p_s - f p_s'\varphi - f p_s\varphi' + \sigma^2[\tfrac{1}{2}(g^2)''\varphi p_s + (g^2)'\varphi' p_s + (g^2)'\varphi p_s' \\ &\quad + \tfrac{1}{2}g^2\varphi'' p_s + g^2\varphi' p_s' + \tfrac{1}{2}g^2\varphi p_s''] \\ &= f p_s\varphi' + \frac{\sigma^2}{2}g^2 p_s\varphi'' \end{aligned} \tag{6.103}$$

since

$$-\partial_x f p_s + \frac{\sigma^2}{2}\partial_{xx} g^2 p_s = 0 \quad \text{and} \quad p_s' = \left(-\frac{2g'}{g} + \frac{2}{\sigma^2}\frac{f}{g^2}\right)p_s .$$

Note that (6.103) is the Kolmogorov backward equation. This implies that the FPE and KBE have the same spectrum and that their eigenfunctions are related to each other by the simple relation (6.102). The right-hand side of (6.103) can be written as $\sigma^2/2\,[g^2 p_s \varphi']'$, so that we obtain the final result:

$$\frac{\sigma^2}{2}[g^2 p_s \varphi']' + \mu p_s \varphi = 0 , \tag{6.104}$$

with the boundary condition (6.98)

$$g^2 p_s \varphi' |_{x=b_1,b_2} = 0 . \tag{6.105}$$

The FPE has thus been transformed into a Sturm-Liouville problem with

$$P(x) = \frac{\sigma^2}{2} g^2(x) p_s(x) , \tag{6.106}$$

$$Q(x,\mu) = \mu p_s(x) , \tag{6.107}$$

$$\alpha = \gamma = 0 , \quad \beta = g^2(b_1)p_s(b_1) , \quad \delta = g^2(b_2)p_s(b_2) . \tag{6.108}$$

It is important for the following to stress again that the eigenfunctions of the Sturm-Liouville equation (6.104) are identical to those of the KBE and not to those of the FPE. The eigenfunctions of the FPE are obtained by multiplication with the stationary probability density, according to (6.102). Since the spectra of KBE and FPE coincide, let us first discuss the eigenvalues of the Sturm-Liouville problem. Using results from the classical Sturm-Liouville theory [6.24, 25], we find the following facts:

i) If both boundaries b_1 and b_2 are finite, then the spectrum is purely discrete, i.e., the eigenvalues form a denumerably infinite discrete set.

ii) If one or both of the boundaries are infinite, a continuous range of eigenvalues may be present.

iii) Since we consider only cases where a unique stationary probability exists, the spectrum always contains at least one discrete nondegenerate eigenvalue, namely $\mu_0 = 0$. It corresponds to the stationary probability density $\psi_0(x) = p_s(x)$ or $\varphi_0(x) = 1$, respectively. This eigenvalue is nondegenerate, i.e., except for a multiplicative constant there is only one nontrivial eigenfunction for $\mu_0 = 0$, since the stationary solution of the FPE is unique.

iv) All eigenvalues are real and nonnegative.

Let us now turn our attention to the eigenfunctions of the KBE and the FPE. The former coincide with those of (6.104), i.e., the $\varphi_\mu(x)$; the latter are given by

$$\psi_\mu(x) = p_s(x)\,\varphi_\mu(x)\,.$$

Again using results from the classical Sturm-Liouville theory, we arrive at the following facts. If both boundaries are finite, the eigenfunctions of the KBE are orthogonal with respect to the weight function $p_s(x)$, i.e.,

$$\int_{b_1}^{b_2} p_s(x)\,\varphi_{\mu_m}(x)\,\varphi_{\mu_n}(x)\,dx = 0 \quad \text{for} \quad n \neq m\,. \tag{6.109}$$

Consequently the eigenfunctions of the FPE are orthogonal with respect to $p_s^{-1}(x)$ $[p_s^{-1}(x) = 1/p_s(x)]$:

$$\int_{b_1}^{b_2} p_s^{-1}(x)\,\psi_{\mu_m}(x)\,\psi_{\mu_n}(x)\,dx = 0\,, \qquad n \neq m\,. \tag{6.110}$$

Furthermore, the eigenfunctions can be normalized, i.e.,

$$\int_{b_1}^{b_2} p_s(x)\,\varphi_{\mu_m}^2(x)\,dx = 1 \tag{6.111}$$

and

$$\int_{b_1}^{b_2} p_s^{-1}(x)\,\psi_{\mu_m}^2(x)\,dx = 1\,, \tag{6.112}$$

respectively.

The classical Sturm-Liouville theory also states that the eigenfunctions are complete in the appropriately weighted $L_2(b_1, b_2)$ space. To be precise, the following two theorems hold.

Theorem I. If the function $u(x)$ is square integrable with respect to the weight function $p_s(x)$ over the interval $[b_1, b_2]$, i.e.,

$$\int_{b_1}^{b_2} p_s(x)\,u(x)^2 dx < \infty\,, \tag{6.113}$$

then $u(x)$ can be written as an expansion in eigenfunctions of the KBE, namely

$$u(x) = \sum_{n=0}^{\infty} u_n \varphi_{\mu_n}(x)\,, \tag{6.114}$$

where

$$u_n = \int_{b_1}^{b_2} p_s(x)\, u(x)\, \varphi_{\mu_n}(x)\, dx$$

$$= \int_{b_1}^{b_2} u(x)\, \psi_{\mu_n}(x)\, dx\,. \tag{6.115}$$

Similarly:

Theorem II. If the function $q(x)$ is square integrable with respct to the weight function $p_s^{-1}(x)$ over the interval $[b_1, b_2]$, i.e.,

$$\int_{b_1}^{b_2} p_s^{-1}(x)\, q^2(x)\, dx < \infty\,, \tag{6.116}$$

then $q(x)$ can be written as an expansion in eigenfunctions of the FPE, namely

$$q(x) = \sum_{n=0}^{\infty} q_n \psi_{\mu_n}(x)\,, \tag{6.117}$$

where

$$q_n = \int_{b_1}^{b_2} p_s^{-1}(x)\, q(x)\, \psi_{\mu_n}(x)\, dx$$

$$= \int_{b_1}^{b_2} q(x)\, \varphi_{\mu_n}(x)\, dx\,. \tag{6.118}$$

Applying these results to the time-dependent solution of the FPE, we have

$$p(x,t\,|x_0) = \sum_{n=0}^{\infty} p_n \psi_{\mu_n}(x)\, \mathrm{e}^{-\mu_n t} \tag{6.119}$$

$$= \sum_{n=0}^{\infty} \varphi_{\mu_n}(x_0)\, \psi_{\mu_n}(x)\, \mathrm{e}^{-\mu_n t}$$

$$= p_s(x) \sum_{n=0}^{\infty} \varphi_{\mu_n}(x_0)\, \varphi_{\mu_n}(x)\, \mathrm{e}^{-\mu_n t}\,, \tag{6.120}$$

if b_1 and b_2 are finite, and *provided*

$$\int_{b_1}^{b_2} p_s^{-1}(x)\, p(x,t\,|x_0)^2\, dx < \infty\,. \tag{6.121}$$

This shows that the eigenvalue problem for the FPE is indeed only closely related to the Sturm-Liouville problem, as mentioned before, but *not* identical. The

Sturm-Liouville equation is generally studied on an appropriate space of square integrable functions. The eigenvalue problem of the FPE, however, has to be solved for the space of integrable functions, as emphasized earlier. In fact, there is no reason why a solution of the FPE should fulfill the inequality (6.121). It only has to have the property that

$$\int_{b_1}^{b_2} p(x,t\,|x_0)\,dx = 1 < \infty\,. \tag{6.122}$$

Thus, a gap exists between the classical Sturm-Liouville problem and the eigenvalue problem of the FPE. The question that has to be answered is whether the eigenfunctions $\{\psi_{\mu_n}\,|\,n=0,1,\dots\}$ are also complete in $L_1[b_1,b_2]$, i.e., if a function $q(x)$ for which

$$\int_{b_1}^{b_2} |q(x)|\,dx < \infty\,, \tag{6.123}$$

can be expanded in terms of the eigenfunctions $\psi_{\mu_n}(x)$:

$$q(x) = \sum_{n=0}^{\infty} q_n \psi_{\mu_n}(x)\,. \tag{6.124}$$

(Note that the same problem arises for the backward equation, where the appropriate function is $C[b_1,b_2]$, also not a L_2 space.) This gap has been bridged by *Elliott* [6.20]. In fact, a slightly more general result is established.

Elliott's Theorem. If *neither* of the boundaries b_1 and b_2 is F natural (cf. p. 107), then the spectrum is purely discrete, identical to the spectrum of the Sturm-Liouville operator, and the eigenfunctions $\varphi_{\mu_n}(x)$ of the KBE and the eigenfunctions $\psi_{\mu_n}(x)$ of the FPE, as obtained from the Sturm-Liouville problem, form a complete system in $C[b_1,b_2]$ and $L_1(b_1,b_2)$, respectively.

It is important to underline that this theorem does not require b_1 or b_2 to be finite. The only condition is that the drift $f(x)$ and the diffusion coefficient $g(x)$ are differentiable on (b_1,b_2) and $g(x)>0$ for $b_1<x<b_2$. As pointed out already repeatedly, we consider only diffusion processes in this monograph which fulfill these conditions.

6.7.2 Examples: The Ornstein-Uhlenbeck Process and the Verhulst Equation

As a first application of the spectral theory of the Fokker-Planck operator we consider Brownian motion in velocity space, i.e., the Ornstein-Uhlenbeck process given by the SDE:

$$dX_t = -\lambda X_t dt + \sigma dW_t\,. \tag{6.125}$$

Here $b_1 = -\infty$, $b_2 = +\infty$ and both boundaries are F natural as is easily verified. This means that the O-U process is not covered by Elliott's theorem. The SDE can be cast in nondimensionalized standard form by rescaling the time $t' = \lambda t$ and the state variable $y = (\sqrt{\lambda}/\sigma)x$. Writing t again for t', we have the SDE

$$dY_t = -Y_t dt + dW_t, \tag{6.126}$$

and the associated FPE

$$\partial_t p(y,t|y_0) = \partial_y y p(y,t|y_0) + \tfrac{1}{2}\partial_{yy} p(y,t|y_0). \tag{6.127}$$

Using the transformation (6.102) we obtain the following eigenvalue problem (for the KBE):

$$-y\partial_y \varphi(y) + \tfrac{1}{2}\partial_{yy}\varphi(y) = -\mu\varphi(y). \tag{6.128}$$

The solutions of this eigenvalue problem are the classical Hermite polynomials

$$\varphi_{\mu_n} = \mathrm{H}_n(y) = (-1)^n \mathrm{e}^{y^2}\frac{d^n}{dy^n}(\mathrm{e}^{-y^2}) \tag{6.129}$$

with the eigenvalues

$$\mu_n = n \tag{6.130}$$

and

$$\int_{\mathbb{R}} \mathrm{e}^{-y^2}\mathrm{H}_n(y)\mathrm{H}_m(y)\,dy = 0 \quad \text{for} \quad n \neq m \tag{6.131}$$

$$\int_{\mathbb{R}} \mathrm{e}^{-y^2}\mathrm{H}_n^2(y)\,dy = \sqrt{\pi}\,2^n n! = C_n^{-1}. \tag{6.132}$$

The properly orthonormalized eigenfunctions are thus

$$\varphi_n = \tilde{\mathrm{H}}_n(y) = \sqrt{C_n}\,\mathrm{H}_n(y). \tag{6.133}$$

Using (6.120), we obtain the spectral representation of the transition probability density:

$$p(y,t|y_0) = \mathrm{e}^{-y^2}\sum_{n=0}^{\infty}\mathrm{e}^{-nt}\tilde{\mathrm{H}}_n(y_0)\,\tilde{\mathrm{H}}_n(y). \tag{6.134}$$

By virtue of Mehler's formula [6.26], the series in (6.134) can be summed and we find:

$$p(y,t|y_0) = [\pi(1-\mathrm{e}^{-2t})]^{-1/2}\exp\left(-\frac{(y-y_0\mathrm{e}^{-t})^2}{1-\mathrm{e}^{-2t}}\right) \tag{6.135}$$

which coincides with (2.121) or (6.93). We can therefore conclude that indeed (6.134) is the correct spectral representation of $p(y,t\,|\,y_0)$ and that (6.130 and 133) form the complete solution of the KBE eigenvalue problem. Taking into account the rescaling of the time and state space variable, we have established the following result:

i) The spectrum of the O-U process (6.125) is purely discrete,

$$\mu_n = n\lambda\,. \tag{6.136}$$

In words, the spectrum consists of all the harmonics of the relaxation frequency λ of the O-U process.

ii) The eigenfunctions of the FPE are

$$\psi_n(x) = \exp\left(-\frac{\lambda}{\sigma^2}x^2\right)\tilde{\mathrm{H}}_n\left(\frac{\lambda^{1/2}}{\sigma}x\right). \tag{6.137}$$

This example illustrates that the condition of Elliott's theorem, namely neither b_1 nor b_2 is F natural, is a sufficient condition for a purely discrete spectrum, but not a necessary one.

As the second example we choose the simplest model exhibiting noise-induced transitions, namely the Verhulst equation. As we have seen in Sect. 6.6.2, the SDE of the Verhulst model can be transformed into the linear SDE ($\nu = 1$):

$$dY_t = \left[\left(\frac{1}{2}\nu^2 - \lambda\right)Y_t + 1\right]dt + \sigma Y_t dW_t\,. \tag{6.138}$$

Rescaling the time $\tau = \sigma^2 t/2$ and the state variable $z = \sigma^2 y/2$ we obtain:

$$dZ_\tau = \left[\left(1 - \frac{2\lambda}{\sigma^2}\right)Z_\tau + 1\right]d\tau + \sqrt{2}Z_\tau dW_\tau\,. \tag{6.139}$$

Defining

$$\alpha = \frac{\lambda}{\sigma^2}\,, \tag{6.140}$$

the associated FPE reads

$$\begin{aligned}\partial_\tau p(z,\tau) = &-\partial_z[1-(2\alpha-1)z]\,p(z,\tau)\\ &+\partial_{zz}z^2 p(z,\tau)\,.\end{aligned} \tag{6.141}$$

Using (6.102) to transform (6.141) to the Sturm-Liouville problem

$$\left[\frac{z^2 z^{-2\alpha-1}}{2}\mathrm{e}^{-z^{-1}}\varphi'(z)\right]' + \frac{\mu}{2}z^{-2\alpha-1}\mathrm{e}^{-z^{-1}}\varphi(z) = 0\,, \tag{6.142}$$

$$\tfrac{1}{2}z^2 z^{-2\alpha-1}e^{-z^{-1}}\varphi'(z)\,|_{z=0,\infty}=0\,, \tag{6.143}$$

Wong [6.25] has shown that the solution of this eigenvalue problem admits $N+1$ discrete eigenvalues ($\alpha-1\leqslant N<\alpha$) and a continuous range of eigenvalues:

$$\mu_n=n(2\alpha-n) \qquad n=0,1,\ldots,N \tag{6.144}$$

and

$$\mu=\alpha^2+\eta^2 \qquad \eta\geqslant 0\,. \tag{6.145}$$

Accordingly the t.p.d. (transition probability density) can be written as

$$\begin{aligned}p(z,\tau|z_0)=z^{-(2\alpha+1)}e^{-z^{-1}}\cdot\Bigg[&\sum_{n=0}^{N}A_n e^{-n(2\alpha-n)\tau}\varphi_n(z_0)\,\varphi_n(z)\\&+\frac{1}{2\pi}\int_0^\infty e^{-(\alpha^2+\eta^2)\tau}A(\eta)\,\varphi_\eta(z_0)\,\varphi_\eta(z)\,d\eta\Bigg]\,.\end{aligned} \tag{6.146}$$

Here the following orthonormalization conditions have been used:

$$\int_{b_1}^{b_2}p_s(z)\,\varphi_m(z)\,\varphi_n(z)A_n dz=\delta_{mn}=\begin{cases}1 & m=n\\ 0 & m\neq n\end{cases} \tag{6.147}$$

for the discrete spectrum and

$$\int_{b_1}^{b_2}p_s(z)\,\varphi_\eta(z)\,\varphi_{\eta'}(z)A(\eta)\,dz=\delta(\eta-\eta') \tag{6.148}$$

for the continuous spectrum.

The $\varphi_n(z)$ are orthogonal polynomials of degree n,

$$\varphi_n(z)=(-1)^n z^{2\alpha+1}e^{z^{-1}}(d^n/dz^n)(z^{2n-2\alpha-1}e^{-z^{-1}}) \tag{6.149}$$

and the normalization factor A_n is given by

$$A_n=\frac{2(\alpha-n)}{\Gamma(2\alpha+1-n)}\cdot\frac{1}{n!}\,. \tag{6.150}$$

The eigenfunctions for the continuous spectrum are

$$\varphi_\eta(z)={}_2F_0(-\alpha-i\eta,\,-\alpha+i\eta,\,-z)\,, \tag{6.151}$$

where ${}_2F_0$ is the generalized hypergeometric series [6.26]. The normalization factor $A(\eta)$ has the form

$$A(\eta)=\frac{\Gamma(-\alpha+\mathrm{i}\eta)\Gamma(-\alpha-\mathrm{i}\eta)}{\Gamma(\mathrm{i}\eta)\Gamma(-\mathrm{i}\eta)}. \tag{6.152}$$

Taking in account the rescaling of time $\tau=\sigma^2 t/2$, and the transformation of the state variable $z=\sigma^2 x^{-1}/2$, the eigenvalues and eigenfunctions of the Verhulst model are easily obtained [6.27]. In particular, for the spectrum we have:

$$\mu_n=n\left(\lambda-n\frac{\sigma^2}{2}\right) \tag{6.153}$$

for $n=0,1,\ldots,N$ where $(\lambda/\sigma^2)-1\leqslant N<\lambda/\sigma^2$ and

$$\mu=\frac{1}{2}\left(\frac{\lambda^2}{\sigma^2}+\sigma^2\eta^2\right). \tag{6.154}$$

Note that the existence of a continuous range of eigenvalues is compatible with Elliott's theorem since $b_1=0$ is an F natural boundary. Note, furthermore, that due to the nonapplicability of this theorem, the delicate question whether the eigenfunctions form a complete system in $C_1[0,\infty)$ or $\psi_\mu=p_s(x)\varphi_n$ in $L_1(0,\infty)$ remains open at this point. We shall come back to it at the end of this section. The results concerning the eigenvalues and eigenfunctions of the FPE for the Verhulst model can be obtained in a different manner [6.28], namely the eigenvalue problem can be cast into the form of Whittaker's equation with appropriate boundary conditions. Indeed writing[6]

$$\psi(x)=x^{(\lambda/\sigma^2)+(1/2)}\mathrm{e}^{-(x/\sigma^2)}v\left(\frac{2}{\sigma^2}x\right), \tag{6.155}$$

it is found, that $v(z)$, where $z=2x/\sigma^2$, obeys Whittaker's equation

$$v''(z)+\left[\left(\frac{2\mu}{\sigma^2}-\frac{\lambda^2}{\sigma^4}+\frac{1}{4}\right)\Big/z^2+\left(\frac{\lambda}{\sigma^2}+\frac{1}{2}\right)\Big/z-1/4\right]v=0. \tag{6.156}$$

The convergence of the solutions for $x\to\infty$, and the normalization conditions

$$\int_0^\infty p_s^{-1}(x)\,\psi_n(x)^2dx=1$$

for the discrete spectrum and

$$\int_0^\infty p_s^{-1}(x)\,\psi_\mu(x)\,\psi_{\mu'}(x)\,dx=\delta(\mu-\mu')$$

[6] We shall again consider only the Stratonovich case i.e., $\nu=1$, here.

for the continuous part select the appropriate eigenvalues and eigenfunctions. The spectrum obtained this way coincides with (6.153, 154). Using this method we have, as in Wong's method, that a function $q(x)$ fulfilling

$$\int_0^\infty p_s^{-1}(x) q(x) dx < \infty ,$$

can be represented as an eigenfunction expansion, i.e.,

$$q(x) = \sum_{n=0}^{N} q_n \psi_{\mu_n}(x) + \int_0^\infty q_\eta \psi_\eta(x) d\eta .$$

Both methods leave the question unanswered whether the eigenfunctions of the KBE or the FPE form a complete system in $C_1[0, \infty)$ or $L_1(0, \infty)$, respectively. To address that question, consider the function (again we shall consider only the Stratonovich version explicitly)

$$\tilde{\psi} = p_s(x) \left(\frac{1}{x} - \frac{1}{\lambda - \sigma^2/2} \right) . \tag{6.157}$$

It is easily verified that for $\lambda > \sigma^2/2$

$$\int_0^\infty |\tilde{\psi}| dx < \infty ,$$

i.e., $\tilde{\psi} \in L_1(0, \infty)$. On the other hand

$$\int_0^\infty p_s^{-1}(x) \tilde{\psi}^2 dx = \int_0^\infty p_s(x) \left(\frac{1}{x} - \frac{1}{\lambda - \sigma^2/2} \right)^2 dx < \infty$$

only for $\lambda > \sigma^2$. In view of (6.102, 103), we obtain, injecting $\tilde{\psi}$ into the FPE;

$$\begin{aligned} &[(\lambda + \sigma^2/2)x - x^2]\, \partial_x \left(\frac{1}{x} - \frac{1}{\lambda - \sigma^2/2} \right) + \frac{\sigma^2}{2} x^2 \partial_{xx} \left(\frac{1}{x} - \frac{1}{\lambda - \sigma^2/2} \right) \\ &\quad = -(\lambda + \sigma^2/2)/x + 1 + \sigma^2/x \\ &\quad = -(\lambda - \sigma^2/2)[1/x - 1/(\lambda - \sigma^2/2)] . \end{aligned}$$

Furthermore, $\tilde{\psi}$ is orthogonal to the eigenfunction corresponding to $\mu_0 = 0$, namely $p_s(x)$. Indeed,

$$\begin{aligned}\int_0^\infty p_s^{-1}(x)p_s(x)\tilde{\psi}\,dx &= \int_0^\infty p_s(x)\left(\frac{1}{x}-\frac{1}{\lambda-\sigma^2/2}\right)dx\\ &= E\left\{\frac{1}{x}\right\}-\frac{1}{\lambda-\sigma^2/2}\\ &= 1/(\lambda-\sigma^2/2)-1/(\lambda-\sigma^2/2)=0\,.\end{aligned}$$

This establishes the result [6.29] that $\tilde{\psi}$, being a member of the space of admissible functions for the FPE, namely $L_1(0,\infty)$, is an eigenfunction to the eigenvalue $\mu_1=\lambda-\sigma^2/2$, as long as $\mu_1>0$. However, the Sturm-Liouville problem or the transformation into Whittaker's equation yield the result that $\mu_1=\lambda-\sigma^2/2$ belongs to the spectrum only if $\lambda>\sigma^2$, since otherwise ψ_1 is not square integrable with respect to the weight function $p_s^{-1}(x)$. This means that the eigenvalue problem, as treated by Wong and by Schenzle and Brand, is closely related to that of the FPE but not identical to it. They obtain eigenfunctions corresponding to a set of eigenvalues which from a complete set in the space of functions being square integrable with respect to $p_s^{-1}(x)$, but not in the space $L_1(0,\infty)$. The Verhulst model is thus an excellent illustration of the value of Elliott's theorem which allows the gap between the two function spaces to be bridged, if neither boundary is F natural. Furthermore, the above discussion shows that Elliott's theorem cannot in general be extended to the case where one or both boundaries are F natural.

6.8 Critical Dynamics of Noise-Induced Transitions

It is a well-known feature of equilibrium as well as of nonequilibrium phase transitions that in the neighborhood of a critical point the dynamical response of the system becomes sluggish. In more precise terms, this means that certain perturbations acquire an extremely long lifetime, or equivalently, that they relax on a macroscopic time scale which becomes slower and slower. This is what is usually understood by the phenomenon of critical slowing down. We saw earlier that noise-induced transitions display essential features known to characterize classical equilibrium and nonequilibrium phase transitions. The noise-induced critical point of the genetic model, for instance, is governed by classical critical exponents. Thus the question arises naturally in how far external noise influences the dynamics of a system and in particular if critical slowing down occurs near noise-induced critical points.

This section will therefore be devoted mainly to a detailed study of the dynamical behavior of the genetic model. However, to begin with, let us first consider the Verhulst model. If we admit only perturbations that are square integrable with respect to $p_s^{-1}(x)$, then $p(x,t)$ can be written as a spectral representation, which involves only the eigenvalues and eigenfunctions obtained by Wong and by Schenzle and Brand. Equations (6.153, 154) reveal that the eigenvalues of the FPE may depend on the noise intensity σ. If σ increases, the eigenvalues decrease. A higher noise level leads thus to a slower decay of a particular perturbation. It should however be stressed, as first pointed out by *Schenzle* and *Brand* [6.28], that for initial perturbations belonging to $L_2(0, \infty, p_s^{-1})$ no eigenvalue tends to zero as the noise-induced transition point is approached. At this point two remarks are in order:

i) As we have seen in Sect. 6.4, at the noise-induced transition point of the Verhulst model the qualitative change in $p_s(x)$ corresponds to jumps. As σ^2 is increased, we have

$$p_s(0) = 0 \quad \text{for} \quad 0 < \sigma^2/2 < \lambda \,,$$

$$p_s(0) = 2/\sigma^2 = 1/\lambda \quad \sigma^2/2 = \lambda \quad \text{double extremum} \,,$$

$$p_s(0) = \infty \quad \sigma^2/2 > \lambda \,.$$

Though $x = 0$ is a double extremum at $\lambda = \sigma^2/2$, a feature typical of a critical point, the change in p_s to this situation is discontinuous, a feature more reminiscent of a hard or first-order transition. The latter type of transition is not associated with any critical slowing in classical equilibrium and nonequilibrium situations. This discussion shows clearly that the noise-induced transition of the Verhulst model [or of the related models, $dX_t = (\lambda X_t - X_t^m)dt + \sigma X_t \circ dW_t$, $m > 2$] defies a neat classification in terms of the classical scheme as a second-order transition, for which critical slowing down occurs, or as a first-order transition, which does not display any critical slowing down. This underlines once more that this transition is indeed a noise effect and has no immediate deterministic analog. The particularity of the above models, which give rise to these features, is that a stationary point and a boundary coincide. This leads to a *split and shift* of the deterministic transition point in lieu of a mere shift. Due to these special properties of the model and of the noise-induced transition point there is no a priori reason to expect any critical slowing in the Verhulst model; it does not display a one hundred per cent pure critical point.

ii) After this lengthy digression on the particularities of the Verhulst model, let us now turn to our second remark, concerning the noise dependency of the FPE spectrum. It must be emphasized that the slowing down observed in the Verhulst case for increasing noise intensity cannot be considered to be a general feature of an FPE. In fact, Hongler's model furnishes a striking counter example. According to Sect. 6.6.2, this model is related to the O-U process with $\lambda = 1$ via a bijective transformation. This implies that the spectra coincide. In other words, the eigenvalues of the FPE for Hongler's model are the nonnegative integers; the spectrum does not display any dependence at all on the intensity of the external noise [6.30].

So far we have considered perturbations belonging only to $L_2(0, \infty, p_s^{-1})$. In this case, the spectrum and the associated system of eigenfunctions are those given by the Sturm-Liouville problem. As mentioned before, this problem, however, is in general only a restricted version of the eigenvalue problem of the FPE, since the space of admissible functions is $L_1(0, \infty)$ and not $L_2(0, \infty, p_s^{-1})$. In full generality the initial perturbations need only be normalizable.

If $p(x,0) \notin L_2(0,\infty,p_s^{-1})$, but

$$\int_0^\infty p(x,0)\,dx < \infty ,$$

then the so-called divergent modes, as e.g., (6.157), have to be considered [6.29]. Note that these modes correspond to eigenfunctions φ of the KBE which are not bounded on $[0,\infty)$; in fact they diverge near 0. If φ were bounded, i.e., $\varphi(x) \leqslant C$ for $x \in \mathbb{R}^+$, then

$$\begin{aligned}\int_0^\infty p_s^{-1}(x)\,\psi^2 dx &= \int_0^\infty p_s^{-1}(x) p_s(x)^2 \varphi(x)^2 dx \\ &= \int_0^\infty p_s(x)\,\varphi(x)^2 dx \\ &\leqslant C^2 ,\end{aligned}$$

and ψ would belong to $L_2(0,\infty,p_s^{-1})$. The unboundedness of φ is thus a necessary feature (but not sufficient) for a divergent mode to occur. To understand fully the significance of these modes, consider the class of models:

$$dX_t = (\lambda X_t - X_t^m)\,dt + \sigma X_t \circ dW_t , \quad m \geqslant 2 . \tag{6.158}$$

According to (6.13), the stationary probability density is given by

$$p_s(x) = N x^{(2\lambda/\sigma^2)-1} \exp\left(-\frac{2}{\sigma^2(m-1)} x^{m-1}\right), \tag{6.159}$$

which is normalizable for $\lambda > 0$ and has a divergence at $x = 0$ for $0 < \lambda < \sigma^2/2$. All models display thus the same noise-induced transition point at $\lambda = \sigma^2/2$ as the Verhulst model. To investigate the divergent modes, it is convenient to find the evolution equation for $E\{X_t^{-m+1}\}$. Setting $y = x^{-m+1}$, we have

$$dY_t = (1-m)(\lambda Y_t - 1)\,dt + \sigma(m-1)\,Y_t \circ dW_t$$

or

$$dY_t = (1-m)\left[\left(\lambda + \frac{1-m}{2}\sigma^2\right) Y_t - 1\right] dt + \sigma(m-1)\,Y_t dW_t$$

and thus

$$E\{\dot{Y}_t\} = (1-m)\left(\lambda + \frac{1-m}{2}\sigma^2\right) E\{Y_t\} + m - 1 . \tag{6.160}$$

The time-dependent solution of (6.160) is obviously

$$E\{Y_t\} = E\{Y_0\} \exp\left[(1-m)\left(\lambda + \frac{1-m}{2}\sigma^2\right)t\right] + 1/\left(\lambda + \frac{1-m}{2}\sigma^2\right). \tag{6.161}$$

Here $E\{X_t^{1-m}\}$ is finite for $t \to \infty$, if $\lambda > (m-1)\sigma^2/2$. As λ approaches $(m-1)\sigma^2/2$, $E\{X_t^{1-m}\}_{t\to\infty}$ tends to infinity and the "mode" $E\{X_t^{1-m}\}$ "decays" slower and slower, since the decay time is $[\lambda + (1-m)\sigma^2/2]^{-1}$, as is clear from (6.161). It is only in the Verhulst model $m = 2$ that the point of "critical slowing down" for this divergent mode coincides with the noise-induced transition point $\lambda = \sigma^2/2$ [6.29]. For all other models, this point occurs for λ well above the noise-induced transition point. The critical slowing down in a divergent mode is *not* due to a qualitative change in the random variable representing the state of the system, i.e., a qualitative change in the stationary probability density, but to the fact that these modes involve unbounded functions. The strong growth of the unbounded functions, in these models near zero, exaggerates the quantitative changes in $p_s(x)$ as σ^2 is varied, and leads to a divergence of the stationary expectation value at a noise level which is in general below that of the transition point. It is this divergence and not any transition phenomena in the system which gives rise to the "critical slowing down" for divergent modes. We conclude therefore, that modes involving unbounded functions should bot be used to decide whether a system displays critical slowing down or not.

The influence of the intensity of external noise on the dynamic behavior of the modified Verhulst model ($m = 3$) has also been studied by using the stationary state correlation function [6.31] and by obtaining exact expressions for the time dependence of the moments via embedding techniques [6.32–34]. It is found that an increase in the noise level leads to a decrease of the decay rates but no critical slowing down was observed in the correlation function or the moments at the noise-induced transition point. Thus analysis of the spectrum, excluding divergent modes, of (bounded) moments or the correlation function fails to yield any indication for noise-induced critical slowing down, though a dependence of the dynamical behavior on the noise level is found.

Apart from the aforementioned fact that the noise-induced transition in the Verhulst model cannot be directly identified as a critical point, this failure is not at all surprising in the light of Sect. 6.3. As was emphasized there, the state of the system is described by the random variable X_t. This is the fundamental quantity we have to deal with and not the moments which might not even uniquely determine the random variable. The cherished belief that moments are all there is to a random variable originates from the analysis of systems with internal fluctuations, fluctuations which are macroscopically small. The unquestioned extension of the notions which were developed to deal with small fluctuations to situations with external noise is dangerous and hampers a real understanding of the phenomena involved. If fluctuations are present in a system, then the only solid starting point is the trivial fact that the state of the system is described by a random variable. Our discussion in Sect. 6.3 of the steady-state case was rigorously based on this one solid fact. A transition occurs if this random variable and not some derived quantity as the moments changes qualitatively. This qualitative change of functional form of the mapping from the sample space into the state space is by virtue of convention (2.15) equivalent to a qualitative change in the probability law. It then becomes a question of practical considerations of how to find the best way to monitor such a qualitative change. As explained in Sect. 6.3, in analogy to the deterministic case, this is best done by studying the behavior of the extrema of $p_s(x)$. (The only exception is the transition from a degenerate to a genuine random variable, where the variance is the most appropriate quantity.) Furthermore, we have seen that the extrema have a particular physical significance. They can be identified with the macroscopic phases of the system and can be used to define the order parameter of the transition, as most clearly illustrated in Sect. 6.5. To cut a long story short, in order to settle the question whether noise-induced critical points display critical slowing down, we have to study the dynamics of the random variable X_t, the relaxation from one functional form to another. For the reasons expounded in Sect. 6.3 and repeated above, this is most appropriately done by studying the dynamics of the extrema. Not surprisingly we shall indeed find that critical slowing down is a characteristic feature of pure noise-induced critical points.

For the sake of clarity, let us first illustrate this phenomenon on Hongler's exactly soluble model. From (6.94) it follows directly that in the course of time, the extrema of $p(x,t|x_0)$ evolve as the roots of

$$\sinh[2\sqrt{2}x_m(t)] - \cosh^2[2\sqrt{2}x_m(t)] \cdot$$

$$\{\sinh[2\sqrt{2}x_m(t)] - e^{-t}\sinh(2\sqrt{2}x_0)\}\frac{4}{\sigma^2}(1-e^{-2t})^{-1} = 0\,. \tag{6.162}$$

Hongler's system is symmetric with respect to $x=0$. Namely $f(x)$ is an odd function and $g(x)$ is even, giving rise to a stationary density $p_s(x)$ which is symmetric around zero. To avoid any spurious transient effects, we start therefore with a symmetric situation around the deterministic steady state. In particular we choose for the sake of simplicity but without restriction of generality a delta peak centered at zero.

Equation (6.162) decomposes then into two factors which yield the following conditions

$$\sinh(2\sqrt{2}x_m(t)) = 0 \quad \text{and} \tag{6.163}$$

$$\cosh^2[2\sqrt{2}x_m(t)] = \frac{\sigma^2}{4}(1-e^{-2t})\,. \tag{6.164}$$

Condition (6.163) implies that $x_m(t)=0$ is always an extremum, independently of the noise intensity and of the time. The left-hand side of (6.164) is always greater or equal to one. This implies that (6.164) can be fulfilled only for times larger than t_c, where

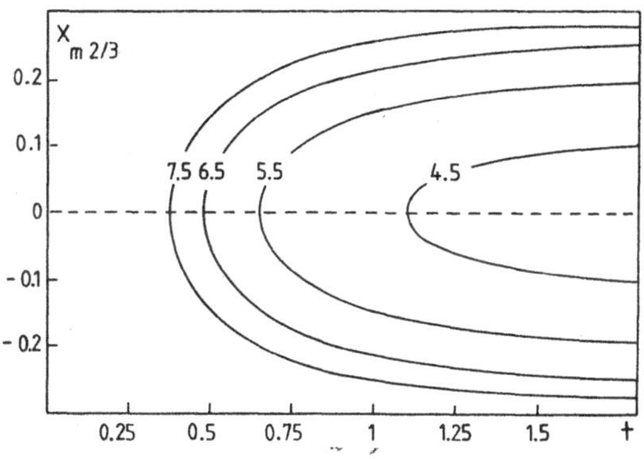

Fig. 6.7. Plot of the time evolution of the extrema $x_{m2/3}$ of Hongler's model for the values of σ^2 indicated. As $\sigma^2 \downarrow \sigma_c^2$ the critical time t_c at which the initial peak at $x=0$ splits into two peaks tends to infinity

$$t_c = -\frac{1}{2}\ln\left(1-\frac{4}{\sigma^2}\right). \tag{6.165}$$

As is shown in Fig. 6.7, t_c is the critical time at which the initial peak at $x_m = 0$ becomes flat, i.e., a double maximum, and then splits into two peaks with maxima at

$$x_{m2/3} = \pm\frac{1}{2\sqrt{2}}\operatorname{arcosh}\left(\frac{\sigma}{2}\sqrt{1-e^{-2t}}\right). \tag{6.166}$$

This splitting of the initially unimodal distribution is the noise-induced analog of the phenomenon of spinodal decomposition, well known in the theory of first-order equilibrium phase transitions. As (6.165) shows, the transition from monomodal to bimodal behavior takes more and more time, i.e., the system stays increasingly in the "unstable" state $x_m = 0$, when σ^2 approaches $\sigma_c^2 = 4$ from above. The dynamical behavior of X_t as reflected by the dynamics of the extrema, i.e., the order parameter, does indeed display a critical slowing down for this noise-induced critical point.

We expect this behavior to be representative for any model that displays a noise-induced critical point. In particular, we shall now show that this behavior is also found in the case of the genetic model, which has a clear physico-chemical or biological interpretation and which possesses a pure noise-induced transition point. In fact, we shall prove a more general result and establish that for a large class of systems, to which the genetic model belongs, the dynamics of X_t as reflected by the extrema undergo a critical slowing down at a noise-induced critical point. These systems are characterized by the following properties:

i) Both boundaries of the state space are F entrance boundaries. That this is the case for the genetic model will be verified in a different context in Chap. 8.

ii) The state space is either a finite interval or $\mathbb{R}$.

iii) The drift $f(x)$ vanishes at the midpoint $\tilde{x}$ of the state space and is antisymmetric around it, i.e., $f(x) = f(\tilde{x}+u) = -f(\tilde{x}-u)$.

iv) The diffusion $g(x)$ is symmetric around the midpoint of the state space, i.e., $g(x) = g(\tilde{x}+u) = g(\tilde{x}-u)$.[7]

The last two conditions imply that any transition probability density that is symmetric around the midpoint will remain so forever. The first condition ensures that Elliott's theorem is applicable. This implies that the spectrum is purely discrete. Furthermore, the eigenvalues and associated eigenfunctions are given by the Sturm-Liouville problem and the eigenfunctions are complete in $L_1(b_1, b_2)$. The following fact is given by the classical Sturm-Liouville theory [6.24]:

The eigenfunction $\varphi_n(x)$, corresponding to the n-th eigenvalue, has exactly n *simple* zeros in (b_1, b_2).

This allows us to conclude that all eigenfunctions $\varphi_{2m}(x)$ are symmetric around the midpoint, while all $\varphi_{2m+1}(x)$ are antisymmetric around it. This means that only the even eigenfunctions can appear in the expansion of the t.p.d. if the system is prepared initially to be in the midpoint $\tilde{x}$:

$$p(x,t\,|\,\tilde{x}) = p_s \sum_{m=0}^{\infty} \varphi_{2m}(\tilde{x})\,\varphi_{2m}(x)\,e^{-\mu_{2m}t}\,. \tag{6.167}$$

Since the eigenvalues form a discrete increasing sequence [6.24],

$$0 = \mu_0 < \mu_1 < \ldots < \mu_n < \ldots, \quad \lim_{n\to\infty} \mu_n = +\infty\,,$$

we can write for the long-term behavior of the system

$$p(x,t\,|\,\tilde{x}) \approx p_s(x)\,[1 + \varphi_2(\tilde{x})\,\varphi_2(x)\,e^{-\mu_2 t}]\,. \tag{6.168}$$

The evolution of the extrema for long times is thus given by

$$\begin{aligned}\partial_x p(x,t\,|\,\tilde{x})\,|_{x=x_m(t)} = \Bigg[\left(-\frac{\nu g'}{g} + \frac{2f}{\sigma^2 g^2}\right) p_s(x)\,[1 + \varphi_2(\tilde{x})\,\varphi_2(x)\,e^{-\mu_2 t}]& \\ + p_s(x)\,\varphi_2(\tilde{x})\,\varphi_2'(x)\,e^{-\mu_2 t}\Bigg]_{x=x_m(t)} &= 0\end{aligned}$$

or

$$\begin{aligned}&\left[f(x_m(t)) - \frac{\nu\sigma^2}{2} g(x_m(t))\,g'(x_m(t))\right][1 + \varphi_2(\tilde{x})\,\varphi_2(x_m(t))\,e^{-\mu_2 t}] \\ &\quad + \frac{\sigma^2}{2} g^2(x_m(t))\,\varphi_2(\tilde{x})\,\varphi_2'(x_m(t))\,e^{-\mu_2 t} = 0\,.\end{aligned} \tag{6.169}$$

[7] Note that these conditions imply that the noise-induced drift $\sigma^2 g' g/2$ has the same properties as f. The conditions hold therefore for the Stratonovich version as well.

Taking into account that f is antisymmetric around the midpoint, i.e., $f(\tilde{x}) = 0$, g and φ_2 symmetric, i.e., $g'(\tilde{x}) = 0$ and $\varphi_2'(\tilde{x}) = 0$, we have

$$f(x) - \frac{\nu\sigma^2}{2} g(x) g'(x) = (x-\tilde{x}) \tilde{k}(x)$$

$$\varphi_2'(x) = (x-\tilde{x}) \tilde{\varphi}_2'(x) \tag{6.170}$$

defining the function $\tilde{k}$ and $\tilde{\varphi}_2'$. In other words, the symmetry properties of the systems ensure that the midpoint $\tilde{x}$ is an extremum of the t.p.d. for all σ^2 and t. Injecting (6.170) into (6.171), we obtain

$$\tilde{k}(x_{\mathrm{m}}(t)) [1 + \varphi_2(\tilde{x}) \varphi_2(x_{\mathrm{m}}(t)) \mathrm{e}^{-\mu_2 t}] + \frac{\sigma^2}{2} g^2(x_{\mathrm{m}}(t)) \varphi_2(\tilde{x}) \tilde{\varphi}_2'(x_{\mathrm{m}}(t)) \mathrm{e}^{-\mu_2 t} = 0 \,. \tag{6.171}$$

For the midpoint to be a critical point at a certain time $t = t_{\mathrm{c}}$, it has to be at least a double root of (6.169) for $t = t_{\mathrm{c}}$, i.e., it must also solve (6.171) for $t = t_{\mathrm{c}}$. Due to the symmetry of the model, it is then also a triple root, that is to say a true critical point. In other words, the critical time t_{c} is given by:

$$t_{\mathrm{c}} = -\frac{1}{\mu_2} \ln \left(\frac{-\tilde{k}(\tilde{x})}{\varphi_2(\tilde{x}) \left[\tilde{k}(\tilde{x}) \varphi_2(\tilde{x}) + \frac{\sigma^2}{2} g^2(\tilde{x}) \tilde{\varphi}_2'(\tilde{x}) \right]} \right) . \tag{6.172}$$

As the noise-induced critical point is approached from above, $\tilde{k}(x) \to 0$, since for $\sigma^2 = \sigma_{\mathrm{c}}^2$ the midpoint $\tilde{x}$ has to be a triple root of $f - \nu\sigma^2 g g'/2 = 0$. Thus the critical time t_{c} tends to infinity, if $g^2(\tilde{x}) \varphi_2(\tilde{x}) \tilde{\varphi}_2'(\tilde{x})$ remains finite and nonzero as $\sigma^2 \downarrow \sigma_{\mathrm{c}}^2$. Let us discuss the three factors separately:

i) $g(x) > 0$ for $x \in (b_1, b_2)$,

ii) $\varphi_2(x)$ has *two simple* zeros in (b_1, b_2) and is symmetric around $\tilde{x}$, which implies $\varphi_2(\tilde{x}) \neq 0$ for all values of σ^2.

iii) It remains thus to prove that the third factor does not tend to zero as $\sigma^2 \downarrow \sigma_{\mathrm{c}}^2$. In other words, we have to show that $\varphi_2(x)$ has only a *simple* extremum at $x = \tilde{x}$. To do so we will employ Prüfer's method for the Sturm-Liouville equation (6.100) [6.24]. This method introduces a type of "polar coordinates" ρ, ϑ by setting

$$y(x,\mu) = \rho(x,\mu) \sin \vartheta(x,\mu) \tag{6.173}$$

and

$$P(x) y'(x,\mu) = \rho(x,\mu) \cos \vartheta(x,\mu) \,, \tag{6.174}$$

where $y(x,\mu)$ is a nontrivial solution of (6.100). The "polar coordinates" obey the following set of equations:

$$\rho'(x,\mu) = \frac{1}{2}\left[\frac{1}{P(x)} - Q(x,\mu)\right]\sin 2\vartheta(x,\mu)\,\rho(x,\mu) \tag{6.175}$$

$$\vartheta'(x,\mu) = \frac{1}{P(x)}\cos^2\vartheta(x,\mu) + Q(x,\mu)\sin^2\vartheta(x,\mu)\,. \tag{6.176}$$

Note that the last equation involves only the “angle” $\vartheta(x,\mu)$. Once it is known, $\rho(x,\mu)$ is easily obtained from (6.175):

$$\ln\rho(x,\mu) = \frac{1}{2}\int_{b_1}^{x}\left[\frac{1}{P(z)} - Q(z,\mu)\right]\sin 2\vartheta(z,\mu)\,dz\,. \tag{6.177}$$

Applying Prüfer's method to our problem we have

$$P(x) = \frac{\sigma^2}{2}g^2(x)p_s(x) \quad \text{and} \quad Q(x,\mu) = \mu p_s(x)\,.$$

Note that $\rho(x,\mu)$ is bounded away from zero. For the eigenfunctions $\varphi_\mu(x)$ we have

$$\varphi'_\mu(x) = \left[\frac{\sigma^2}{2}g^2(x)p_s(x)\right]^{-1}\rho(x,\mu)\cos\vartheta(x,\mu) \tag{6.178}$$

for $x\in(b_1,b_2)$, since g and $p_s(x)$ vanish only at the boundaries. Furthermore, g and p_s are bounded away from infinity. Together with the fact that $\rho(x,\mu)$ is bounded away from zero, this implies

$$\varphi'_\mu(x) = 0$$

if and only if

$$\cos\vartheta(x,\mu) = 0\,. \tag{6.179}$$

Therefore $\varphi'_{\mu_2}(\tilde{x}) = 0$ if and only if

$$\vartheta(\tilde{x},\mu_2) = \frac{\pi}{2}, \frac{3}{2}\pi, \ldots\,. \tag{6.180}$$

In view of this fact, it remains to prove that $\tilde{x}$ is a simple zero of $\cos\vartheta(x,\mu_2)$. We have:

$$\frac{d}{dx}[\cos\vartheta(x,\mu_2)] = -\sin\vartheta(x,\mu_2)\,\vartheta'(x,\mu_2) \tag{6.181}$$

and

$$\sin \vartheta(\tilde{x}, \mu_2) = \pm 1 . \tag{6.182}$$

Equation (6.176) then yields:

$$\begin{aligned}\vartheta'(\tilde{x}, \mu_2) &= \left[\frac{\sigma^2}{2} g^2(\tilde{x}) p_s(\tilde{x})\right]^{-1} \cos^2 \vartheta(\tilde{x}, \mu_2) \\ &\quad + \mu_2 p_s(\tilde{x}) \sin^2 \vartheta(\tilde{x}, \mu_2) \\ &= \mu_2 p_s(\tilde{x}) .\end{aligned} \tag{6.183}$$

Since $\mu_2 > \mu_1 > 0$ and $p_s(\tilde{x}) > 0$, the derivative of $\vartheta(x, \mu_2)$ does not vanish at the midpoint. Together with (6.182 and 181) this implies that $\tilde{x}$ is a simple zero of $\cos \vartheta(x, \mu_2)$ and of $\varphi'_{\mu_2}(x)$. We have thus established that

$$\tilde{\varphi}_2'(\tilde{x}) \neq 0$$

for all values of σ^2. As the noise-induced critical point is approached from above, $\sigma^2 \downarrow \sigma_c^2$, $\tilde{k}(\tilde{x})$ tends to zero and $g^2(\tilde{x})\,\varphi_2(\tilde{x})\,\tilde{\varphi}_2'(x)$ to a nonvanishing finite value, leading to a divergence in the critical time t_c. This establishes that a broad class of systems with (pure) noise-induced critical points displays a critical slowing down as its classical counterparts. The time for the "peak splitting" or "spinodal decomposition" to occur tends logarithmically to infinity as the distance from the critical point is decreased.

Recently, *Hongler* [6.35, 36] studied an exactly solvable class of models describing motion in nonharmonic attractive potentials. Depending on the value of a deterministic parameter, the potential has one or two wells, i.e., these models exhibit a deterministic bifurcation. The noise is additive in these models and thus does not modify the deterministic transition phenomenon. Hongler showed that the transition from monomodal to bimodal behavior occurs at a time t_c that depends logarithmically on the deterministic bifurcation parameter. These exact results concerning the pure noise-induced spinodal decomposition and the corresponding deterministic phenomenon establish that both share the same features.

Note that the existence of critical slowing down for the above-defined class of models was established by exploiting only some general properties, namely symmetry around the critical point and the existence of a discrete (branch of the) spectrum. The latter property is intimately connected with the character of the boundaries, as is clear from Elliott's theorem. Roughly speaking, the boundaries should not play too important a role, in the sense that they are not stationary points of the system. In the light of these facts, it is reasonable to expect that critical slowing down is a rather general feature of noise-induced critical points. Indeed, noise-induced critical points represent local modification of the probability density and the latter is always, at least locally, symmetric around the critical point $\tilde{x}$. Thus the above results should be applicable to most systems displaying noise-induced critical points.

7. Noise-Induced Transitions in Physics, Chemistry, and Biology

The purpose of the present chapter is twofold: first, to review the experiments carried out so far to test the theoretical predictions concerning noise-induced transitions. Second, we shall study a number of model systems which provide a satisfactory and well-established description of the real systems they address. Up to now these systems have not been tested under the influence of an external noise. The results of this chapter, however, furnish ample motivations for such studies. They demonstrate that further experimental evidence on noise-induced transitions and on aspects of these phenomena not yet covered in experiments up to now, could in these systems be the object of qualitative and quantitative verification.

7.1 Noise-Induced Transitions in a Parametric Oscillator

Two concrete and simple model systems were studied in Chap. 6 to illustrate the two possible types of noise-induced transitions: the shift of the deterministic bifurcation diagram as encountered in the Verhulst model and the appearance of critical points which are impossible under deterministic external constraints as in the genetic model. The theoretical framework within which the effect of external noise on nonlinear systems is described, has of course, as every physical theory, been constructed using several, albeit plausible, assumptions and idealizations all of which have been discussed in detail. Let us briefly recall them:

i) the system is at the thermodynamic limit and spatially homogeneous, i.e., internal fluctuations can be neglected;

ii) the state of the system can be satisfactorily described by one variable;

iii) the external fluctuations have an extremely short correlation time and the white-noise idealization can be used to model them.

These assumptions all seem to be reasonable. However, they lead to theoretical predictions that are quite surprising, not to say counterintuitive: noise, a factor of disorganization, can create new macroscopic states if the coupling between the system and the environment is multiplicative. Experimental confirmation of these predictions is certainly most desirable. In keeping with the spirit of Chap. 6, we shall focus our attention first on simple experimental setups. They are also preferable for obvious methodological reasons: a clear-cut confirmation, or rejection, of the theoretically predicted existence of noise-

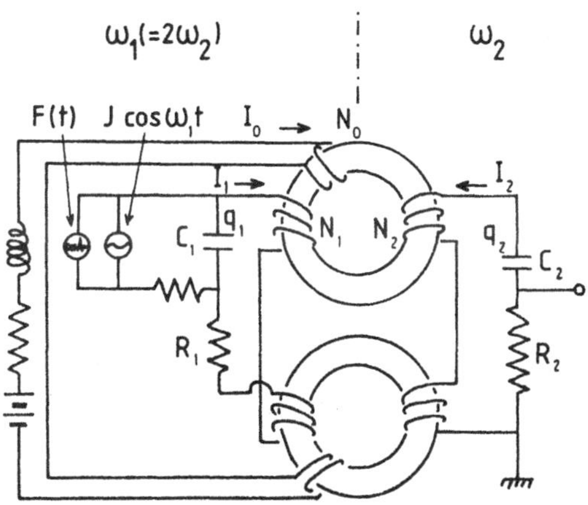

Fig. 7.1. Oscillating electrical circuit of *Kabashima* and co-workers [7.2]. The external noise $F(t)$ perturbs the ac current input $J\cos\omega_1 t$ supplied to the primary circuit

induced transitions is most easily achieved on simple experimental systems. They should possess the following features. Their evolution mechanism is well known for deterministic external conditions. Their experimental manipulation poses no great technical difficulties and the state variables of the system as well as the characteristics of the external noise are easily monitored. Looking for such systems, one quickly hits on electrical circuits as the ideal choice. They present a rich variety of nonlinear behavior and conform perfectly to the above requirements for experimental feasibility. Most conveniently for our purposes they also fulfill assumptions (i) and (ii) extremely well. As to the third assumption, electronic devices exist that generate noise the power spectrum of which is flat up to some cutoff frequency where it drops rapidly to zero. If this cutoff frequency is much larger than all relevant frequencies of the system, such a noise will be called quasi-white. A poor man's means to generate quasi-white noise is to amplify the thermal noise of a resistance. This noise has a flat spectrum for a very large range of frequencies. We conclude that electric circuits are admirably well suited for an experimental test of the existence of noise-induced transitions. Not surprisingly, such systems were the first on which experimental evidence for noise-induced transitions was obtained.

The experimental study of noise-induced transitions was initiated by *Kabashima*, *Kawakubo* and co-workers [7.1, 2]. This group previously studied the properties of nonequilibrium phase transitions in electrical circuits. To explore the influence of external noise on the behavior of an electrical circuit *Kabashima* et al. [7.2] studied the degenerate parametric oscillator, represented in Fig. 7.1. Its main elements are a primary and a secondary circuit loop coupled by two ferrite toroidals. The currents in the primary and secondary circuits are respectively I_1 and I_2. The dc current I_0 in the additional bias circuit serves the purpose of shifting the operating point of the parametric oscillator into the domain where second-order nonlinearities come into play in the relation between the magnetic flux of the coils and the currents. The ac current $J\cos\omega_1 t$ supplied to the primary circuit acts as a pump to excite subharmonic oscillations, i.e., $\omega_2 = \omega_1/2$, in the secondary circuit. The specific value used by Kabashima et al. is $\omega_1 = 50$ kHz. In addition to the pumping current, a random current ζ_t [denoted

$F(t)$ in Fig. 7.1] having a flat spectrum between 0.01 and 100 kHz can be supplied to the primary circuit. The state variable describing the system is the *slowly* varying *amplitude* b of the secondary current. Writing the current in the primary and secondary circuit as

$$I_1(t) = a(t)\sin[\omega_1 t + \phi(t)] \tag{7.1}$$

and

$$I_2(t) = b(t)\sin[\omega_2 t + \psi(t)]\,, \tag{7.2}$$

Kabashima et al. remarked that the variables characterizing the primary circuit can be adiabatically eliminated. This means that they vary much more rapidly than the corresponding variables of the secondary circuit and can therefore be replaced by their stationary values. Furthermore, under these conditions phase locking occurs in the secondary circuit. The phase ψ quickly reaches its steady-state value 0 or π and remains then fixed at this value. Thus one is left only with the following equation for the slowly varying amplitude $b(t)$ of $I_2(t)$:

$$\dot{b}(t) = \frac{2\mathfrak{N}}{\gamma_1}(f - f_{\text{th}})b - \frac{2\mathfrak{N}^2}{\gamma_1}b^3\,, \tag{7.3}$$ [1]

where $\mathfrak{N} > 0$ is the negative of the coupling constant between the two circuits, γ_1 and γ_2 are the damping constants of the primary and secondary circuit, $f = J/(2\omega_1 C_1)$ is the pump strength and $f_{\text{th}} = \gamma_1\gamma_2/(2\mathfrak{N})$. This equation closely resembles the Verhulst model. The only difference is the power of the self-restriction term. This changes the form of the switching curve but does not modify in any essential way the bifurcation diagram. Similar to the Verhulst equation this equation exhibits a second-order transition point, namely at $f = f_{\text{th}}$. This can be seen from the steady-state solution, for which we straightforwardly obtain

$$b_1 = 0 \tag{7.4}$$

and

$$b_2 = \pm\,\mathfrak{N}^{-1/2}(f - f_{\text{th}})^{1/2}\,. \tag{7.5}$$

Kabashima et al. experimentally showed that in the noiseless case the steady-state behavior as well as the dynamics are in close agreement with the theoretical predictions based on (7.3). This confirms that (7.3) satisfactorily describes the parametric oscillator. The experimental results for the steady-state values of the amplitude b of $I_2(t)$ in the noiseless case are represented by the open circles in Fig. 7.2. The theoretical values are given by (7.4, 5). The latter is represented in the form

$$\bar{I}_2 = c(J - J_{\text{th}})^{1/2} \tag{7.6}$$

[1] The derivation of (7.3) is rather tedious and not very useful for our purposes. For details see [7.2].

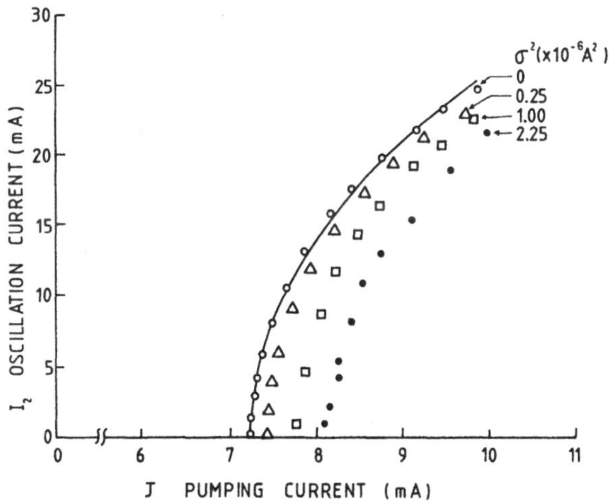

Fig. 7.2. Steady-state values of the amplitude of the current I_2 as a function of the pumping current and for the values of σ^2 indicated in the figure. The solid curve $\bar{I}_2 = 0.499\ (J - J_{th})^{1/2}$ fits experimental behavior (open circles) in the absence of noise

by the solid line with an appropriate numerical value for the constant c. These results show that an oscillation current is observed in the secondary circuit if the amplitude of the pumping current increases beyond a certain threshold value J_{th}. If, however, the pumping current has an amplitude which is too small to excite the subharmonic oscillations, the current in the second circuit fluctuates around zero. It is Gaussian distributed except in the immediate vicinity of the transition point. Its variance and correlation time diverge like $(J_{th} - J)^{-1}$ if the threshold is approached from below.

In what ways is this behavior of the parametric oscillator modified, if in addition to the pumping current a random current is supplied to the primary circuit? As is clear from (7.3), the coupling between the two circuits is multiplicative. This is basically due to the fact that the amplitude of the dc bias current is chosen of sufficient strength to bring out the second-order nonlinearities in the toroidals. Hence the necessary prerequisite for noise-induced transitions is fulfilled. In view of the close resemblance of (7.3) to the Verhulst equation, we expect to encounter qualitatively the same kind of noise-induced transitions phenomena. In fact, (7.3) belongs to the class of generalized Verhulst models discussed in Sect. 6.8, cf. (6.158):

$$dX_t = (\lambda X_t - X_t^n)\,dt + \sigma X_t \circ dW_t\,. \tag{7.7}$$

(If n is odd, then the model remains meaningful, if instead of $[0, \infty)$ the real line $(-\infty, \infty)$ is chosen as state space. However $x = 0$ is an intrinsic boundary. Therefore the two halves do not communicate, unless at least a small amount of additive noise is present).

It was found that all these models have qualitatively the same transition phenomena: namely a transition point at $\lambda = 0$, where a genuine stationary probability density appears which still, however, has zero as its most probable state, and a second transition point at

$$\lambda = \sigma^2/2\,, \tag{7.8}$$

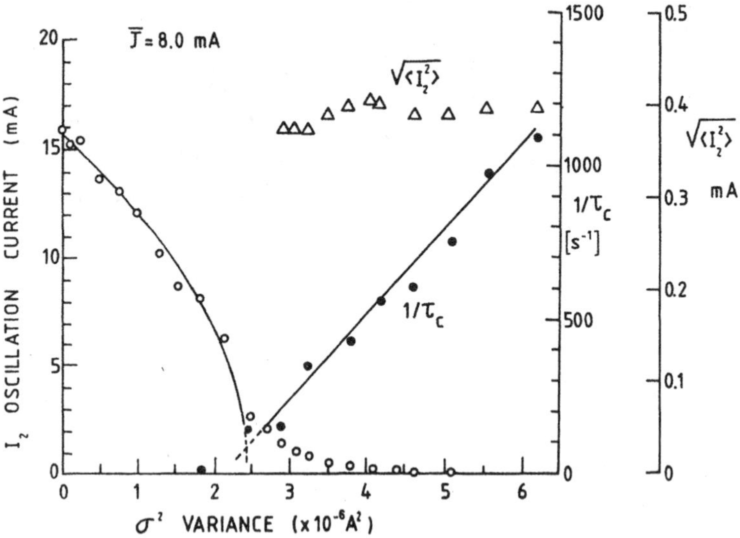

Fig. 7.3. Influence of noise on the oscillatory to nonoscillatory transition. (Open circles: oscillation current; closed circles: reciprocal of the correlation time; triangles: square root of the variance of the fluctuating current.)

where the most probable state acquires a nonvanishing value. The experimental results of Kabashima et al. for the case where the noise current is added to the pumping current agree completely with these theoretical predictions. If the fluctuating secondary current does not contain a systematic oscillating contribution, the peak value of its spectrum will be at zero frequency. On the other hand, if the most probable state corresponds to a current with an oscillating part, the peak value will be located at this oscillation frequency. In Fig. 7.2 the effective value of the oscillation current, as determined from the peak value of the spectrum of the output current, is shown for several noise intensities. As predicted, the threshold for the onset of oscillations is found to be shifted to higher values. Indeed, it increases linearly with σ^2 in complete agreement with (7.8). If the transition point is approached from above, i.e., the noise intensity is decreased from above to this threshold value, no divergence in the variance of the oscillation current is observed, in contradistinction to the noiseless case. The correlation time however increases as shown in Fig. 7.3. The experimental results seem to suggest that critical slowing down occurs in the correlation function, contrary to the theoretical predictions of *Schenzle* and *Brand* [6.28], *Graham* and *Schenzle* [6.34], *Fujisaka* and *Grossmann* [6.31], and *Brening* and *Banai* [6.33] discussed in Sect. 6.8. It is at present impossible to pinpoint the exact origin of the discrepancy. It might be that (7.3), though it satisfactorily describes the stationary behavior in the deterministic and the noise case, fails to represent accurately certain dynamical aspects of the parametric oscillator. If the noise intensity exceeds the threshold value, the oscillations in the secondary circuit disappear and the current becomes noisy. However, in contrast to the subthreshold fluctuating current of the noiseless case it behaves in a non-Gaussian way. An example of the sta-

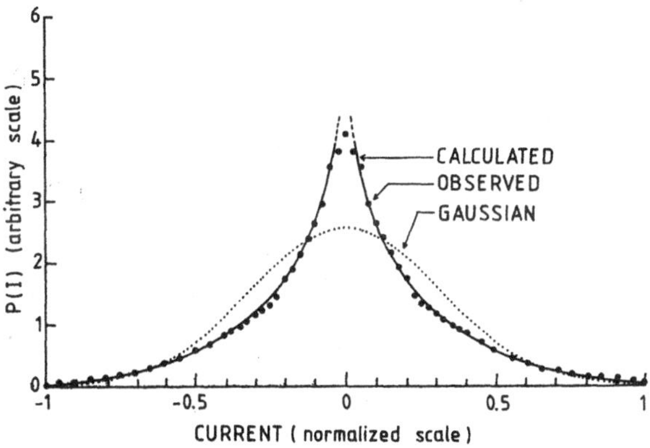

Fig. 7.4. Stationary probability density of the fluctuating current. Experimental values: $\bar{J} = 8\,\text{mA}$, $\sigma^2 = 3.22 \times 10^{-6}\,\text{A}^2$. Calculated values: $(f - f_{\text{th}})/\mathfrak{N} = 0.245$, $\mathfrak{N}(\lambda/\sigma^2) = 2.45$

tionary probability density for the current in this case is plotted in Fig. 7.4. For easy comparison a Gaussian probability density with mean value zero and the same variance is also represented in this figure. The solid curve is the theoretical prediction for the stationary probability density of the current in the secondary circuit. It is given by the following formula

$$p(I) = \int_I^\infty \frac{db}{\pi} \frac{p_s(b)}{(b^2 - I^2)^{1/2}}, \tag{7.9}$$

where according to (6.159)

$$p_s(b) = N b^{([2\lambda/\sigma^2] - 1)} \exp(-b^2/\sigma^2), \tag{7.10}$$

with $\lambda = (f - f_{\text{th}})/\mathfrak{N}$ and σ is the intensity of random current, divided by the coupling constant $\mathfrak{N}$. The derivation of formula (7.9) for the probability density is as follows. The current I_2 in the secondary circuit is given by

$$I_2 = b_t \sin \omega_2 t, \tag{7.11}$$

where b_t is a diffusion process with the stationary probability density (7.10). Since the second factor rapidly varies, it is permissible to consider the instant of time t a random variable with state space $[-T/4, T/4]$ and a uniform probability density on it, i.e.,

$$p(t)\,dt = \frac{\omega_2}{\pi}\,dt \quad \text{for} \quad t \in [-T/4, T/4], \tag{7.12}$$

where T is the period of oscillations $T = 2\pi/\omega_2$. Since the probability density of I_2 is obviously symmetric around 0, it is sufficient to determine it for the non-negative half line. We can write for $I \geq 0$:

$$p_s(I) = \int_0^\infty p(b_t \sin \omega_2 t = I \mid b_t = b)\, p_s(b)\, db. \tag{7.13}$$

Obviously

$$p(b_t \sin \omega_2 t = I | b_t = b) = 0 \quad \text{for} \quad b < I \tag{7.14}$$

and we can choose as the lower boundary of the integral in (7.13) the value I instead of 0. To determine the first factor of the integrand, we have to calculate the probability density of the random variable

$$x = a \sin \omega t \,, \tag{7.15}$$

where a is a constant and t a random variable with (7.12). On the interval $[-T/4, T/4]$ the sine function is bijective and thus the probability is conserved under the transformation (7.15):

$$p(x)\,dx = p(t)\,dt \,. \tag{7.16}$$

It follows:

$$p(x) = p(t(x)) \frac{dt(x)}{dx} \,, \tag{7.17}$$

where $t(x)$ is the inverse function of (7.15), namely:

$$t = \frac{1}{\omega} \arcsin(x/a) \qquad x \in [-a, a] \,. \tag{7.18}$$

The Jacobian is straightforwardly calculated to be

$$\frac{dt}{dx} = \frac{1}{\omega}(a^2 - x^2)^{-1/2} \,, \tag{7.19}$$

and we finally obtain

$$p(x) = \frac{1}{\pi}(a^2 - x^2)^{-1/2} \,. \tag{7.20}$$

Thus we have

$$p(b_t \sin \omega_2 t = I | b_t = b) = \frac{1}{\pi}(b^2 - I^2)^{-1/2} \quad \text{for} \quad b \geqslant I \tag{7.21}$$

and (7.9) follows.

As can be seen from Fig. 7.4, the experimentally obtained stationary probability density deviates slightly from the theoretical one, given by (7.9). This is easily explained by a small amount of additive external noise [6.28]. To sum up, the good agreement between experimental results and theoretical predictions for the parametric oscillator establishes that noise-induced transitions are a real phenomenon and not an artefact generated by the white-noise idealization.

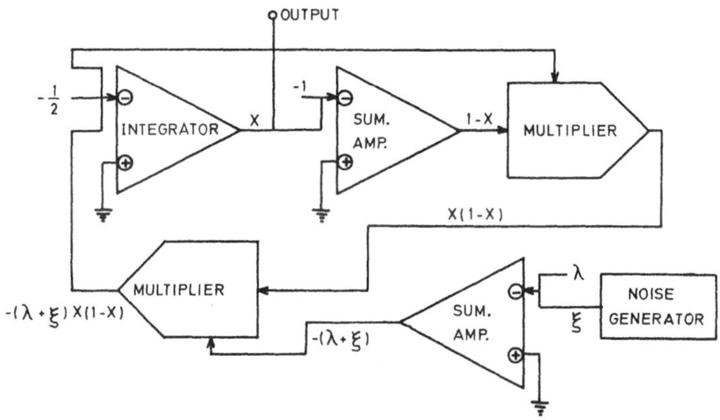

Fig. 7.5. An electrical circuit simulating the genetic model [7.3]

In order to study experimentally the second kind of noise-induced transitions, for which the genetic model is a typical representative, *Kabashima* constructed an electrical circuit with multipliers and summation amplifiers [7.3]. The circuit diagram is shown in Fig. 7.5. The output of the integrator is

$$X_t = -\omega \int [X_t - 1/2 + (\lambda + \xi_t) X_t (1 - X_t)] \, dt \,, \tag{7.22}$$

where the time constant of the circuit was chosen to be $\omega = 314 \, s^{-1}$. The fluctuating input ξ_t is produced by a noise generator which furnishes a Gaussian noise whose power spectrum is flat up to 500 Hz. The data accumulation is

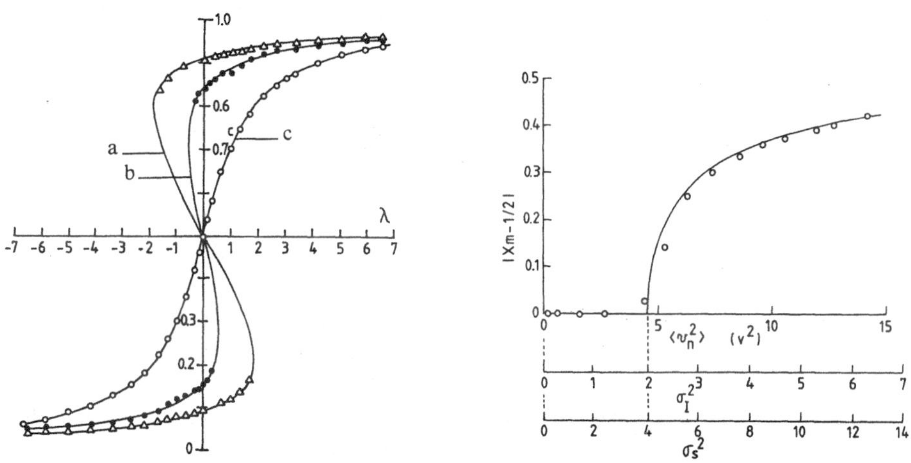

Fig. 7.6a **Fig. 7.6b**

Fig. 7.6. (**a**) Stationary values of the extrema of the probability density for fixed intensities of the noise as a function of λ. (*Curve a*) $\sigma_S^2 = 12.5$; (*Curve b*) $\sigma_S^2 = 7.6$; (*Curve c*) $\sigma_S^2 = 0$ (the values of σ_S^2 correspond to the Stratonovich interpretation).
(**b**) Plot of the extrema x_m of $p_s(x)$ as a function of intensity of the noise generator $\langle v_n^2 \rangle$ (Fig. 7.5) for $\lambda = 0$. The theoretical values σ_I^2 and σ_S^2 of the noise intensity corresponding to the Ito and Stratonovich interpretations are determined from the scaling indicated in this figure

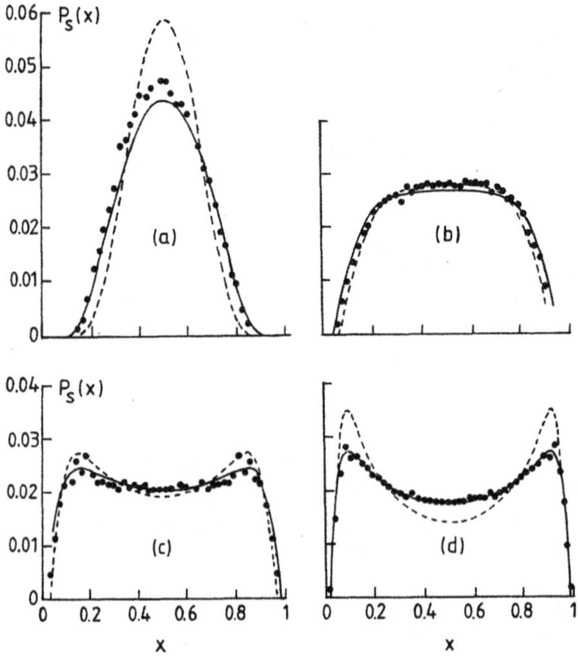

Fig. 7.7. Stationary density distribution for increasing values of the noise intensity. The broken and solid curves correspond to the Ito and Stratonovich interpretations, respectively (**a**): $\sigma_S^2 = 1.11$; (**b**) $\sigma_S^2 = 3.88$; (**c**) $\sigma_S^2 = 7.60$; (**d**) $\sigma_S^2 = 12.5$

effected by a computer, into which the output of the integration is fed via an analog-digital converter. The sampling rate is 4096 Hz for the measurements of the probability densities. The experimental results are summarized in Fig. 7.6, 7 and agree very well with the theoretical predictions.

7.2 Noise-Induced Transitions in an Open Chemical System: The Briggs-Rauscher Reaction

Besides quantum optical and hydrodynamical systems, chemical reactions play an important role in the experimental investigation of nonequilibrium phenomena. Systems capable of self-sustained oscillations or limit cycles of purely chemical origin (as opposed to a feedback via temperature) have attracted widespread interest. The prototype of this kind of chemical behavior is the Belousov-Zhabotinskii reaction [7.4–7]. One can say that there is nearly no nonequilibrium phenomenon that has not been found in this reaction – multiple steady states, limit cycles, chemical waves, and most recently chaotic behavior [1.35].

Oscillating chemical reactions are of manifest interest for an experimental test of the existence of noise-induced transitions in realistic physico-chemical systems. In the following we shall describe the experiments of *De Kepper* and *Horsthemke* [7.8, 9] on the Briggs-Rauscher (BR) reaction. This reaction displays the same rich behavior under nonequilibrium conditions as the Belousov-Zhabotinskii reaction. Though less popular than the latter, the BR reaction has also been extensively studied [7.10] and from our viewpoint has some rather advantageous features. The BR reaction involves the following chemicals as major reactants: KIO_3 (potassium iodate), H_2O_2 (hydrogenperoxide), $HClO_4$ (perchloric acid), $MnSO_4$ (magnesium sulfate), $CH_2(COOH)_2$ (malonic acid).

The theoretical treatment of noise-induced transitions is formulated for spatially homogeneous systems. Thus experiments on the BR reaction in a so-called continuous flow stirred tank reactor (CSTR) are of particular interest. In such experiments the reactants mentioned above are pumped at a steady flow rate from reserve vessels containing the chemical with a specified concentration into the reactor. Mixing is achieved by mechanical means, in general a rapidly revolving stirrer. The reactor is equipped with an overflow, so that the reaction volume remains constant and temperature is kept constant by thermostatic control. The state of the chemical system is recorded by measuring state variables like its redox potential or the optical density at a given wavelength.

In a CSTR the BR reaction exhibits multiple steady-state behavior and transitions from nonoscillating to oscillating states depending on the values of external constraints such as the concentrations of the chemicals, the flow rate, the temperature. The most advantageous feature of the BR reaction for our purposes is that it is extremely photosensitive to visible light. It absorbs strongly around 460 nm. Light at this wavelength splits the I_2 molecules produced in the reactor into two extremely reactive radicals I, accounting for the very photosensitive nature of the BR reaction. From an experimental point of view, this is a true boon: i) Variations in the incident light intensity affect the reactor as a whole and not only via some boundary effect. It is thus a situation to which our theoretical analysis applies well. Furthermore, it is intuitively clear that fluctuations in parameters that affect the bulk properties of the system will have a more drastic effect than mere surface quantities. Hence they hold greater promise for a successful experiment. ii) It is fairly easy to contruct a source of fluctuating light intensity with a correlation time that is very short compared to the macroscopic time scale of the system.

For these reasons, it is extremely attractive to study the influence of rapid light intensity fluctuations on the BR reaction. The experimental setup is shown in Fig. 7.8. Here S is a white-light source, namely the bulb of a projector lamp, and P denotes a pair of polaroids used to vary the light intensity. The noise generator is B. It consists of a box containing small polystyrene balls dancing in a turbulent airstream. This device produces a noise which is (almost) Gaussian

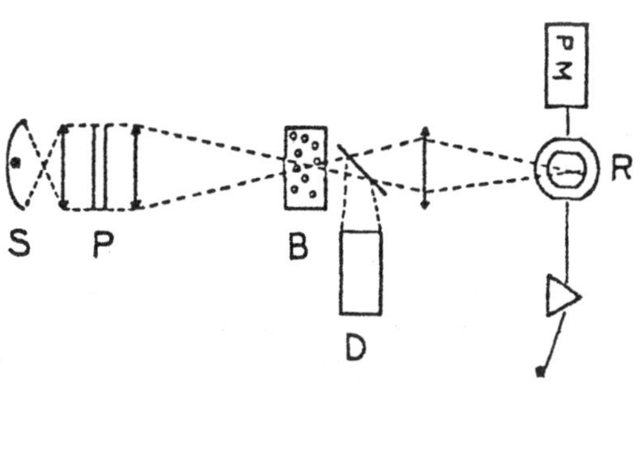

Fig. 7.8. Experimental setup used to study the influence of a fluctuating illumination on the Briggs-Rauscher reaction

distributed with a relative variance of 13%. The intensity fluctuations have an exponentially decreasing correlation function with $\tau_{cor} \simeq 0.05$ s. This corresponds to about one-twentieth of the typical macroscopic time for the experimental conditions used. This external noise can thus be considered to be quasi-white. Further, D is a light detector with which the mean light intensity is monitored, and R is the continuous flow stirred tank reactor. Lenses are used to focus the beam in box B and reactor R. The state of the reactor is monitored by measuring the redox potential, the optical density at 460 nm permitting the concentration of I_2 in the reactor to be determined, and the temperature. The temperature of the reactor was held constant to better than 0.1 °C. This excludes that any of the transitions found in the experiments are due to temperature changes caused by the incident light. Under the experimental conditions used a change in the reactor temperature of 10° to 20°C is necessary to induce a transition. The concentrations of the reactants and the flow rate is given in Fig. 7.9.

Basically two different sets of experiments are performed, according to whether one is below or above the deterministic critical point corresponding to the occurrence of a bistability region. In the first series of experiments, a bistable situation as a function of the light intensity is explored. The main results are displayed in Fig. 7.9. The full-drawn curves describe the situation when the noise generator is switched off, whereas the broken curves describe the fluctuating situation (in this case I denotes the mean incident light intensity).

Let us first consider the result in the case without external noise. The values of the other constraints, i.e., the concentrations of the major reactants, the temperature T of the reactor and the renewal time τ of the reactor, a measure for the flow rate, are indicated in the figure. It is found that the system displays bistable behavior as a function of the intensity of the incident light. Indeed, as the light

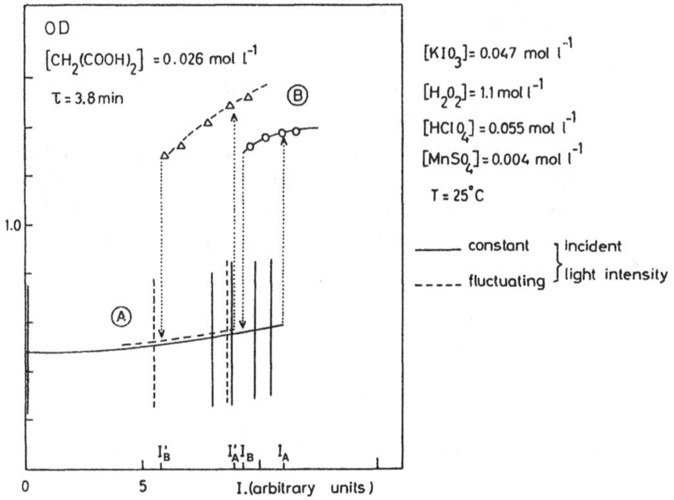

Fig. 7.9. Optical density (OD) at 460 nm as a function of the light intensity I (nonfluctuating condition: solid curves and fluctuating conditions: broken-line curves). (Constraint values as indicated. A: low concentration of I_2; B: steady state of high concentration)

intensity is first increased and then decreased again, the state of the BR system describes a hysteresis loop. For a certain range of intensities, two branches of steady states exist simultaneously. Note that on the lower branch the steady states are unstable for the external constraints considered and are surrounded by a stable limit cycle corresponding to homogeneous oscillations of the chemicals in time. Hence the bistability is between an oscillating state A of low concentration of I_2 and a steady-state B of high concentration (Fig. 7.9). These two phases can be distinguished by looking at the reactor vessel. In state A, the reacting mixture is brownish yellow, in state B, blue violet. As the figure shows, under deterministic external conditions[2] the chemical system is in the oscillating state A for no or little illumination of the reactor[3]. The large vertical bars in Fig. 7.9 denote the amplitude of the limit cycle. As the incident light intensity is increased, the mean value of the oscillations increases slightly, the amplitude remains almost unchanged, whereas the period increases till the intensity I_A is reached where the system undergoes a spontaneous transition to state B; the reactor vessel changes suddenly from yellow to violet. It goes without saying that after every change in intensity I, the system is given ample time to relax to its new state. If the incident intensity is now decreased, the system remains in state B of high concentration, the reactor is dark violet. Thus the system is in a region of bistability till intensity I_B is reached, when a spontaneous transition to the lower state A occurs. If the noise generator is switched on, we observe the situation represented by the broken curves. In the case of the fluctuating intensity, the region of bistability is shifted to lower (mean) intensities to such an extent that there is *no* overlap with the hysteresis loop of the nonfluctuating case. Furthermore, the width of the hysteresis loop I'_B, I'_A is enlarged by about a factor 2. The fact that the two regions of bistability are separated from each other implies the existence of noise-induced transitions for values of light intensity between I'_A and I_B. Keeping the mean intensity constant, the system undergoes a transition if the noise is switched on or off. Indeed, for $I \in (I'_A, I_B)$ and $\sigma^2 = 0$, only state A exists, whereas for $I \in (I'_A, I_B)$ and $\sigma^2 = 13\%$ only state B exists. This is the first experimental evidence for noise-induced transitions in complex chemical systems. Note also that the shift and broadening of the hysteresis loop is a nontrivial result. Arguing along intuitive lines of reasoning one would expect the following behavior. If the mean intensity I is near but below I_A, the fluctuating light intensity will be larger than I_A part of the time and under fluctuating conditions the transition from A to B should occur sooner. This is indeed observed. However, the same argument applies to the transition from B to A. If I is near but above I_B, due to the external noise the light intensity will spend an appreciable fraction of the time below I_B, thus triggering the transition. One expects therefore that the width of the hysteresis loop should be reduced and that in the fluctuating case the loop is completely contained in the deterministic hysteresis loop. This is, however, not

---

[2] The intrinsic intensity fluctuations of the light bulb have an amplitude too small to play any role. They can be neglected.

[3] The experiment is performed in a dark room. Otherwise the results are perturbed by variations in the background illumination.

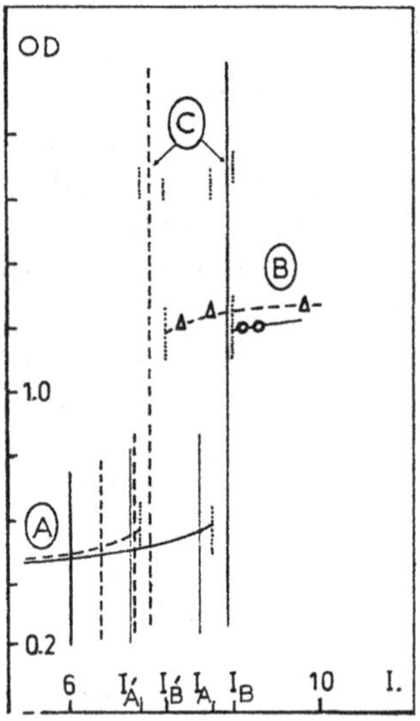

Fig. 7.10. Optical density at 460 nm of completely reversible first-order transitions as a function of light intensity *I*. Same constraints as Fig. 7.9 except $(CH_2(COOH)_2)_0 = 0.026$ mole l^{-1} and $\tau = 3.3$ min. (A: simple oscillating state; B: steady state; C: complex oscillating state)

the outcome of the experiment, showing that the noise-induced transitions observed in the BR reaction are a nontrivial result of the interplay between the nonlinearity of the system and the external noise.

The second part of the experiments was performed for external constraints corresponding to a situation beyond the deterministic critical point of bistability as shown in Fig. 7.10. Here for a given value of the incident light intensity only one state of the system is stable. As a function of the illumination, the simple oscillating state A changes to the steady-state B through a narrow zone of complex oscillations C. State C is in a certain sense a composite state, made up of states A and B. In state C the system oscillates, so to speak, between the oscillating state A and the steady-state B. Under fluctuating illumination the same switching curve is observed, however again shifted by a certain amount to lower (mean) intensities. As can be seen from Fig. 7.10, for values of I included between $I_{B'}$, and I_A a noise induced transition from the oscillating state A to a steady-state B and vice versa is expected. This is experimentally verified for $I = 8$ units (Fig. 7.10) by switching the noise generator on or off and keeping the mean light intensity constant. Some minutes after the switch a spontaneous transition to the other state occurred. (Of course, the waiting time for the transition to occur is random).

These experimental results, establishing the existence of noice-induced nonequilibrium phase transitions in a complex chemical system, are of particular importance in view of the fact that the theoretical results of Chap. 6 were obtained using certain idealizations as already mentioned above. Let us comment on some of these points in more detail.

i) The phenomena observed in the Briggs-Rauscher reaction demonstrate the importance of external noise for the macroscopic behavior of complex real systems. We think that these results are of considerable interest in biological systems, a point to which we shall return later.

ii) In the experiments a "real noise" was of course generated. Though the correlation time of the intensity fluctuations was sufficiently short, so that the external noise can be considered to be quasi-white, it is obviously not a genuine white noise. The experimental results give credence to the arguments employed in Chap. 3 to justify the replacement of an external noise with an extremely short correlation time by Gaussian white noise. The fact that the results of the white-noise analysis are indeed robust and hold qualitatively in the neighborhood of white noise will be established in an analytical way in the following chapter.

iii) The experimental results on the Briggs-Rauscher reaction hold one surprise. As is indicated in Figs. 7.9, 10, the state of the system, subjected to a fluctuating light intensity, is as well defined as the state of the deterministic system. At least for the state variable, concentration of I_2, the probability density has sharp narrow peaks, in contradistinction to the model systems studied theoretically in Chap. 6 and to the electrical circuits studied experimentally by Kabashima and co-workers. Furthermore, during the entire time span of the experiment, about one hour for every value of the mean intensity, no spontaneous transition between the two states in the region of bistability is observed. In our opinion, a plausible explanation for this phenomenon is the existence of a nucleation mechanism also in the case of noise-induced bistability. Due to the thorough stirring any "droplets", being in the other state, are broken up in the reacting mixture before they can grow to the finite critical size necessary to tip the system over into the other state. Thus stirring stabilizes the two states and leads to narrow peaks in the probability density. This nucleation mechanism is not operational in the electrical circuits studied by Kabashima and co-workers, since those system are effectively zero-dimensional systems. From a theoretical point of view, these results show that a quantitative description of noise-induced phenomena in nonzero-dimensional systems requires the inclusion of spatial dependence in the equations, even if the system remains homogeneous on the average. For such systems we expect that in general the mean-field theory, based on SDE's which do not take into account the possibility of spatial inhomogeneities, qualitatively gives a good description of noise-induced transitions. However, fluctuations away from spatial homogeneity might quantitatively change the mean-field predictions as in the theory of classical equilibrium phase transitions.

7.3 Optical Bistability

Among the systems extensively studied theoretically as well as experimentally from the viewpoint of nonequilibrium phase transitions are optical systems such as the laser [1.46, 49, 54, 7.11 – 14]. Laser-like systems are also ideal for the purpose of studying noise-induced transitions in spatially homogeneous media, since in general the space dependence of the electric field in the laser cavity can be safely neglected. Such systems can indeed be described satisfactorily by global variables. In this section, we discuss in particular optical devices that can show bistable behavior. The first experimental observation of optical bistability was made on the transmission of a laser beam through a Fabry-Perot cavity filled with sodium vapor [7.15]. It was found that under certain conditions the plot of

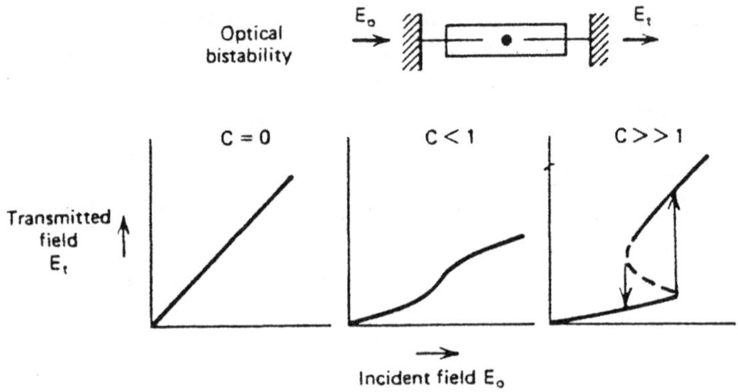

Fig. 7.11. Transmitted light versus incident light in an optical bistable device for three different values of the cooperativity parameter c, cf. (7.29). (From [7.14])

the transmitted light versus the incident light exhibits a hysteresis cycle (Fig. 7.11). From a theoretical point of view two mechanisms can give rise to such bistable behavior, namely nonlinear absorption and nonlinear dispersion [7.16]. In the following we shall consider only the first mechanism, namely absorptive optical bistability. It should be noted that experimentally it requires great care to prepare a purely absorptive bistable system. In most experiments, the effects of nonlinear dispersion dominate those of absorption.

The now standard model for absorptive optical bistability was first derived by *Bonifacio* and *Lugiato* [7.17]. It is a simple generalization of the one-mode laser model to include the effects of an external coherent field that continuously drives the atoms, i.e., an injected laser beam. Though different authors formulated descriptions for optical bistability taking fluctuations into account (see [7.14] for a review), the first investigation of noise-induced transitions in this system was carried out by *Bulsara* et al. [7.18].

The standard model of absorptive optical bistability considers a gas of two-level atoms in a Fabry-Perot cavity. The atoms are driven incoherently by optical pumping and coherently by an injected laser beam. It is assumed that the length L of the pencil-shaped optical cavity is an integer multiple of the wavelength of the injected field, i.e., the cavity is perfectly tuned to the incident laser beam. Furthermore, it is supposed that the frequency of the laser beam coincides exactly with the resonance frequency of the two-level atoms. This assumption of "on resonance" between the injected field and the cavity as well as between the injected field and the atoms implies the absence of any dispersion effects. Following the standard approximations for the one-mode laser model, the atomic system can be adiabatically eliminated and a set of equations for the *total* cavity field alone (its amplitude and phase) is obtained. This relies mainly on the assumption of a good cavity, i.e., the incoherent free atomic decay rates are much faster than the empty cavity field decay. Under this condition, the atoms follow the field almost instantaneously, the atoms are "slaved" by the field [7.13] and the system is described by the cavity field alone. The evolution equation for the field amplitude X is given by [7.19]

$$\dot{X} = Y\cos(\varphi - \varphi_0) - X - 2c\frac{X}{1+X^2} \tag{7.23}$$

and for its phase

$$\dot{\varphi} = -\frac{Y}{X}\sin(\varphi - \varphi_0) . \tag{7.24}$$

Here Y is the amplitude of the injected laser beam and φ_0 its phase. The cooperativity parameter c is proportional to the negative of the population inversion of the atoms: $c > 0$ implies that (number of excited atoms – number of ground-state atoms) < 0 and corresponds thus to absorption. Further, $c < 0$ is a positive inversion, i.e., the number of excited atoms exceeds those in the ground state, and corresponds to stimulated emission. The following potential

$$V(X,\varphi) = \tfrac{1}{4}[X^2 - Y^2 - 2XY\cos(\varphi - \varphi_0) + 2c\ln(1+X^2)]$$

can be associated with (7.23, 24) in the sense that

$$\dot{X} = -2\frac{\partial V}{\partial X}, \tag{7.25}$$

$$\dot{\varphi} = -\frac{2}{X^2}\frac{\partial V}{\partial \varphi}. \tag{7.26}$$

For $Y > 0$, the deterministic steady states are given by

$$Y = X + 2c\frac{X}{1+X^2} \tag{7.27}$$

and

$$\varphi = \varphi_0 . \tag{7.28}$$

Thus the injected laser beam confers a preferred phase to the cavity. Moving away from the line $\varphi = \varphi_0$, the potential rises very steeply. This indicates that the phase motion is on a much faster time scale than the amplitude motion. The phase of the cavity field relaxes quickly to the phase of the injected laser beam, i.e., phase locking occurs. Therefore the phase variable φ can be adiabatically eliminated and we end up with the one-variable problem:

$$\dot{X} = Y - X - 2c\frac{X}{1+X^2} . \tag{7.29}$$

In the following we shall consider only the case of no or weak optical pumping, i.e., $c \geqslant 0$. As far as the steady-state behavior of (7.29) is concerned, it is easily verified that critical behavior occurs for $c=4$. If the cooperativity parameter is smaller, i.e., $c<4$, only one (stable) steady state exists: for every value of the incident intensity there is exactly one value of the transmitted intensity. If the cooperativity parameter is larger than the critical value, i.e., $c>4$, the phenomenon of hysteresis occurs. There is a range of values for the incident intensity, for which three steady states, two stable and one unstable, are possible for the transmitted intensity.

Obviously, there are different sources of external noise for the problem, as discussed in [7.14, 19, 20], in particular fluctuations in the amplitude Y of the incident laser beam, and in the cooperativity parameter c. Since the former correspond to additive noise and therefore will not modify the deterministic bifurcation diagram, we shall consider here only fluctuations of the latter type. Fluctuations in the parameter c are unavoidable, since spontaneous emission, a quantum noise effect, will lead to a random motion for the atomic population inversion. If this is the only source of external noise present, then σ is very small, in the units chosen here $\sigma \sim 10^{-3}$, and a modification of the deterministic stationary behavior will occur only in a very small neighborhood around the deterministic critical point [7.19].

There are, however, two further sources of external noise, namely fluctuations in the optical pumping and fluctuations in the atomic density in the cavity. These fluctuations can be controlled to a good degree in the laboratory. In other words, external noise with given characteristics can be applied to the optical bistable system. In view of the advantageous features of quantum optical systems, enumerated at the beginning of this section, the absorptive optical bistability is thus particularly well suited to study noise-induced nonequilibrium phase transitions. In the following we shall assume, in the spirit of our theoretical approach so far, that rapid external noise is applied to the system causing fluctuations in the cooperativity parameter which can be described satisfactorily by Gaussian white noise. Optical bistability can then be modeled by the diffusion process which is the solution of the following Stratonovich SDE:

$$dX_t = \left(Y - X_t - 2c\frac{X_t}{1+X_t^2}\right)dt + \sigma\frac{X_t}{1+X_t^2}\circ dW_t\,. \tag{7.30}$$

We suppose here that external noise does not destroy the phase locking and therefore neglect fluctuations in the phase of the field. In the case we are interested in here, namely multiplicative noise due to random changes in the cooperativity parameter c and no additive noise, this procedure is legitimate since $\varphi = \varphi_0$ is a stable stationary point of the SDEs for the two variable process (X_t, φ_t).

Equation (7.29) is also encountered in biological applications. It has been used to model some of the features of the respiratory process in a Klebsiella aeorgenes bacterial culture. The effect of external multiplicative white noise in this context was studied by *de la Rubia* and *Velarde* [7.21].

To highlight the essential properties of noise-induced transitions in this system, we pass immediately to the equation for the extrema of the stationary probability density, which is easily verified to exist, and find:

$$Y - X_m - 2c\frac{X_m^2}{1+X_m^2} - \frac{\sigma^2}{2}\frac{X_m}{1+X_m^2}\left(\frac{1}{1+X_m^2} - \frac{2X_m^2}{(1+X_m^2)^2}\right) = 0\,, \tag{7.31}$$

or

$$Y - X_m - 2c\frac{X_m}{1+X_m^2} - \frac{\sigma^2}{2}\frac{X_m - X_m^3}{(1+X_m^2)^3} = 0\,. \tag{7.32}$$

To elucidate the characteristic features of the noise-induced transitions, consider the curve of the extrema given by the functional relation

$$Y = X + 2c\frac{X}{1+X^2} + \frac{\sigma^2}{2}\frac{X - X^3}{(1+X^2)^3}\,. \tag{7.33}$$

For a region of bistability to occur $Y(X)$ must not be a monotone function. For the slope of the curve given by (7.33), we have

$$Y' = 1 + 2c\frac{1-X^2}{(1+X^2)^2} + \frac{\sigma^2}{2}\frac{1-8X^2+3X^4}{(1+X^2)^4}\,, \tag{7.34}$$

which can be decomposed into a part stemming from the phenomenological kinetics and a part due to external noise:

$$Y'_{\text{det}} = 1 + 2c\frac{1-X^2}{(1+X^2)^2}\,, \tag{7.35}$$

$$Y'_{\text{stoch}} = \frac{\sigma^2}{2}\frac{1-8X^2+3X^4}{(1+X^2)^4}\,. \tag{7.36}$$

For $X = 0$ we have:

$$Y'(0) = 1 + 2c + \frac{\sigma^2}{2} > 0 \quad \text{for all} \quad \sigma^2\,, \tag{7.37}$$

if we restrict ourselves to the weak pumping case such that $c > -1/2$. (This excludes the laser transition. For $Y = 0$, $c = -1/2$ corresponds to the laser threshold. If the pumping is such that $c < -1/2$, the Fabry-Perot emits coherent light.) According to (7.34), bistability occurs if an interval exists for which $Y' < 0$. The deterministic critical point is given by $c = 4$, $X_c = \sqrt{3}$, i.e., $Y'_{\text{det}}(\sqrt{3}) = 0$ for $c = 4$.

The zeros of Y'_{stoch} are easily determined to be $X_1^s = 0.362$ and $X_2^s = 1.592$. It follows that $Y'_{\text{stoch}}(\sqrt{3}) > 0$ for all σ^2. This implies that arbitrarily small external noise suppresses the deterministic critical point at $c = 4$. The deterministic bistability can of course no longer be destroyed by external noise for those values of c where $Y'_{\text{det}}(X_{1,2}^s) < 0$, since this ensures the existence of an interval where $Y' < 0$, independent of the noise intensity. Whereas for $X_1^s = 0.362$ we have

$$Y'(X_1^s) = Y'_{\text{det}}(X_1^s) = 1 + 2c \cdot 0.6783 > 0 \qquad (\text{for } c > -1/2)\,, \tag{7.38a}$$

we obtain for the other zero

$$Y'(X_2^s) = Y'_{\text{det}}(X_2^s) = 1 + 2c(-0.1228)\,. \tag{7.38b}$$

This expression vanishes, and then becomes negative, for $c \geqslant 4.0704$, implying that bistable behavior occurs for arbitrary values of the noise intensity, if $c \geqslant 4.0704$. As is clear from the fact that the noise intensity plays no role in the existence of the hysteresis phenomenon, this bistable behavior corresponds to that already present under deterministic conditions. External noise shifts its occurrence to slightly higher mean values of the cooperativity parameter c.

However, the system displays also a pure noise-induced transition. To see this, note that for $X = 1$

$$Y'(1) = 1 - \sigma^2/8 \tag{7.39}$$

independently of c. We have thus the following *sufficient* condition for noise-induced bistability to occur. If the intensity of the external noise is sufficiently large, namely $\sigma^2 > 8$, then there are two maxima for the transmitted field. The existence of this bistable behavior is independent of the value of the cooperativity parameter. It is observed even for $c < 4$ and is a pure noise-induced effect. Its mechanism is different from the deterministic hysteresis phenomenon. It comes about through the interplay of phenomenological nonlinear kinetics and external noise. To summarize, optical bistability displays both types of noise-induced transitions. The deterministic critical point is suppressed by arbitrarily small noise and shifted to higher value of c. At the same time, the mechanism of a pure noise-induced critical point operates and causes bistable behavior for *all* values of the cooperativity parameter.

7.4 Noise-Induced Transitions and the Extinction Problem in Predator-Prey Systems

In population dynamics, the problem often arises to determine the amount of predation which a prey population can sustain without endangering its survival. The answer to this question is fundamental for a good management of biological resources, namely to find the best strategy for optimal harvesting or fishing. The question may also be posed with the opposite purpose in mind, i.e., to drive a prey population to extinction by sufficient predation. Under this angle the extinction problem does not only find applications in pest control but also in the medical sciences. For instance, in the epidemiology of infectious diseases it is necessary to understant the predator-prey type of equilibrium which exists between the therapeutic treatment and the persistence of kernels of infection. At the individual organismic level qualitatively the same type of question arises in the study of the immune defense mechanism against nonself and abnormal antigenes.

In all the above-mentioned cases, one deals with systems subjected to random environmental variations which sometimes may be quite big in magnitude. It is thus natural in this context to inquire about the impact of environmental fluctuations [1.81, 7.22–27]. We have already seen that this impact may be drastic. Even in the case of the simple Verhulst model (6.37) it introduces essential changes in the transition mechanism of the population from growth to extinction. In this section we shall again consider the effect of noise-induced transitions on the persistence of populations, concentrating primarily on the biological implications of these phenomena. We shall consider cases where the prey is subjected to a fluctuating rate of predation which can be described by a scalar equation of the form:

$$\dot{X} = \alpha + \lambda X\left(1 - \frac{X}{K}\right) - \beta g(X) . \tag{7.40}$$

Here X represents the density of the prey in a given territory. This density is supposed to be homogeneous so that spatial effects need not be taken into account. The constant α takes care of a constant source of X, e.g., by immigration. The second term on the rhs of (7.40) is a Fisher logistic growth term with birth rate λ and carrying capacity K. Equations of the form (7.40) apply to situations where the prey population evolves on a slower time scale than the predator population. Often the rate of predation $\beta g(X)$ is then found to be a saturable function of the prey population X. The maximum rate at which predation takes place when the prey is much more abundant than the predator, i.e., $X \to \infty$, is β. A typical ecological example of such a situation, for which a mass of field data exists is the spruce budworm/forest interaction in eastern North America. A study of the properties of this system reveals that an appropriate form for $g(X)$ is $g(X) = X^2/(1+X^2)$. The system has been analyzed under the hypothesis of deterministic environmental conditions by *Ludwig* et al. [7.28].

In the following we investigate in detail the case where $g(X) = X/(1+X)$. It corresponds to a predator-prey system where the total number of predators is constant and where the predators have essentially only two states of activity: "hunting" and "resting". This two-state predator model also finds applications outside of the field of ecology; it furnishes an accurate description of the basic reaction steps between immune cytotoxic cells and tumoral cells [7.29, 30]. We shall illustrate its properties within this context.

7.4.1 Two-State Predator Model

We consider a population of predators living in a given territory and feeding on a population of prey X. The assumptions are:

i) In the absence of the predator the environment of the prey is constant so that its growth is simply logistic, i.e., of the form $\lambda X(1-X/K)$ (K: carrying capacity).

ii) The characteristic time interval over which appreciable variations occur in the populations of predator and prey are very different so that the number of predators is, on the average at least, strictly constant over the generation time of the prey population.

iii) The predator divides its time in two activities: either "hunting" for the prey or "resting". The number of predators in each state is Y and Z, respectively. The predator spends on the average a time τ_H hunting and a time τ_R resting. As is generally the case, these time spans are taken to be short compared to the generation time of the prey, i.e., $\tau_H, \tau_R \ll \lambda^{-1}$.

Under these conditions the predator-prey interactions can be represented by a reversible loop describing the switching of the predator between its two activities:

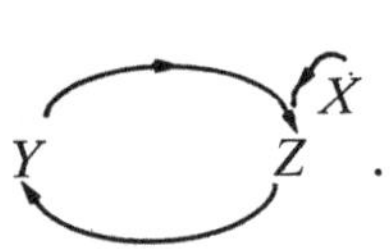

(7.41)

The essential feature captured in model (7.41) is that formally the activity of the predators on the territory resembles the mode of action of enzymes or catalysts in chemical reaction: without disappearing themselves in the reaction, catalysts continuously transform substrates. This is exactly what the constant population of predators on the territory does by feeding on the prey. We may make this analogy more quantitative and admit, as is often done, that the interaction between predator and prey obeys a set of chemical kinetic type of evolution equations

$$\dot{X} = A + \lambda X(1 - X/K) - \frac{1}{\tau_H} XY, \tag{7.42}$$

$$\dot{Y} = -\frac{1}{\tau_H} XY + \frac{1}{\tau_R} Z \quad \text{with} \quad E \equiv Y(t) + Z(t) = \text{const.} \tag{7.43}$$

In (7.42) the first two terms on the rhs represent, respectively, the flux of prey due to immigration and their logistic growth; the last term $-XY/\tau_H$ describes the rate at which the prey is killed. We consider the case that $\lambda^{-1} \gg \tau_H, \tau_R$ and assume further that the total population of predators E is small compared to the prey population X; more precisely we put

$$\tau_H = \varepsilon \tau_H^*, \qquad \tau_R = \varepsilon \tau_R^*, \qquad Y = \varepsilon Y^*, \qquad Z = \varepsilon Z^*, \tag{7.44}$$

where ε is a small quantity, τ_H^*, τ_R^* are quantities of order λ^{-1} and Y^*, Z^* are quantities of order X. Replacing (7.44) in (7.42, 43) yields

$$\dot{X} = A + \lambda X(1 - X/K) - \frac{1}{\tau_H^*} XY^*, \tag{7.45}$$

$$\varepsilon \dot{Y}^* = -\frac{1}{\tau_H^*} XY^* + \frac{1}{\tau_R^*} Z^*. \tag{7.46}$$

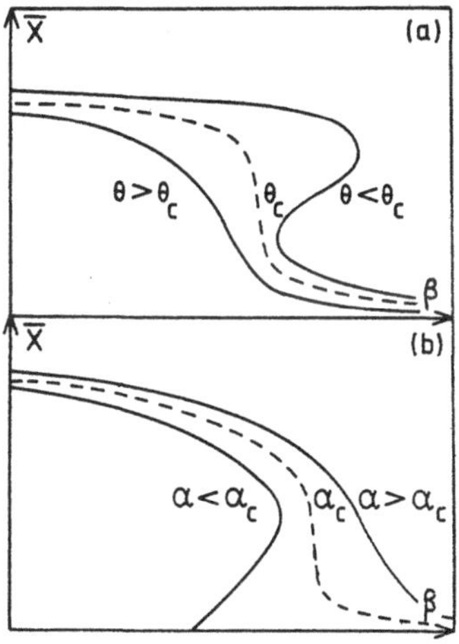

Fig. 7.12. Steady states of model (7.47) as a function of β **(a)** α fixed and values of θ below, at and above the critical point. **(b)** θ fixed ($\theta < 1$) and values of α below, at and above α_c

Taking the limit $\varepsilon \to 0$ and eliminating Y^* from (7.45), one obtains finally an equation of evolution for X of the form (7.40)

$$\dot{X}^* = \alpha + X^*(1 - \theta X^*) - \beta \frac{X^*}{1 + X^*} \tag{7.47}$$

with

$$\alpha = \frac{A \tau_R^*}{\lambda \tau_H^*}, \quad \beta = \frac{E}{\lambda \cdot \tau_H}, \quad \theta = \frac{\tau_H}{\tau_R K}, \quad X^* = \frac{\tau_R^*}{\tau_H^*} X. \tag{7.48}$$

In the following, for simplicity, we drop the * everywhere. The remarkable feature of the deterministic steady-state solutions of (7.47) is that they may display a cusp type of catastrophe. The corresponding critical point (α_c, β_c, X_c) is given by

$$\alpha_c = \frac{(1-\theta)^3}{27\,\theta^2}, \quad \beta_c = \frac{(1+2\,\theta)^3}{27\,\theta^2}, \quad X_c = \frac{1-\theta}{3\,\theta}. \tag{7.49}$$

It is clear from (7.49) that a necessary condition for this critical point to be physically meaningful, i.e., for α_c, X_c to be positive, is that $\theta < 1$. Thus as a function of β the steady states curves $\bar{X}$ will always present a bistability region for small values of θ (Fig. 7.12a). This domain can always be reached by sufficiently increasing the carrying capacity of the territory K or by decreasing the ratio τ_H/τ_R; the better hunter the predator is, the more likely it is that a bistable situation occurs. Or viewed another way, if the population E of a very efficient predator increases on a given territory, the population of the prey will drop in a continuous manner only up to the value of β corresponding to the turning point of the hysteresis loop. Beyond that value a jump takes place and $\bar{X}$ drops abruptly on the lower branch of steady states. If θ is fixed at a value smaller than one, the occurrence of a bistability region depends on the rate of immigration α (Fig. 7.12b). When $\alpha = 0$, $\bar{X}$ has a branching point at $\beta = 1$ with the line $\bar{X} = 0$ which

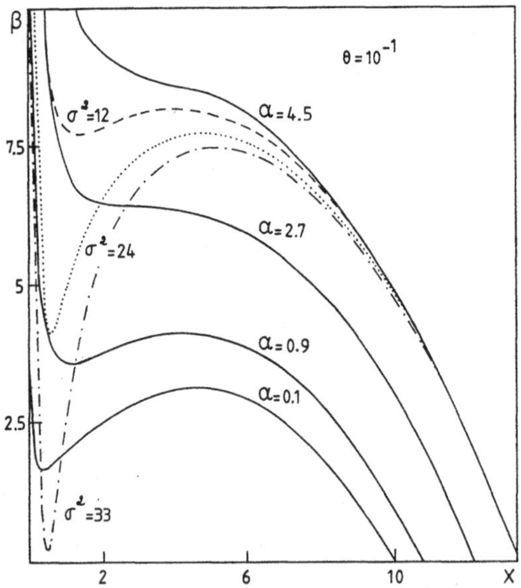

Fig. 7.13. Deterministic stationary states of (7.47) as a function of β and different values of α (solid curves). The broken curves plot the extrema of $p_s(x)$ corresponding to $\alpha = 4.5 > \alpha_c$ and values of σ^2 as indicated

in that case always is a steady-state solution; $\bar{X} = 0$ is stable for $\beta > 1$ and unstable otherwise. Immigration to the territory suppresses this branching point and eventually, when it becomes large, it suppresses the occurrence of a bistable transition region.

Let us now consider the behavior of this system, when β fluctuates arounds its mean value sufficiently rapidly for the white-noise idealization to be plausible, i.e., $\beta_t = \beta + \sigma \xi_t$ with ξ_t corresponding to Gaussian white noise [7.27, 31]. The deterministic kinetic equation is replaced by the SDE

$$dX_t = \left[\alpha + (1 - \theta X_t) X_t - \beta \frac{X_t}{1 + X_t} \right] dt + \sigma \frac{X_t}{1 + X_t} dW_t \,. \tag{7.50}$$

Interpreting it in the Ito sense one obtains the stationary probability density by the now familiar procedure. Its extrema are displayed in Fig. 7.13 as a function of β for different values of the variance σ^2. The external noise has two main effects:

i) The domain of bistability is shifted to lower values of β;

ii) The sigmoidicity of the curves increases. For values of the parameters equal to or above the deterministic critical point (7.49), these effects induce bistability. Then $p_s(x)$ may become double peaked when σ increases, even under conditions where the deterministic stationary curves are single valued for all β. Clearly the noise induces a shift of the deterministic phase transition; a behavior which seems to resemble closely the experimental findings in the BR reaction. In Fig. 7.14 the behavior of $p_s(x)$ is represented for increasing values of σ^2. For $\sigma^2 = 3$ the distribution is single peaked with its maximum near the deterministic steady state. When σ^2 increases, a new peak appears and grows on the side of the

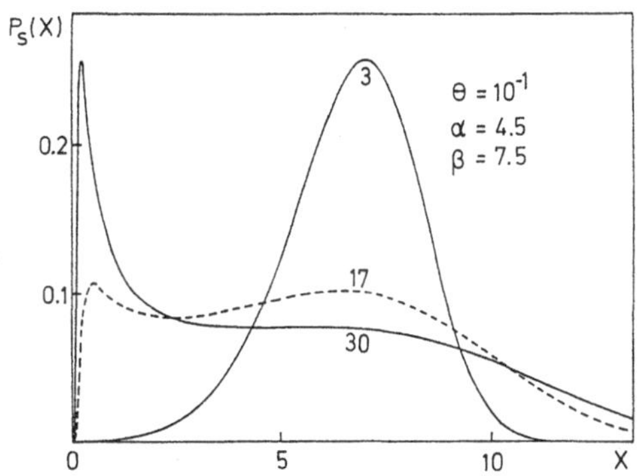

Fig. 7.14. Probability density for three values of the variance σ^2

low values of x while the upper peak disappears. Amazingly, this effect which favors the extinction of the prey population is observed without changing the average predation parameter. It is interesting to note that this phenomenon occurs despite the obvious and unavoidable drawback of the Gaussian white-noise idealization that ξ_t takes negative values; an effect which undoubtedly favors the growth of the X population rather than its extinction. It is therefore expected that in a real environment, i.e., such that β_t is positive, extinction is realized more quickly than suggested by the values of σ^2 considered here.

7.4.2 Cell-Mediated Immune Surveillance: An Example of Two-State Predator Systems

It is a well-established fact that most tumoral cells bear antigens which are recognized as foreign by the immune system. The latter develops a response against these antigens which is mediated by immune cells such as T lymphocytes. Other cells not directly dependent on the immune system, e.g., macrophages or natural killer cells, may also participate in this response. It proceeds via the infiltration of the tumor by these cells which subsequently develop in situ a cytotoxic activity againts the tumoral cells. The dynamics of this process as a whole are extremely complex and will not be described here (for a more detailed discussion see [7.29, 30, 32]). Rather we shall focus our attention on situations where over extended time intervals, i.e., much longer than the average time between successive tumoral cell replications, the immune system can be considered to be in a quasi-stationary state. It is then meaningful to represent the cytotoxic reactions between the cytotoxic cells which have infiltrated the tumor and the tumoral cells by a two-step process of the form (7.41). A population of cytotoxic cells is denoted by Y (predators), X is the target tumoral population (prey) and Z is a complex formed by the binding of Y to X. The cytolysis cycle (7.41) can accurately be described by the evolution equations (7.42, 43). In Table 7.1 we report some representative orders of magnitude for the constants $k_1 = \tau_H^{-1}$ and $k_2 = \tau_R^{-1}$, and the corresponding values for the parameter θ which here is equal to k_1/k_2N, N is the maximum number of cells per unit volume [7.29, 30, 32].

Table 7.1. Approximated values of cytotoxic parameters for each kind of cytotoxic cell

Cytotoxic cell	$k_1 N\,[\mathrm{day}^{-1}]$	$k_2\,[\mathrm{day}^{-1}]$	θ
Allosensitized T cell[a]	18	18	1
Immune T cells[b]	0.425	0.85	$1-2$
Syngeneic NK cells[c]	$1-6$	$0.6-3$	$0.1-3$
Syngeneic activated macrophages[c]	$0.1-0.4$	$0.2-0.7$	$0.5-5$

[a] Allosensitized T cells are T lymphocytes which have been produced in an animal bearing a graft of tumoral cells coming from an animal genetically very different, e.g., of another species. The histocompatibility antigens of the tumoral cells being completely different from those of the host, the latter develops a very strong immune response which is manifested here by the fact that the values of the kinetic constants are much larger than for any other type of cytotoxic cells.

[b] Immune T cells are T lymphocytes obtained from the primary host or by grafting a tumor to an animal genetically of the same strain.

[c] Syngeneic NK (natural killer) cells and syngeneic macrophages are obtained by grafting tumors to animals of the same species.

Equation (7.47) may thus be considered as the appropriate phenomenological equation to describe the cytotoxic reactions taking place inside a tumor. In the following we shall assume that cellular replication is the only source of tumoral cells, i.e., $\alpha = 0$.

As shown by Table 7.1 there is a great variability in the kinetic parameters according to the cytotoxic cell considered; furthermore, the density of the latter inside the tumor may undergo large variations. It is therefore reasonable to assume that as a result of this variability, the parameter describing the efficiency of the cytotoxic process is a fluctuating quantity. This raises the question of how this variability does affect the rejection of the tumor. Clearly the extinction of the X population could be achieved with a fluctuating rate of cytotoxic predation for values of β which under deterministic conditions would necessarily correspond to a progressive tumor (for a more detailed discussion see [7.27]). Indeed, the stationary state probability density $p_s(x)$ corresponding to the Ito interpretation of (7.50) with $\alpha = 0$[4] behaves for small x as:

$$p_s(x) = x^{\{[2(1-\beta)/\sigma^2]-2\}} . \tag{7.51}$$

Its properties are as follows:

i) If $\beta < 1 - \sigma^2$, the distribution is integrable over $[0, \infty)$ and the stationary point $x_0 = 0$ is a natural boundary. The most probable value of x corresponds to a finite size of the neoplastic population. The distribution is monomodal independently of the value of θ.

[4] There is no immigration and we neglect other possible sources of X such as the transformation of normal cells into neoplastic ones by mutation or by chemical, physical or biological agents.

ii) $\beta = 1 - \sigma^2$ is a transition point at which the nature of the probability distribution changes abruptly. For $1 - \sigma^2 < \beta < 1 - \sigma^2/2$ $p_s(x)$ is still normalizable, but diverges at x_0 which becomes the most probable value. If $\theta < 1$, the distribution may, however, retain another peak corresponding to a finite value of x. The condition for the stationary point x_0 to be stable from the right (these form the only perturbation we have considered) is that $J = \int_0^\delta \exp\{\int_u^\delta [2f(z)/g^2(z)]\, dz\} du$ is finite. This implies that the stationary point x_0 is unstable for $\beta < 1 - \sigma^2/2$ and entails that the zero boundary is natural for this range of parameters.

iii) The point $\beta = 1 - \sigma^2/2$ is a soft transition point. For $\beta > 1 - \sigma^2/2$, the distribution is no longer integrable over $[0, \infty)$ and the stationary point x_0 is stable. "The stationary density" is completely concentrated at zero as a Dirac δ-function. Thus although β might be smaller than one, the extinction of the X population is certain.

In summary, though bistability is possible in this model, its properties as far as extinction is concerned resemble the behavior of the Verhulst model. This is due to the fact that the boundary 0 has the same properties here as in the Verhulst case. Under the influence of the fluctuations of the environment, extinction of the population occurs via the same sequence of transitions which take place at the boundary zero of the support of $p_s(x)$. These transitions are independent of the critical point behavior which the model may also present. Again we have an effect connected with the existence of a point of bifurcation between two branches of steady states at $\beta = 1$ in the deterministic case. The transition at $\beta = 1 - \sigma^2$, where a divergence of $p_s(x)$ at $x = 0$ replaces abruptly a peak for finite values of x, is evidently a shift of this deterministic transition point due to the noise.

7.5 Illuminated Chemical Systems

Photochemical reactions play a primordial role in the biosphere on earth. This ranges from simple photochemical processes in the upper atmosphere, e.g., $O_3 + h\nu \rightarrow O_2 + O$, to photosynthesis in plants. The ever-changing environmental conditions, as for instance variations in the cloud cover, cause fluctuations in the incident light intensity. It is therefore important to gain an understanding of the possible effects of such fluctuations for photochemical systems. Fortunately, photochemical processes are also extremely convenient for an experimental study of the influence of external noise. The incident light intensity is an external parameter which the experimenter can, to a large degree, easily manipulate at will. Furthermore, sensitivity to light provides an ideal way to perturb from the outside the bulk of a chemical system in a nearly homogeneous manner. If the sample under study is not too thick, it is reasonable to assume that the light intensity is constant throughout. On the other hand, the intensity of the light source is a multiplicative parameter in Beer-Lambert's absorption law: the intensity fluctuations may therefore be expected to exert a strong influence. An experimental confirmation of this fact has already been provided in the case of the Briggs-Rauscher reaction. The latter is however a very complicated system and its reaction mechanism is not known in complete detail. The experimental system considered in Sect. 7.5.1 has the advantage over the Briggs-Rauscher reaction in that its molecular mechanism is well understood. The process is the thermoluminescence of fluorescein in boric acid glass; this process furnishes an experimental model representative of the important class of photochemical reactions which are biphotonic. These processes are essential building blocks in photobiology; in particular they are the basis of photosynthesis. The system is also advantageous because its nonequilibrium properties are entirely controlled by the energy flux of the incident light: in other words it is a closed system which exchanges no matter with its environment. In the next sections, we study a class of photothermal systems in which under deterministic conditions bistability can arise from a coupling between a light absorption process, a chemical reaction and a heat transport process. This bistability phenomenon can appear with the simplest chemical reactions imaginable, namely monomolecular isomerization reactions. Furthermore, like the biphotonic reaction mentioned above, it is also a closed nonequilibrium chemical system. This extreme simplicity allows us to carry out a detailed theoretical analysis. It demonstrates that besides the noise-induced shifts of deterministic critical points which these systems have in common, e. g., with the systems considered in Sect. 7.4, they present also the possibility of several pure noise-induced transitions which in particular may lead to a tristability phenomenon.

7.5.1 Sensitivity of Biphotonic Systems to Light Intensity Fluctuations

We consider the biphotonic process represented in Fig. 7.15. The singlet ground state is S_0, S_1 the singlet excited state, T the long-lived tripled state and X the photoionized state. By absorption of a first photon the system reaches S_1 from S_0, where it either deactivates back to S_0 or by intersystem crossing reaches the triplet state T. Once in the triplet state it can absorb a second photon to reach the photoionized state X. Both triplet and photoionized state are able to decay back to the ground state S_0. The rates of absorption of the first and second photon are I_s and I_T.

A hysteresis loop appears in the stationary curves of X as a function of the incident light intensity I_0 when the lifetime $\tau_X = 1/k_r$ of the photoionized species increases, while the lifetime $\tau_T = 1/k_p$ of the triplet diminishes. This phenomenon

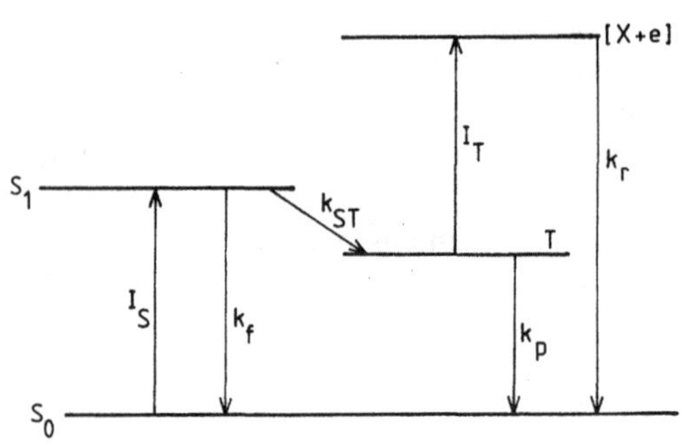

Fig. 7.15. Energetic diagram of a biphotonic model system

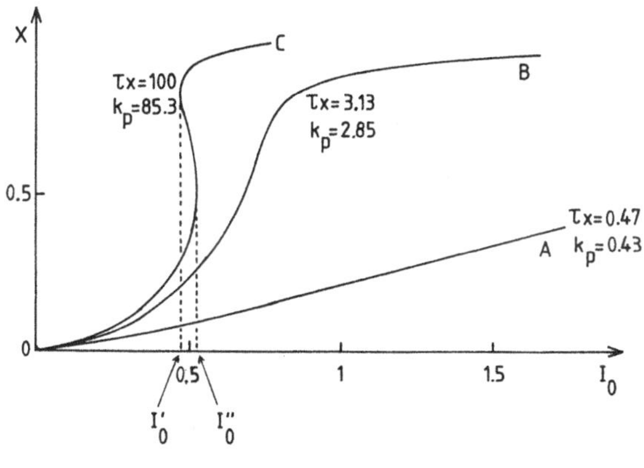

Fig. 7.16. Plot of the stationary states of the biphotonic model as a function of the incident intensity I_0

is illustrated in Fig. 7.16. When τ_X is small, saturation can be achieved only at unrealistically high values of I_0 (Curve A). As τ_X increases, the stationary state curves become more and more sigmoidal; the nonlinear behavior typical of biphotonic systems becomes more obvious (Curve B). Finally, if τ_X is very large the curves are S shaped (Curve C). In the transition region $I_0' < I_0 < I_0''$, two stable stationary states coexist with an unstable stationary state inbetween (see [7.33, 34] for a more quantitative study).

The thermoluminescence of fluorescein in boric acid glass provides a convenient experimental model system for the type of biphotonic reaction mechanism described above. This phenomenon indeed involves the absorption of two consecutive photons [7.35] and can easily be realized at room temperature using a conventional light source [7.36]. Thermoluminescence is a short-lived flash produced by heating a previously irradiated sample. The results of measurements carried out under constant conditions of illumination are reported in Figs. 7.17 – 19. Figure 7.17 shows a typical curve representing the variation of the emitted light after a linear temperature increase. The relative thermoluminescence efficiency E is determined by direct measurement of the area under the glow curve. After extended irradiation, this value reaches a steady state either by growing from zero (Curve a of Fig. 7.18) or by decaying if a higher intensity was initially applied (Curve b). Further, is a monotonously increasing function of I_0 (Fig. 7.19). As expected, for I_0 small the steady-state curve rises parabolically while at large I_0 it saturates to one. This nonlinear behavior is consistent with the commonly postulated reaction mechanism of fluorescein thermoluminescence in boric acid glass (Fig. 7.20). The similarity with the biphotonic process represented in Fig. 7.15 is evident. During irradiation, dye molecules release long-lived tripled states which are the starting point of the biphotonic process. Photoionization then produces free electrons which leads to the formation of atomic hydrogen by a reaction on boric acid protons. Thermal recombinations of the photoionized species with hydrogen atoms directly produce excited singlet states as light emitting species. The value of the area under the glow curve is proportional to the light emission during the heating process. Reactants for the chemiluminescent recombination reaction are obviously issued from the photoionization process. It can thus be assumed that for given experimental conditions the light

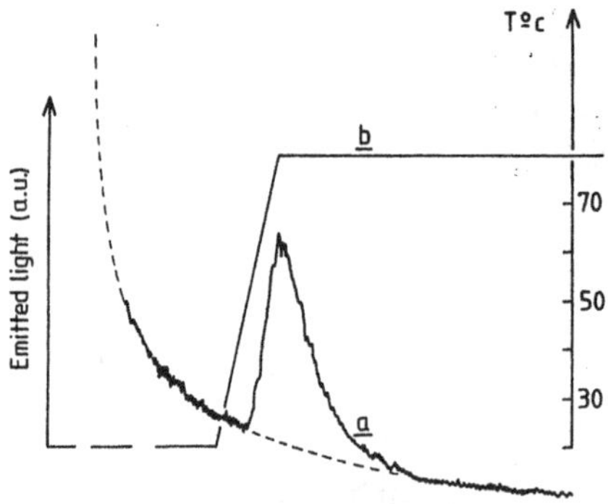

Fig. 7.17. (*Curve a*) Typical glow curve of fluorescein in boric acid glass. (*Curve b*) Temperature change. The area under the glow curve is proportional to the thermoluminescence efficiency E

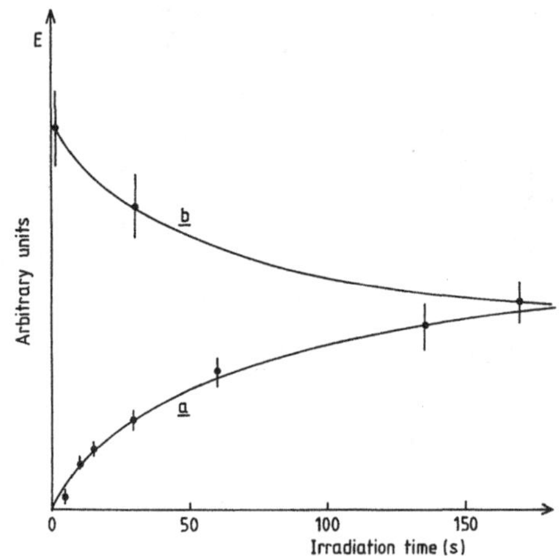

Fig. 7.18. Thermoluminescence efficiency E versus irradiation time, $I_0 = 1.64 \times 10^6$ photons/s at 490 nm. (*Curve a*) initial condition no irradiation; (*Curve b*) initial condition: 1 min irradiation at $I_0 \approx 8.2 \times 10^{16}$ photons/s at 490 nm

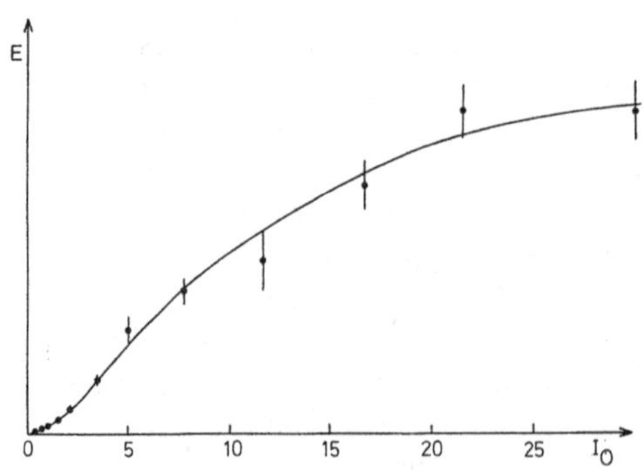

Fig. 7.19

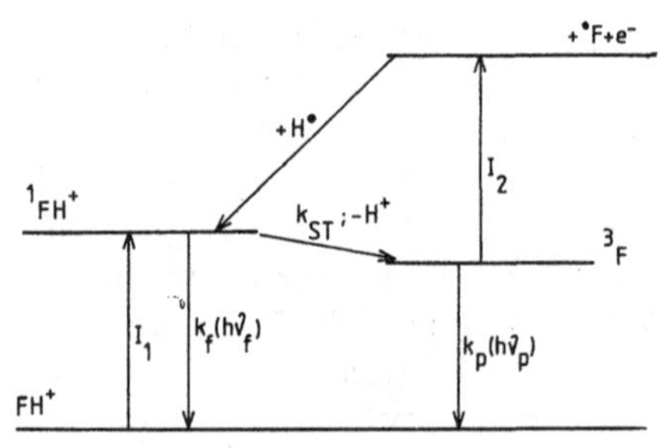

Fig. 7.20

Fig. 7.19. Steady-state plot of E versus I_0

Fig. 7.20. Energetic diagram of the thermoluminescence of fluorescein in boric acid glass

emission is proportional to the extent of photoionization corresponding to the fluorescein radical-cation concentration. The only difference between the biphotonic model and the fluorescein thermoluminescence is that the recombination reaction $F\cdot^{+} + H\cdot \rightarrow {}^1FH^{+}$ has been incorporated as a first-order process.

To investigate the response of this system to environmental fluctuations affecting the incident light intensity an experimental set up quite similar to the one described in Sect. 7.2 to study Briggs-Rauscher reactions has been used (see [7.34] for details). Heavy opaque particles contained in a water filter jacket and dancing in a turbulent water stream chopped up the incident light beam, thus generating the desired optical noise.

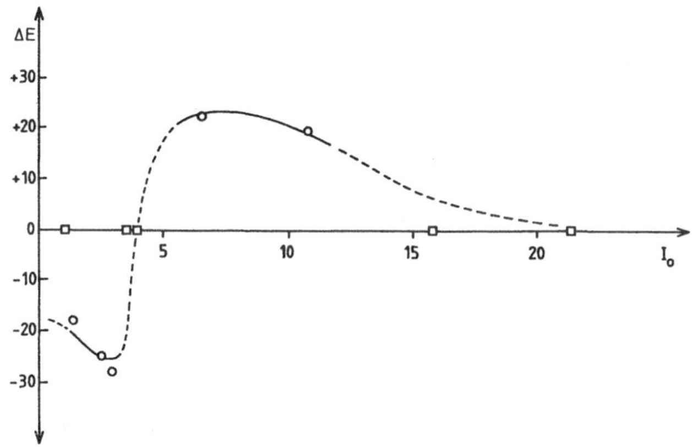

Fig. 7.21. Plot of ΔE versus I_0

The principal experimental results are reported in Fig. 7.21. It represents the relative bias

$$\Delta E = (E_f - E_d)/E_d$$

as a function of the average light intensity. Here E_f and E_d are respectively the thermoluminescence emitted after fluctuating and constant illumination. An important effect is seen for values of the incident intensity I_0 corresponding (under deterministic conditions) to the transition region of the steady-state curve, i.e., the vicinity of its inflection point. The deviations of the location of the extrema under fluctuating light conditions from the values of the deterministic steady states may amount to more than 20%. For small values of I_0 (below 3 a.u.) the thermoluminescence efficiency of fluctuating systems is lower while above 6 a.u. the effect is reserved. In other words, the main effect of the fluctuations is to steepen the transition region. Particularly, the saturation into the energy-rich species (i.e., photoionized dye) is reached for lower mean light intensities. The behavior can be viewed as a distortion of the steady-state curve which appears to rotate around its inflection point. This result could be of great interest in practical photochemistry and photobiology where nonlinear responses to variations of light intensity are frequent. Various applications are indeed conceivable. Let us mention:

i) The yield of thermosolar power plants depends nonlinearly on the solar flux. It is likely to undergo macroscopic changes under the influence of light intensity fluctuations. The more nonlinear the overall yield curve the greater the expected effect will be.

ii) Recent theoretical analysis of solar light conversion systems by water photolysis [7.37] have shown that nonlinear behavior can be expected. One can predict under these conditions a nonnegligible influence of the exciting light intensity fluctuations on the macroscopic properties of such solar gears. The effect of external noise is not necessarily parasitic, e.g., systems whose working point is beyond the inflection point could see their efficiency enhanced by fluctuations of the light; on the contrary, systems designed to work under fairly weak intensities would see their efficiency decreased by fluctuations.

iii) It is particularly remarkable that photosynthesis which operates under very fluctuating conditions seems to withstand external noise by exhibiting quasi-linear behavior in the low-intensity range. Under such conditions the effect of fluctuations in the light intensity is highly minimized. In this respect external noise could be considered as a factor of evolutive pressure.

7.5.2 Illuminated Photothermal Systems

The basic ingredient of the process studied in this section is an isomerization reaction or a simple dissociation reaction in which one of the components absorbs light of a given frequency and restores this energy to the system in the form of heat [7.38–40]. When the formation of the light-absorbing component is endothermic, the rise of the system's temperature and the light-absorption process become coupled (via a positive feedback): the more component present, the more light can be absorbed, the more the temperature rises, and the more the chemical equilibrium is displaced in favor of the active component. We shall base our presentation of the deterministic properties of these systems, i.e., under constant illumination, on the original paper of *Nitzan* and *Ross* [7.38]. (For an experimental realization see [7.40].)

One considers a closed chemical system in which an isomerization reaction of the form

$$\mathrm{A} \underset{k_2}{\overset{k_1}{\rightleftarrows}} \mathrm{B} \tag{7.52}$$

is coupled with a light-absorption and radiationless deexcitation process

$$\mathrm{A} + h\nu \rightarrow \mathrm{A}^* \rightarrow \mathrm{A} + \text{heat} \tag{7.53}$$

involving only component A. The following assumptions are made:

i) The time scale of (7.53) is fast compared to (7.52), i.e., the excitation of A by absorption of a photon and the deexcitation of A* equilibrate instantaneously to the concentration of A present at a given instant of time.

ii) The kinetic constants k_1 and k_2 depend on the illumination only through the temperature variations which the latter impose on the system. This is justified if the radiationless relaxation processes are the fastest processes in the systems.

iii) The kinetic constants depend on the temperature in the Arrhenius form. The equilibrium constant of (7.52) is

$$k_1/k_2 = \exp\left(-\frac{\Delta H}{RT} + \frac{\Delta S}{R}\right) = \varepsilon K(T)\,. \tag{7.54}$$

iv) The system is homogeneous in space.

v) The heat flux between the system and its environment depends linearly on the temperature difference.

The concentration of A and the temperature T inside the system then obey the set of kinetic equations:

$$\dot{A} = -(k_1 + k_2)A + k_2 a \,, \tag{7.55}$$

$$\dot{T} = \alpha I A - \beta(T - T_e) - \lambda \dot{A} \,, \tag{7.56}$$

with $a = A + B = \text{const}$. The external temperature is T_e. The first two terms in (7.56) describe the temperature variations due to the energy fluxes in the form of light and heat exchanged with the environment. The light intensity is I. The third term in (7.56) describes the heat generated by the chemical reaction itself. Here α, β, λ are proportionality constants, characteristic of the reaction and of the experimental setup considered. In the following, we assume that (7.55) equilibrates rapidly to (7.56) so that the concentration variable A may be expressed in terms of the temperature. Introducing the dimensionless variables

$$X = T/T_e \,, \quad \bar{\alpha} = (\alpha I a)/(\beta T_e) \,, \quad \Delta r = -\Delta H/R T_e \,, \quad \tau = \beta t \tag{7.57}$$

we can reduce the system of (7.55, 56) to the one-variable equation:

$$\frac{dX}{d\tau} = \frac{\bar{\alpha}}{1 + \varepsilon \exp(\Delta r/X)} - X + 1 \,. \tag{7.58}$$

Its steady-state solutions $\bar{X}$ are sketched as a function of $\bar{\alpha}$ in Fig. 7.22. For $\Delta r < 0$ (i.e., for exothermic reaction), the steady-state curves $\bar{\alpha} = \bar{\alpha}(\bar{X})$ are monotone functions of $\bar{X}$. When the light intensity increases, the slope of the curves increases continuously from $1 + \varepsilon \exp \Delta r$ (for $\bar{X} = 1$) towards $1 + \varepsilon$ (for $\bar{X} \to \infty$). Whatever the values of $\bar{\alpha}$ and T_e, one has $d\bar{\alpha}/d\bar{X} > 0$, $d^2\bar{\alpha}/d\bar{X}^2 > 0$ (for $\bar{X} > 1$). The value $\Delta r = 0$ is a transition point beyond which there always exists

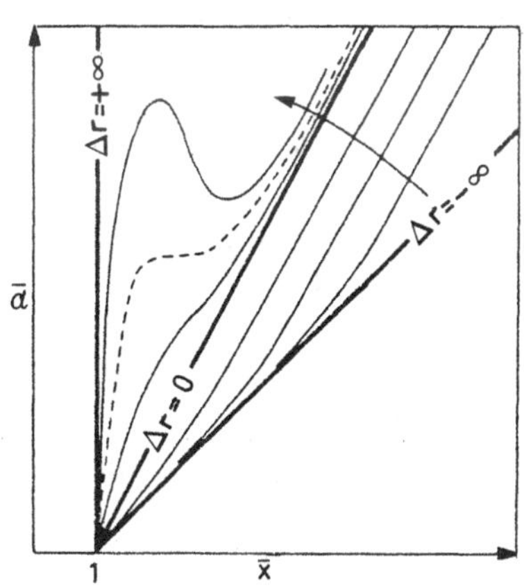

Fig. 7.22. Sketch of the steady state $\bar{\alpha} = \bar{\alpha}(\bar{x})$ for ε fixed and Δr increasing as indicated by the arrow. The occurrence of hysteresis is restricted to values of Δr belonging to the range $0 < \Delta r_c < \Delta r < \infty$. The minimal value of Δr_c is 4. The broken line corresponds to Δr_c

in the interval $[1, \infty)$ a value of X at which $d^2\bar{\alpha}/d\bar{X}^2$ changes sign. At this point, a simultaneous change of sign occurs for $d\bar{\alpha}/d\bar{X}$ provided Δr and ε verify the equality

$$\varepsilon_c = [\Delta r_c \exp(2 - \Delta r_c)]/(\Delta r_c - 4) , \tag{7.59}$$

$$X_c = \Delta r_c/(\Delta r_c - 2) , \quad \bar{\alpha}_c = 4/(\Delta r_c - 4) . \tag{7.60}$$

Once $\Delta H < 0$, this critical point can be reached by decreasing the external temperature T_e. The important point in this deterministic analysis of the steady-state properties is that a necessary condition for bistability, whatever the value of ε, is that Δr be larger than 4 (see (7.59, 60)).

7.5.3 Steady-State Properties for a Fluctuating Light Source

Let us now assume that the light intensity fluctuates with a vanishingly small correlation time so that the white-noise idealization is justified [7.41]. We interpret the SDE corresponding to (7.58) in the Ito sense[5],

$$dX_\tau = \{\bar{\alpha}/[1 + \varepsilon \exp(\Delta r/X_\tau)] - X_\tau + 1\} d\tau + \bar{\sigma}/[1 + \varepsilon \exp(\Delta r/X_\tau)] \, dW_\tau .$$

Looking for the stationary solution of the associated FPE, we first notice that the realization of the process X_τ must be restricted to the semiclosed interval $[1, \infty)$. It is easily verified that the analytical condition for ∞ to be a natural boundary is satisfied. On the contrary, $X = 1$ is not a natural boundary. This is a consequence of the fact that it is not an intrinsic boundary of the problem and that the white-noise idealization contains negative realizations for I_t which are nonphysical. We already encountered this undesirable feature, due to the Gaussian property of the white noise, in the cytotoxic reaction model discussed in Sect. 7.4. (Note that in all other models negative values of the external parameter are not unphysical and no problems arise.) However, in that model the support of $p_s(x)$ remained limited to the set of physically acceptable values for the state variable X: though the fluctuating parameter could become negative, the process X_t nevertheless remained restricted to the physically acceptable interval $[0, \infty)$. Here the problem is more severe: letting X_τ take values smaller than 1 implies that the support of $p_s(x)$ is no longer identical to the set of physically acceptable values for the state variable. A nonzero probability appears that as a result of the light intensity fluctuations the temperature inside the system is smaller than the temperature of the external heat reservoir. This is unacceptable and some remedy has to be found to this situation. This can only be achieved by choosing the "lesser evil" of two courses of action, dictated by conflicting demands. The first course of action that comes to mind is to abandon the Gaussian approximation. A probability

[5] Qualitatively identical results would be obtained in the Stratonovich case: remember that it would amount only to the replacement of σ^2 by $\sigma^2/2$ everywhere it appears in the equation for the extrema of $p_s(x)$.

density which vanishes identically for negative values would not give rise to any unphysical results. However, this is a big leap which one would take only with utmost reluctance. As we have already stated repeatedly, it is to be expected by virtue of the central limit theorem that in most natural systems external fluctuations are to a very good approximation Gaussian distributed. For small and moderate variances with respect to the mean value of the external parameter, negative values occur only with a small though nonzero probability. Therefore fluctuations in external parameters which are intrinsically positive can satisfactorily be modelled by a Gaussian distribution. It is only for very large variances that this approximation becomes questionable and the role of unphysical values may become nonnegligible. However, this problem indeed arises here since we deal with white noise which, loosely speaking, has an infinite variance. This fact suggests that we better abandon the white-noise idealization. However, this entails the loss of the Markov property for the process X_τ. Therefore we would be even more reluctant to choose this course of action. If we want to save both the Gaussian character of the noise and the Markovianity of X_τ and at the same time exclude unphysical results, only one way remains open to us, namely to impose the condition that the solution X_τ of the SDE is restricted to the physically acceptable interval $[1, \infty)$. The way to achieve this is to make the process X_τ undergo a delayed reflection on the boundary $x=1$. This means X_τ spends a positive length of time on this boundary. An estimation of the validity of this approximation will be given by the value of the probability density accumulating on $x=1$. As long as this value is small with respect to the probability density at the extrema belonging to the support $[1, \infty)$, X_τ represents a satisfactory model of the physical process and we are sure that the observed phenomena are not due to the particular boundary condition imposed.

With this boundary condition the stationary distribution function $F(x)$ reads [5.12, 7.41]:

$$F(x) = \begin{cases} 0 & \text{if} \quad x<1 \\ N^{-1}\left[1+2f(1)\int\limits_1^x \dfrac{\phi(z)}{g^2(z)}\,dz\right] & \text{if} \quad x \geqslant 1\,, \end{cases} \tag{7.61}$$

where

$$\phi(z) = \exp \int\limits_1^z \frac{2f(z')}{g^2(z')}\,dz'\,,$$

$$g(z) = \bar{\sigma}/[1+\varepsilon \exp(\Delta r/z)]\,, \tag{7.62}$$

and

$$N = 1 + 2f(1)\int\limits_1^\infty \frac{\phi(z)}{g^2(z)}\,dz\,. \tag{7.63}$$

The probability density of (7.61) is

$$p_s(x) = \frac{\delta(x-1)}{N} + 2\,\frac{f(1)}{N}\,\frac{\phi(x)}{g^2(x)}\,. \tag{7.64}$$

The first term is a δ-function weighted by the factor $1/N$ and represents the probability density of finding the process on the boundary $x=1$. The extrema x_m of (7.64) in the interval $[1,\infty)$ can be calculated as usual from (6.36). We obtain, after expressing $\bar{\alpha}$ in terms of x_m,

$$\bar{\alpha}=[1+\varepsilon\exp(\Delta r/x_m)](x_m-1)+\bar{\sigma}^2\frac{\Delta r}{x_m^2}\frac{\varepsilon\exp(\Delta r/x_m)}{[1+\varepsilon\exp(\Delta r/x_m)]^2}. \tag{7.65}$$

Figures 7.23, 24 report the conditions for which the curves, given by relation (7.65), admit a critical point, i.e., values of the parameters Δr, ε and $\bar{\sigma}^2$ for which there exists on the curve $\bar{\alpha}=\bar{\alpha}(x_m)$ a physically acceptable value at which $d\bar{\alpha}/dx_m$ and $d^2\bar{\alpha}/dx_m^2$ vanish simultaneously. The curves in Fig. 7.23 represent the values of Δr and ε for which the condition is realized. The $(\Delta r-\varepsilon)$ phase space subdivides essentially into four regions:

i) In area (a), $p_s(x)$ admits only one extremum (besides the peak at $x=1$, which in any case is always present) whatever the magnitude of the external fluctuations, i.e., the value of σ^2.

ii) In (b), despite that fact that $\Delta r<0$, the probability density (7.64) is double peaked over a finite range of values of $\bar{\alpha}$ and $\bar{\sigma}^2$. This situation is illustrated in more detail in Fig. 7.25 in which the curves of the extrema, as given by (7.65), have been plotted for a particular choice of the pair of parameters $(\Delta r,\varepsilon)$ and various values of $\bar{\sigma}^2$. Hysteresis occurs for $\bar{\sigma}^2>0.2$. When $\bar{\sigma}$ increases beyond this value, this hysteresis phenomenon becomes more and more important and simultaneously shifts towards lower values of $\bar{\alpha}$. Let us point out that the new

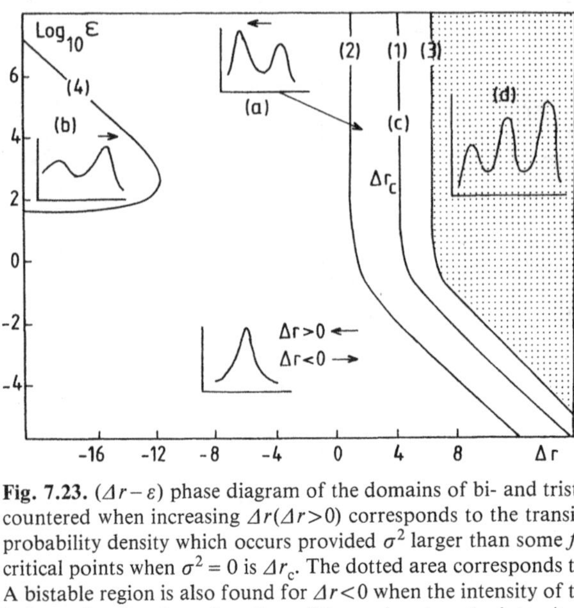

Fig. 7.23. $(\Delta r-\varepsilon)$ phase diagram of the domains of bi- and tristability. For ε fixed, the first line encountered when increasing $\Delta r(\Delta r>0)$ corresponds to the transition from a monomodal to bimodal probability density which occurs provided σ^2 larger than some *finite* value. The line of deterministic critical points when $\sigma^2=0$ is Δr_c. The dotted area corresponds to the region where $p_s(x)$ is trimodal. A bistable region is also found for $\Delta r<0$ when the intensity of the noise is large enough. The arrows indicate the direction of motion of the peaks when the intensity of the noise increases

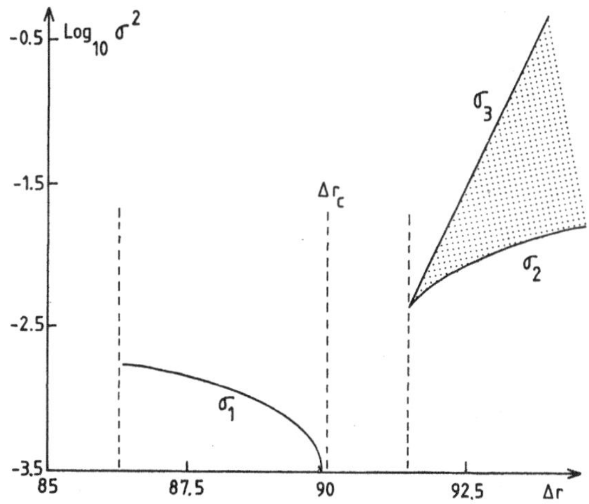

Fig. 7.24. Values of the critical noise intensities $\bar{\sigma}_c^2$ for the transition from monomodal to bimodal probability densities (*curve* σ_1) and from bimodal to trimodal probability densities (dotted area delimitated by the *curves* σ_2 and σ_3). In this particular example $\varepsilon = 6.34 \times 10^{-39}$

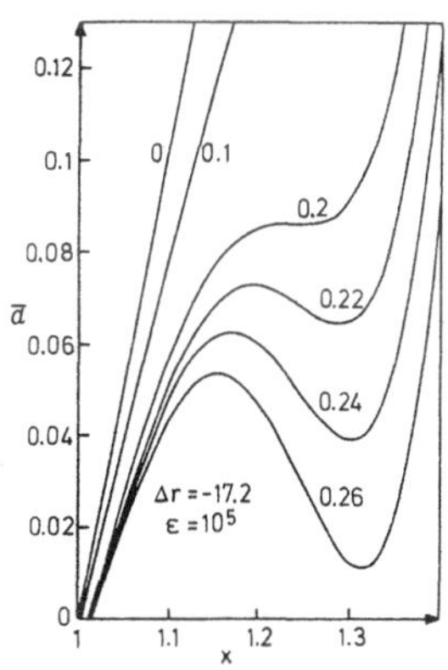

Fig. 7.25. Location of the extrema of $p_s(x)$ for an example chosen in the bistable domain with $\Delta r<0$. The curves are plotted for various values of $\bar{\sigma}^2$

peak of the distribution, induced by the fluctuations of I_t, corresponds to internal temperatures that are higher than in the deterministic case.

iii) In (c), (7.65) exhibits hysteresis. However, contrary to what happens in (ii), this hysteresis amounts to the appearance of a peak near $x = 1$. Since this domain contains the deterministic critical curve given by (7.60), it subdivides into two regions. To the left of Curve 1 (Fig. 7.23) there is no hysteresis under deterministic conditions. When the value of ε is fixed and Δr increases from Curve 2 to Curve 1, hysteresis appears provided that $\bar{\sigma}^2$ is greater than some finite critical value (see σ_1-curve in Figure 7.24). This value tends to zero as Δr approaches Curve 1. (Further, σ_1 vanishes as a power of the distance from the deterministic critical point.) To the right of Curve 1, the effect of fluctuations amplifies a hysteresis phenomenon already existing for $\bar{\sigma}^2 = 0$ and shifts the hysteresis loop towards larger values of $\bar{\alpha}$.

iv) The domain marked (d) has properties that are specific for the fluctuating system and do not show up in any deterministic treatment. Its particularity is to admit a range of values $\sigma_2 < \bar{\sigma}^2 < \sigma_3$ for which $p_s(x)$ may be trimodal (Fig. 7.24)

Table 7.2. Parameters of the probability density for different values of $\bar{\sigma}^2$. ($\Delta x = 10^{-5}$, $\varepsilon = 6.34 \times 10^{-39}$, $\Delta r = 92.2$, $\bar{\alpha} = 0.3$)

$\bar{\sigma}^2$	$p_s(1)\Delta x$	$p_s(x_{m1})\Delta x$	$p_s(x_{m3})\Delta x$	$E\{X\}$
0.008	9.35×10^{-13}	1.22×10^{-3}	6.74	1.297
0.01	7.58×10^{-9}	1.15×10^{-1}	5.63	1.297
0.013	9.84×10^{-7}	1.29	4.85	1.292
0.0198	2.50×10^{-4}	17.4	3.14	1.236
0.05	3.69×10^{-2}	75.1	3.83×10^{-1}	1.049

depending on the values of $\bar{\alpha}$. This trimodal behavior is unlikely to be an artefact due to the white-noise idealization and its Gaussian property. Indeed, comparing the probability at $x = 1$ to that at the two maxima by choosing an extremely narrow window $\Delta x = 10^{-5}$, which in the $NO_2 - N_2O_4$ system corresponds to $\Delta T \sim 0.001$ K, we find that the former is negligible with respect to the latter for small and moderate noise intensities as expected (Table 7.2). This indicates that in experiments where the external noise takes only positive values similar results will be observed, if the distribution is approximately bell-shaped over the positive axis and the correlation time of the real noise is short enough.

To summarize, three kinds of critical points are induced by the noise in this system. The one corresponding to Line (2) in Fig. 7.23 is a shift of the deterministic critical point. The other two corresponding to Lines (3) and (4) are pure noise-induced transitions, similar to the pure noise-induced transition of the genetic model.

8. External Colored Noise

The appeal of the white-noise idealization used in the preceding chapters stems from the fact that it permits us to model the system's temporal evolution by a diffusion process. This places the powerful arsenal of the theory of Markov processes at our disposition, which is extremely advantageous: for instance, it allows us to obtain exact analytical results for the stationary behavior of open systems subjected to noise. Thus white noise is a theoretician's boon, though on the mathematical side it has to be handled with some circumspection as repeatedly emphasized. Furthermore, on the physical side, the correlation time of a real noise, however small it may be, is never strictly equal to zero. While it is of course eminently plausible that rapid external noise can be satisfactorily described by white noise and while this is further supported by the experimental results reported in Chap. 7, where an excellent qualitative agreement with white-noise predictions is found, one would like to put this intuition and these findings on solid mathematical grounds. This is one of the objects of this chapter.

The main object of Chap. 3 was basically the analysis of how to take the limit $\tau_{\text{cor}} \to 0$ so as to avoid a noiseless limit and an investigation of the properties of white noise. In other words, it is necessary at this stage in our development of a theory of noise-induced transitions to take up once more the analysis of the passage from a real noise to the white-noise idealization. There are essentially two aspects which are now ripe for a detailed investigation:

i) The precise determination of the conditions under which one recovers the predictions of the white-noise analysis in the limit of vanishingly small correlations of the external noise. This analysis addresses essentially the question of the validity of the white-noise idealization. Namely, are the results on the various noise-induced transition phenomena obtained in Chaps. 6, 7 robust, i.e., do they hold qualitatively also in a neighborhood of white noise.

ii) The evaluation of the quantitative modifications to which small correlations in the external noise give rise.

Besides these questions, which concern the immediate neighborhood of white noise, it is of course desirable to extend our approach beyond the case of extremely rapidly fluctuating environments. Though this case covers a broad class of applications, there are obviously situations where no clear-cut separation between the time scales of the noise and system exists. In order to be able to treat these cases also and to achieve a general description of noise-induced transitions, a systematic exploration of the effect of correlations has to be carried out. This will be the main topic of this chapter and Chap. 9.

8.1 Modeling of Environmental Fluctuations Revisited

Any attempt at a theoretical treatment of nonlinear systems, subjected to real external noise, i.e., noise with a nonvanishing correlation time, encounters at first sight two main difficulties. First, the temporal evolution of the system is no longer Markovian. Indeed, it was established in Chaps. 4 and 5 that the evolution of a system subjected to external noise is Markovian if and only if the noise is white. While the loss of the Markov property holds some advantages, e.g., non-Markovian processes have smoother realizations than Markov processes and no generalized stochastic processes or special calculi as the Ito calculus have to be used, the disadvantages weigh far heavier. The powerful tools of the theory of Markov processes are lost, making it far more difficult, if not impossible, to obtain exact explicit results. Second, while there are essentially only two types of white noise, namely Gaussian and Poissonian white noise, between which a choice can be made, once one deals with real noises the variety of possibilities looks incomparably vaster.

Headway can be made in this complex situation by realizing that the non-Markovian behavior of the system does not of course exclude the possibility that the environment is Markovian, though nonwhite. And indeed, this is the case to a very good degree of approximation for a large class of natural systems. In the following we shall impose this mild restriction on the external noise, namely that it has to be Markovian. Admittedly, this is not the most general case of real noise, but we expect it to exhaust the physics of the kind of situations encountered in applications for the following two reasons: (i) As we shall see a little further on, it allows the system to have differentiable realizations; ii) the system is non-Markovian and correlated with the environment. Hence we see that the essential features of real systems can be satisfactorily taken into account already if we choose our models for the fluctuating environment among the class of Markovian noise processes. A convenient feature of this modeling procedure is that it kills two birds with one stone. It has the advantage that it permits us to continue in the framework of the Markov theory and thus overcoming the first difficulty. As far as the choice of models for the noise is concerned, it restricts drastically the possibilities and in most situations selects the Ornstein-Uhlenbeck process as the appropriate model, as we shall see below.

Let us comment on these points in more detail. As already remarked above, if we consider only the process that describes the evolution of the system, this process is non-Markovian. Knowing the present state of the system we cannot make the best possible prediction of its future stochastic evolution. This is due to the fact that the present state of the environment is not known, since only the state variable of the system is considered. However, if the external noise is non-white, then the history of the system and the present state of the noise are correlated. Hence any information about the past of the system will reduce our ignorance on the state of the environment and improve our predictions of the future evolution of the system. In other words, even if the present state of the system is known, its past and future are not stochastically independent. They are correlated via the memory of the environment. This verbal description of the origin of non-

Markovian behavior shows in a transparent way how to stay in the framework of a Markovian description. Consider the two-dimensional process which consists of the system variable X_t and of the fluctuating parameter ζ_t. Though its first component taken by itself is non-Markovian, the complete two-dimensional process is Markovian, under some mild condition on the initial values, e.g., that the initial state of the system is independent of the noise [8.1]. Intuitively, this can be seen as follows: knowing the present state of the two-dimensional process implies that the present state of the system *and* the present state of the environment are known. The future evolution of the system depends only on its present state and the present state of the environment; the environment itself is Markovian. These facts imply that any prediction of the future stochastic evolution of the two-dimensional process based on its present state is the best possible. Information about its past behavior will not improve this prediction. We will consider here moreover that the environment, besides being Markovian, is also ergodic. Such an environment is then called a *colored*-noise environment.

Keeping this discussion in mind, we now come back to the second difficulty mentioned above. As already remarked in Chap. 1, it is often the case that environmental fluctuations are the cumulative effect of weakly coupled environmental factors. The central limit theorem implies then that the fluctuations in the external parameter are Gaussian distributed. This property of external noise has been used throughout the preceding chapters and we shall continue to use it here. It is a nice feature that these characteristics of the environmental fluctuations, namely ergodicity, Markovianity and Gaussian behavior, reduce the vast choice of possible model processes to *exactly* one. Indeed, Doob's theorem, which was already used in Chap. 4, states that a regular stationary Gaussian-Markov process is necessarily an O-U process.

Thus we can conclude that in most applications the appropriate choice to model a colored-noise environment will be the O-U process. For this reason, and for the sake of concretness, we use the O-U process for the discussion of the influence of colored noise in this chapter. However, the various results we derive in this chapter hold for more general noises, either exploiting only the Gaussian character of the real noise (Sect. 8.2) or only its Markovian character (Sect. 8.4). We shall remark at the appropriate places on the extension of these results to these other classes of noises.

Even for the simple case of an O-U process one is left with technical difficulties which in general make it impossible to evaluate exactly the stationary probability density of the system. We now have to deal with a pair of Ito SDEs[1]

$$dX_t = f(X_t)\,dt + g(X_t)\,\zeta_t dt\,, \tag{8.1}$$

$$d\zeta_t = -\gamma\zeta_t dt + \sigma dW_t \tag{8.2}$$

for the pair process (X_t, ζ_t). Being a solution of an Ito SDE, the pair process is a diffusion process and its transition probability density (tpd) is the fundamental

[1] Note that the diffusion matrix is constant and there is no difference between the Ito and Stratonovich interpretation in this case.

solution of the corresponding FPE:

$$\partial_t p(x,z,t|x_0,z_0) = -\partial_x(f+zg)p(x,z,t|x_0,z_0) + \partial_z \gamma z p(x,z,t|x_0,z_0) + \frac{\sigma^2}{2}\partial_{zz}p(x,z,t|x_0,z_0)\,. \tag{8.3}$$

As is well known, the stationary pd of a FPE of two or more variables can in general be found only under restricted conditions known as detailed balance [8.2–4]. Equation (8.3) does not obey detailed balance since the diffusion matrix is degenerate. However, we shall show that a class of models exists that does not obey detailed balance where the stationary density and even the time dependent tpd can be obtained exactly. This class of systems will be discussed in Sect. 8.3. They will serve as a first stepping stone to gain insight into the behavior of systems subjected to colored noise. However, their physical meaning is rather tenuous. Therefore, in order to approach the problem in general and to answer the questions raised at the beginning of this chapter it is necessary to resort to approximate procedures. This will be done in Sects. 8.4, 5 and 6 for two limiting cases: (i) the white-noise neighborhood which can be investigated when the noise is sufficiently fast compared to the typical evolution time of the system; (ii) the opposite limit, which can be taken when the correlation time of the noise is much longer than the characteristic time of the system. Sect. 8.7 will deal with the case of nonlinear external noise.

8.2 Some General Remarks on Stochastic Differential Equations with Colored Noise

Before we start the investigation of the effects of colored noise, it is necessary to gain some familiarity with the theoretical tools employed to describe such situations. A few remarks on SDE's with colored noise are thus in order. Let us begin by comparing the SDE for $\tilde{X}_t$ with a nonwhite noise with the corresponding SDE for $\hat{X}_t$ in the white-noise case:

$$\frac{d\tilde{X}}{dt} = f(\tilde{X}_t) + \sigma g(\tilde{X}_t)\zeta_t \rightarrow d\tilde{X}_t = f(\tilde{X}_t)dt + \sigma g(\tilde{X}_t)\zeta_t dt\,, \tag{8.4}$$

$$\frac{d\hat{X}}{dt} = f(\hat{X}_t) + \sigma g(\hat{X}_t)\xi_t \rightarrow d\hat{X}_t = f(\hat{X}_t)dt + \sigma g(\hat{X}_t)dW_t\,. \tag{8.5}$$

The difference between the two types of SDE's is most transparent in their equivalent integral forms:

$$\tilde{X}_t = \tilde{X}_0 + \int_{t_0}^{t} f(\tilde{X}_s)ds + \sigma\int_{t_0}^{t} g(\tilde{X}_s)\zeta_s ds\,, \tag{8.6}$$

$$\hat{X}_t = \hat{X}_0 + \int_{t_0}^{t} f(\hat{X}_s)\,ds + \sigma \int_{t_0}^{t} g(\hat{X}_s)\,dW_s\,. \tag{8.7}$$

The first integral in the rhs of both equations presents no difficulties; it can be interpreted as an ordinary Riemann integral. Recall that the second integral in (8.7) cannot be interpreted as an ordinary integral. Ito or Stratonovich calculus is needed to give a precise meaning to it. In contradistinction, the second integral of the rhs of (8.6) can be interpreted as an ordinary Riemann integral. This is due to the fact that the nonvanishing correlations in the colored external noise imply "smooth" realizations, as for instance continuous realizations for the noise.

White noise is a generalized stochastic process with independent values at every instant of time. If one wants to picture it in the language of ordinary processes, the white noise jumps wildly around, having infinite discontinuities at every instant of time. The integration in (8.7) leads to a smoothing. The solution $\hat{X}_t$ is an ordinary stochastic process. Furthermore, as we saw in Chap. 5, the solution of an Ito (or Stratonovich) SDE has a.s. continuous realizations. However, the realizations of $\hat{X}_t$ inherit the properties of the Wiener process W_t, i.e., in particular they are a.s. continuous but nowhere differentiable. (Recall that this nonanalytic behavior goes hand in hand with the Markov property). This implies, of course, that (8.5) cannot be interpreted in any straightforward manner as an ensemble of ordinary differential equations whose solutions are the realizations of the process $\hat{X}_t$, i.e., a set of DE's indexed by $\omega \in \Omega$. The situation is completely different for the colored-noise case. If ζ_t in (8.4) is a real noise and contrary to the white noise ξ_t in (8.5) has a.s. continuous realizations, we expect that the solution (8.6), due to the smoothing effect of the integration, will correspond to a process $\tilde{X}_t$ with a.s. differentiable realizations. Thus (8.4) may plausibly be interpreted as an ensemble of ordinary DE's for the realizations. Let us formulate this in more precise terms. A stochastic process X_t is said to be the R solution (realizationwise solution) of the differential equation with a random right-hand side

$$\frac{dX_t}{dt} = f(X_t, t)\,, \tag{8.8}$$

if almost surely the realizations of X_t obeys the ordinary DE [Ref. 8.5, p. 13]

$$\frac{dX_t(\omega)}{dt} = f(X_t(\omega), t, \omega)\,. \tag{8.9}$$

Note that if a stochastic differential equation of the form (8.4) is solved realizationwise, then the ensemble of realizations is not necessarily a stochastic process. More explicitly, if we solve the ensemble of DEs

$$\frac{dX_t(\omega)}{dt} = f(X_t(\omega)) + \sigma g(X_t(\omega))\,\zeta_t(\omega) \tag{8.10}$$

for every ω, then it still remains to be verified that at every instant of time t, the so defined function X_t from the sample space Ω into the real numbers, X_t: $\Omega \rightarrow \mathbb{R}$, is a random variable, i.e., fulfills the defining condition (2.8). If not, there is no R solution and no probability can be assigned to the event that X_t takes values in a certain interval. Hence it would not be meaningful on this basis to say that the system is in a given state with a certain probability. However, the fact that no R solution exists, does not of course exclude that a stochastic process X_t exists which solves the SDE in a different sense. This is, for example, illustrated in the white-noise case where obviously no R solution exists since (8.4) cannot even be interpreted as an ordinary SDE, but where the Ito or Stratonovich calculus gives a precise meaning to the SDE (8.4) and to the notion that X_t is a solution of (8.4). Note that the R solution to an ordinary SDE is the solution in the strongest sense, but that several weaker definitions of a solution of a SDE, e.g., in mean square or in probability [8.5] are possible, analogous to the different possibilities for the definition of the continuity and differentiability of a stochastic process. With these remarks we conclude the brief presentation of some theoretical aspects of DEs with a random rhs and now turn our attention again to questions concerning the influence of external noise on nonlinear systems.

8.3 Real External Noise: A Class of Soluble Models

Consider the class of systems described by the ordinary SDE

$$dX_t = -\alpha g(X_t)H(X_t)dt + \sigma g(X_t)\zeta_t dt\,, \tag{8.11}$$

where ζ_t is an ergodic, not necessarily Markovian, random process with mean value zero and

$$H(x) = \int^x dz/g(z)\,. \tag{8.12}$$

The particularity of these systems is that via the transformation

$$u = H(x) \tag{8.13}$$

we obtain a linear SDE

$$dU_t = -\alpha U_t dt + \sigma\zeta_t dt\,. \tag{8.14}$$

If ζ_t has almost surely continuous sample paths, then

$$\begin{aligned} U_t(\omega) &= U_0(\omega)\exp[-\alpha(t-t_0)] + \int_{t_0}^{t} \zeta_s(\omega)\exp[-\alpha(t-s)]\,ds \\ &\equiv U_0(\omega)\exp[-\alpha(t-t_0)] + Y_t(\omega) \end{aligned} \tag{8.15}$$

is the unique R solution of (8.14) to the initial condition U_0 at time t_0 [Ref. 8.5, p. 39].

From this solution, it is in principle possible to obtain the hierarchy of probability densities for $U_t, p(u_1 t_1, \ldots, u_n, t_n)$, and hence via the inverse transformation of (8.13), the hierarchy for the original process X_t. In particular, the time-dependent probability density $p(x,t)$ with $p(x, t=t_0) = p_0(x)$ can be determined. In practice, it is rarely feasible to carry out this program. One encounters insuperable technical difficulties. There is however one class of real-noise processes, where it is child's play to calculate the hierarchy for U_t and X_t, respectively. It should come as no surprise that this is the class of Gaussian processes. The fact that in this case the calculations can easily be carried out stems from two features particular to Gaussian processes:

i) Gaussian processes are already completely characterized by the expectation and correlation function;

ii) random variables, which result from a linear transformation of jointly Gaussian random variables, are also jointly Gaussian distributed.

The second property is at the basis of the fact exploited earlier (p. 50) that the integral Y_t is a Gaussian process if ζ_t is a Gaussian process. Hence, according to (i), we only have to calculate $E\{Y_t\} = m_Y(t)$ and $E\{(Y_{t_1} - m(t_1))(Y_{t_2} - m(t_2))\} = C_Y(t_1, t_2)$ to characterize the random process Y_t completely.

In the following, we shall essentially be interested in the situation that at the initial time t_0 the system is prepared in such a way as to be in a particular state x_0. This implies that u_0 is a constant and thus stochastically independent of Y_t. Hence U_t is also a Gaussian process and

$$E\{U_t\} = u_0 \exp[-\alpha(t-t_0)] + E\{Y_t\}, \tag{8.16}$$

$$C_U(t_1, t_2) = E\{(U_{t_1} - E\{U_{t_2}\})(U_{t_2} - E\{U_{t_2}\})\}. \tag{8.17}$$

For the Gaussian process Y_t we have

$$E\{Y_t\} = \int_{t_0}^{t} E\{\zeta_s\} \exp[-\alpha(t-s)]\, ds = 0, \tag{8.18}$$

since $E\{\zeta_s\} = 0$ and

$$C_Y(t_1, t_2) = \int_{t_0}^{t_1} \int_{t_0}^{t_2} C_\zeta(\tau_1, \tau_2) \exp[-\alpha(t_1 - \tau_1)] \exp[-\alpha(t_2 - \tau_2)]\, d\tau_1 d\tau_2 . \tag{8.19}$$

Note that if the correlation function $C_\zeta(\tau_1, \tau_2)$ of the external noise is known, the time-dependent probability density, in fact the complete hierarchy, of U_t and hence of X_t can be determined. This exploits only the Gaussian character of the external noise and does not assume in any way that the noise is Markovian. The results apply thus to any real Gaussian noise. We can choose without restriction of generality $t_0 = 0$. Recall that we consider only situations where the initial condition u_0 is *nonrandom*. Then the time-dependent pd reads:

$$p(u,t;u_0) = [2\pi C_Y(t,t)]^{-1/2} \exp\left(-\frac{1}{2}\frac{(u-u_0\mathrm{e}^{-\alpha t})^2}{C_Y(t,t)}\right), \tag{8.20}$$

and using the inverse transformation of (8.13), we obtain for the original variable x:

$$p(x,t;x_0) = [2\pi C_Y(t,t)]^{-1/2}[g(x)]^{-1} \exp\left(-\frac{1}{2}\frac{[H(x)-H(x_0)\mathrm{e}^{-\alpha t}]^2}{C_Y(t,t)}\right) \tag{8.21}$$

where $1/g(x)$ is the Jacobian of the transformation (8.13). Note that since neither U_t nor X_t are Markov processes, (8.20, 21) are not transition probability densities, but are the one-time probability density describing the evolution of the system that had been prepared to be initially in state $x_0(u_0)$.

For the sake of concreteness, and in view of its wide occurrence in applications, as mentioned above, we shall now treat the case that the external noise ζ_t is a colored noise given by a stationary O-U process with $N(0,\mu^2/2\gamma)$. Thus the correlation function $C_\zeta(\tau_1,\tau_2)$ of the noise exponentially decreases:

$$C_\zeta(\tau_1,\tau_2) = \frac{\mu^2}{2\gamma}\exp(-\gamma|\tau_1-\tau_2|). \tag{8.22}$$

Taking $t_1 \geqslant t_2$, the correlation function $C_Y(t_1,t_2)$ can be straightforwardly evaluated from (8.19):

$$C_Y(t_1,t_2) = \frac{\mu^2}{2\gamma}\frac{\exp[-\alpha(t_1+t_2)]}{2\alpha(\alpha^2-\gamma^2)}\cdot[-2\gamma\mathrm{e}^{2\alpha t_2}+2\gamma-2\alpha\mathrm{e}^{(\alpha-\gamma)t_2}+2\alpha$$
$$+2\alpha\mathrm{e}^{(\alpha-\gamma)t_1}\mathrm{e}^{(\alpha+\gamma)t_2}-2\alpha\mathrm{e}^{(\alpha-\gamma)t_1}], \quad \alpha\neq\gamma \tag{8.23}$$

and

$$C_Y(t_1,t_2) = \frac{\mu^2}{2\gamma}\exp[-\alpha(t_1+t_2)]$$
$$\times\left(\frac{\mathrm{e}^{2\alpha t_2}}{2\alpha^2}-\frac{1}{2\alpha^2}+\frac{(t_1-t_2)}{2\alpha}\mathrm{e}^{2\alpha t_2}-\frac{t_1+t_2}{2\alpha}\right), \quad \alpha=\gamma. \tag{8.24}$$

Considering $t_1 = t_2 = t$, we obtain for the variance of Y_t:

$$C_Y(t,t) = \frac{\mu^2}{2\gamma}\left(\frac{1}{\alpha(\alpha+\gamma)}+\frac{\mathrm{e}^{-2\alpha t}}{\alpha(\alpha-\gamma)}-\frac{2\mathrm{e}^{-(\alpha+\gamma)t}}{\alpha^2-\gamma^2}\right), \quad \alpha\neq\gamma \tag{8.25}$$

and

$$C_Y(t,t) = \frac{\mu^2}{2\gamma}\left(\frac{1}{2\alpha^2}-\frac{\mathrm{e}^{-2\alpha t}}{2\alpha^2}-\frac{t\mathrm{e}^{-2\alpha t}}{\alpha}\right), \quad \alpha=\gamma. \tag{8.26}$$

Let us end the general presentation of the class of systems defined by (8.11), which are exactly soluble for real external noise, and turn our attention to a particular member of this class which is specially interesting from the viewpoint of noise-induced transitions, namely Hongler's model, already discussed in detail in the white-noise case. Consider

$$\dot{X}_t = -\frac{1}{2\sqrt{2}}\tanh(2\sqrt{2}X_t) + \frac{\zeta_t}{4\cosh(2\sqrt{2}X_t)}, \tag{8.27}$$

where ζ_t is the stationary O-U process given by the Ito SDE

$$d\zeta_t = -\gamma\zeta_t dt + \mu dW_t. \tag{8.28}$$

The transformation (8.13) corresponds here to $u(x) = \sinh(2\sqrt{2}x)$ and yields the linear equation:

$$dU_t = -U_t dt + \frac{\sqrt{2}}{2}\zeta_t dt. \tag{8.29}$$

It obviously has the same structure as the corresponding white-noise equation (6.92) of Chap. 6. According to (8.21) the one-time probability density of the solution of (8.27) is given by:

$$p(x,t;x_0) = \frac{2\sqrt{2}\cosh(2\sqrt{2}x)}{[2\pi C_Y(t,t)]^{1/2}}\exp\left(-\frac{1}{2}\frac{[\sinh(2\sqrt{2}x)-\sinh(2\sqrt{2}x_0)e^{-t}]^2}{C_Y(t,t)}\right) \tag{8.30}$$

with

$$C_Y(t,t) = \frac{\mu^2}{4\gamma}\left(\frac{1}{1+\gamma}+\frac{e^{-2t}}{1-\gamma}-\frac{2e^{-(1+\gamma)t}}{1-\gamma^2}\right), \quad \gamma \neq 1 \tag{8.31}$$

or

$$C_Y(t,t) = \frac{\mu^2}{4}\left(\frac{1}{2}-\frac{e^{-2t}}{2}-te^{-2t}\right), \quad \gamma = 1. \tag{8.32}$$

Here $p(x,t;x_0)$ has the same functional dependence on x as $p(x,t\,|\,x_0)$ in the white-noise case (6.94). The difference between the two expressions lies in the correlation functions. For the stationary probability density,

$$p_s(x) = \frac{2\sqrt{2}\cosh(2\sqrt{2}x)}{\left(2\pi\frac{\mu^2}{4\gamma(1+\gamma)}\right)^{1/2}}\exp\left[-\frac{1}{2}\frac{\sinh^2(2\sqrt{2}x)}{\frac{\mu^2}{4\gamma(1+\gamma)}}\right], \tag{8.33}$$

the difference from the white-noise case reduces to the fact that the white-noise intensity σ^2 has been replaced by $\mu^2/\gamma(1+\gamma)$. As in the white-noise case, Hongler's model displays a noise-induced critical point at which the stationary pd switches from monomodal to bimodal behavior. Recall that in the white-noise case the critical variance at which this phenomenon occurs is $\sigma^2 = \sigma_c^2 = 4$. In the colored-noise case, we have obviously

$$\frac{\mu^2}{\gamma(1+\gamma)} = 4 \rightarrow \mu^2 = 4\,\gamma(1+\gamma)\,. \tag{8.34}$$

Recall further that the white-noise limit corresponds to $\mu \rightarrow \infty$, $\gamma \rightarrow \infty$ such that μ^2/γ^2 is finite and equal to σ^2. Hence the white-noise result can be rewritten in the form

$$\sigma_c^2 = (\mu/\gamma)_c^2 = 4\,, \tag{8.35}$$

which has to be compared with the value for the colored-noise case, namely

$$(\mu/\gamma)_c^2 = 4 + (4/\gamma)\,. \tag{8.36}$$

In this particular model the effect of nonvashing correlations in the noise, i.e., $\gamma < \infty$, increases the intensity which is necessary to induce the transition.

Let us now investigate the dynamics of the extrema. In order to compare the temporal behavior of Hongler's model in the colored-noise case with that in the white-noise case, we start with the initial condition $x_0 = 0$. From (8.30) we obtain the following equation for the temporal evolution of the extrema $x_m(t)$ of the probability density:

$$\sinh\,[2\sqrt{2}x_m(t)]\left(1 - \frac{\cosh^2[2\sqrt{2}x_m(t)]}{C_Y(t,t)}\right) = 0\,. \tag{8.37}$$

Obviously, $x_m(t) = 0$ is an extremum for all times t. Initially, it is a maximum. If $\mu^2 > 4\,\gamma(1+\gamma)$, the state $x = 0$ is unstable and the final state $x_m = 0$ corresponds to a minimum. As in the white-noise case, a spinodal decomposition occurs. The probability density becomes "flat" at a certain instant of time t_c and displays a critical point. Then two maxima emerge from zero which tend to their steady-state values for $t \rightarrow \infty$. At the critical time t_c, $x_m = 0$ is a triple root. This occurs when the variance of Y_t becomes larger than one. Hence the critical time t_c is given by

$$C_Y(t_c, t_c) = 1\,. \tag{8.38}$$

8.4 Perturbation Expansion in the Bandwidth Parameter for the Probability Density

In Sect. 8.3 we obtained some preliminary results on the influence of nonvanishing correlations on noise-induced transitions. As far as the nature of the external noise is concerned, these results are very general. They hold for any kind of real noise, as long as it is a Gaussian process. As we have repeatedly pointed out by invoking the central limit theorem, this covers a broad class of applications. Let us once more emphasize that the noise is not required to be Markovian. This generality in the nature of the noise is however offset by the very particular kind of systems studied in Sect. 8.3. Indeed, the physical meaning of (8.11) is rather tenuous. We know of no physico-chemical system that belongs to this class. While the above results represent a first insight into the complex problem of the influence of colored noise, no general conclusions can be inferred from them, since they apply only to a rather particular class of systems.

A more general approach to the questions raised at the beginning of this chapter is therefore called for. In this section we shall establish the robustness of the white-noise analysis for a general one-variable system and determine the quantitative modifications in noise-induced transition phenomena due to a non-zero but short correlation time of the noise. In other words, we shall explore the immediate neighborhood of white noise. Here we consider only colored noise, in particular the O-U process. The nice feature of the O-U process is that it is not only in a vast number of applications the most appropriate model for correlated environmental fluctuations but also that the approach developed below is most clearly illustrated and most practically employed for the O-U process. Let us, however, emphasize that in principle everything that follows can be carried out for any Markovian ergodic noise.

To explore the behavior of nonlinear systems in the vicinity of Gaussian white noise, we consider the following set of SDEs

$$\begin{aligned} dX_t^\varepsilon &= [h(X_t^\varepsilon) + \lambda g(X_t^\varepsilon)]\,dt + \frac{\zeta_t}{\varepsilon} g(X_t^\varepsilon)\,dt \\ &= f(X_t^\varepsilon)\,dt + \frac{\zeta_t}{\varepsilon} g(X_t^\varepsilon)\,dt\,, \qquad x \in [b_1, b_2]\,, \end{aligned} \tag{8.39}$$

$$d\zeta_t = -\frac{\zeta_t}{\varepsilon^2}dt + \frac{\sigma}{\varepsilon}dW_t\,, \qquad \zeta_0 N\left(0, \frac{\sigma^2}{2}\right), \tag{8.40}$$

where we suppose that the functions $h(x)$ and $g(x)$ are independent of ε. Obviously, (8.40) defines a stationary O-U process with the probability density:

$$p_s(z) = \left(2\pi\frac{\sigma^2}{2}\right)^{-1/2} \exp\left(-\frac{1}{2}\,\frac{z^2}{\sigma^2/2}\right), \tag{8.41}$$

i.e., independent of ε, and a correlation time

$$\tau_{\text{cor}} = \varepsilon^2\,. \tag{8.42}$$

As hinted by (8.42), scaling with ε was introduced as a convenient way to measure the "distance" from the white-noise situation. Indeed, it is easily verified that the limit $\varepsilon \to 0$ corresponds to the white-noise limit as introduced in Chap. 3. To see this, consider the spectral density of the O-U process ζ_t, which is according to (3.19) given by the Lorentzian:

$$S(\nu)=\frac{\sigma^2}{2\pi(\varepsilon^2\nu^2+\varepsilon^{-2})}. \tag{8.43}$$

In the limit $\varepsilon\to 0$, the correlation time $\tau_{\rm cor}$ tends to zero and the bandwidth $\nu_{\rm b}=\varepsilon^{-2}$ (3.17) goes to infinity. Therefore the spectrum becomes flat, but obviously the spectral density vanishes for all finite frequencies. This is the noiseless limit, already encountered in Chap. 3, which results by simply letting the correlation time go to zero and not paying proper attention to the intensity of the fluctuations. To avoid ending up with a noiseless limit and to obtain the correct white-noise limit, the intensity of the external fluctuations has to be appropriately scaled in (8.39), which accounts for the factor ε^{-1} in the random force term of (8.39). Indeed, the spectral density of $\varepsilon^{-1}\zeta_t$ is given by

$$S^{\varepsilon}(\nu)=\frac{\sigma^2}{2\pi(\varepsilon^4\nu^2+1)}, \tag{8.44}$$

which converges to $\sigma^2/2\pi$ in the limit $\varepsilon\to 0$. It is thus clear that the scaling adopted for (8.39, 40) is appropriate to explore the vicinity of Gaussian white noise. Written in this form, the white-noise idealization of (8.39) is obtained by simply letting ε go to zero. There is a different way to understand the above scaling, which sheds additional light on it. Consider the system

$$dX_t=f(X_t)\,dt+\bar{\zeta}_t g(X_t)\,dt\,,$$

where $\bar{\zeta}_t$ is an O-U noise with a correlation time equal to the time unit:

$$d\bar{\zeta}_t=-\bar{\zeta}_t dt+\sigma dW_t\,,\qquad \bar{\zeta}_0 N(0,\sigma^2/2)\,.$$

To investigate the white-noise neighborhood, i.e., the environmental fluctuations are rapid on the macroscopic time scale, it is natural to speed up the noise by considering it on the time scale $t\to t/\varepsilon^2$, i.e., $\zeta_t=\bar{\zeta}_{t/\varepsilon^2}$. Furthermore, the amplitude has to be appropriately increased by a factor ε^{-1}, since

$$S_{\zeta}(\nu)=\int e^{it\nu}E\{\zeta_t\zeta_0\}dt=\varepsilon^2 S_{\bar{\zeta}}(\varepsilon^2\nu)\,.$$

In fact, a general prescription for the exploration of the white-noise neighborhood can be deduced from the above considerations [5.9]. Consider the general system

$$dX_t=f(X_t)\,dt+g(X_t,\bar{\zeta}_t)\,dt\,,$$

where g is possibly nonlinear in the colored noise $\bar{\zeta}_t$. Suppose that for fixed x,

$$E\{g(x,\bar{\zeta}_t)\}=0\,.$$

This means that any systematic effect of the external noise, i.e., a possibly nonzero mean value of the nonlinear noise, has been incorporated in $f(X_t)$; g is a pure noise term. Then the appropriate system to describe the white-noise neighborhood is

$$dX_t^\varepsilon = f(X_t^\varepsilon)\,dt + \frac{1}{\varepsilon} g(X_t^\varepsilon, \zeta_t)\,dt \tag{8.39a}$$

with $\zeta_t = \bar{\zeta}_{t/\varepsilon^2}$. The white-noise limit corresponds in this scaling simply to $\varepsilon \to 0$. We shall take up again the case of nonlinear external noise in Sect. 8.7. Let us now come back to the pair process $(X_t^\varepsilon, \zeta_t)$ defined by (8.39, 40).

Since it is defined by a set of Ito SDE's, it is a diffusion process (if X_0^ε and ζ_0 are independent of W_t). Its transition probability density is the fundamental solution of the following FPE:

$$\partial_t p^\varepsilon(x,z,t) = \left(\frac{F_1}{\varepsilon^2} + \frac{F_2}{\varepsilon} + F_3\right) p^\varepsilon(x,z,t) \tag{8.45}$$

with

$$F_1 = \partial_z z + \frac{\sigma^2}{2}\partial_{zz}\,, \tag{8.46}$$

$$F_2 = -z\partial_x g(x)\,, \tag{8.47}$$

$$F_3 = -\partial_x f(x)\,. \tag{8.48}$$

As usual, the first task is to analyze the steady-state behavior of the system. Since $(X_t^\varepsilon, \zeta_t)$ is a two-variable diffusion process with a degenerate diffusion matrix, i.e., (8.39) for X_t^ε does not contain any dW_t term, it is in general impossible to obtain an exact analytical expression for the stationary solution of the FPE (8.45) or the marginal density $p_s^\varepsilon(x)$. The latter probability density is of course the main object we have to determine; it is this quantity that describes the steady-state behavior of the system. Facing the impossibility of obtaining a general explicit expression for the steady-state solution of (8.45), we have to resort to approximation procedures to explore the neighborhood of white-noise. Our problem contains an obvious smallness parameter, namely the scaling parameter ε, i.e., the distance from the white-noise situation. Furthermore, the form of the Fokker-Planck operator suggests the following expansion of the transition probability density:

$$p^\varepsilon(x,z,t) = p_0(x,z,t) + \varepsilon p_1(x,z,t) + \varepsilon^2 p_2(x,z,t) + \dots \tag{8.49}$$

and specializing to the stationary probability density

$$p_s^\varepsilon(x,z) = p_0(x,z) + \varepsilon p_1(x,z) + \varepsilon^2 p_2(x,z) + \dots\,. \tag{8.50}$$

This is an expansion in the square root of the correlation time or the square root of the inverse bandwidth. The stationary probability density $p_s^\varepsilon(x,z)$ has to be normalized to all orders in ε; this implies that:

$$\int_{z=-\infty}^{+\infty} \int_{b_1}^{b_2} p_0(x,z)\,dx\,dz = 1 \tag{8.51}$$

and

$$\int_{z=-\infty}^{+\infty} \int_{b_1}^{b_2} p_k(x,z)\,dx\,dz = 0\,. \tag{8.52}$$

Obviously, we have

$$\int_{b_1}^{b_2} p^\varepsilon(x,z)\,dx = p_s(z)\,, \qquad \forall\varepsilon \tag{8.53}$$

since the stationary pd of the O-U process ζ_t is independent of ε (8.41), which implies the stronger condition

$$\int_{b_1}^{b_2} p_k(x,z)\,dx = 0\,, \qquad k = 1,2\ldots. \tag{8.54}$$

Injecting (8.50) into the stationary form of (8.45) and equating coefficients of equal power in ε we obtain:

$$\varepsilon^{-2}:\ F_1 p_0(x,z) = 0\,, \tag{8.55}$$

$$\varepsilon^{-1}:\ F_1 p_1(x,z) = -F_2 p_0(x,z)\,, \tag{8.56}$$

$$\varepsilon^{k-2}:\ F_1 p_k(x,z) = -F_2 p_{k-1}(x,z) - F_3 p_{k-2}(x,z)\,, \qquad k = 2,3,\ldots. \tag{8.57}$$

Note that the operator F_1 is the Fokker-Planck operator of the O-U process ζ_t and acts of course only on the second variable z. This implies that the probability density $p_s^\varepsilon(x,z)$ factorizes to the lowest order of the perturbation scheme. In other words, in the lowest order in ε the system variable and the fluctuating parameter are stochastically independent at the same instant of time:

$$p_0(x,z) = p_0(x)\,p_s(z)\,. \tag{8.58}$$

This is obviously the situation that prevails in the white-noise case. Here $p_s(z)$ is the stationary probability density of ζ_t as given by (8.41). As is clear from above, $p_0(x)$ has to be a probability density, implying in particular that

$$\int_{b_1}^{b_2} p_0(x)\,dx = 1\,. \tag{8.59}$$

This condition obviously is far from sufficient to determine $p_0(x)$. This does not however mean that our ansatz (8.50) for $p_s^\varepsilon(x,z)$ does not work and that ε is not the right expansion parameter. As we shall see shortly, $p_0(x)$ is obtained from a solvability condition on the higher-order equations (8.56, 57). This is not an uncommon situation in series expansions and is, for instance, frequently encountered in bifurcation analysis. It is convenient to write the $p_k(x,t)$ in the form

$$p_k(x,z) = p_s(z)\, r_k(x,z)\,, \qquad k = 0,1,2,\dots; \qquad r_0(x,z) = r_0(x) = p_0(x)\,; \quad (8.60)$$

(8.56, 57) transform into

$$F_1^+ r_1(x,z) = z\,\partial_x g(x)\, r_0(x) = I_1(x,z)\,, \quad (8.61)$$

$$\begin{aligned} F_1^+ r_k(x,z) &= [z\,\partial_x g(x)\, r_{k-1}(x,z) + \partial_x f(x)\, r_{k-2}(x,z)] \\ &= I_k(x,z)\,, \qquad k = 2,3,\dots\,. \end{aligned} \quad (8.62)$$

To obtain (8.61, 62), we have exploited the fact that $p_s(z)$ is the stationary solution of the FPE (4.53) of the O-U process, i.e., $F_1 p_s(z) = 0$. Further, F_1^+ is the Kolmogorov backward operator of the O-U process

$$F_1^+ = -z\,\partial_z + \frac{\sigma^2}{2}\,\partial_{zz}\,. \quad (8.63)$$

The eigenvalues of F_1^+ and F_1 coincide and in particular they have the eigenvalue zero. Thus F_1^+ can not be inverted to obtain straightforwardly the solutions of (8.61, 62) as $r_k(x,z) = F_1^{+\,-1} I_k(x,z)$. Therefore, (8.61, 62) have to fulfill a solvability condition known as the Fredholm alternative. To see this and to obtain this condition, we take the scalar product on both sides of (8.61, 62) with $p_s(z)$:

$$\int_{\mathbb{R}} p_s(z)\, F_1^+ r_k(x,z)\, dz = \int_{\mathbb{R}} p_s(z)\, I_k(x,z)\, dz\,, \qquad k = 1,2,\dots\,. \quad (8.64)$$

The Kolmogorov backward operator is the adjoint operator of the Fokker-Planck operator, if the diffusion process has natural boundaries. Taking the adjoint operator corresponds in the present context to an integration by parts. The O-U process has indeed natural boundaries and we have:

$$\begin{aligned} \int_{\mathbb{R}} p_s(z)\, F_1^+ r_k(x,z)\, dz &= \int_{\mathbb{R}} r_k(x,z)\, F_1 p_s(z)\, dz \\ &= 0, \text{ since } F_1 p_s(z) = 0\,. \end{aligned} \quad (8.65)$$

This implies together with (8.64) that the inhomogeneous parts $I_k(x,z)$ have to be orthogonal to the null space of the backward operator F_1^+, meaning that

$$\int_{\mathbb{R}} p_s(z)\, I_k(x,z)\, dz = 0\,, \qquad k = 1,2,\dots\,. \quad (8.66)$$

This is known as the Fredholm alternative and is the condition that (8.61, 62) possess a solution.

We can now proceed to evaluate $p_s^\varepsilon(x)$ in a systematic way. At the order ε^{-1}, the Fredholm alternative (8.66) is trivially satisfied:

$$\int_{\mathbb{R}} p_s(z) z \partial_x g(x) r_0(x) dz = E\{\zeta_t\} \partial_x g(x) r_0(x) = 0 . \tag{8.67}$$

From the form of (8.61) it follows that the general form of $r_1(x,z)$ is the sum of the homogeneous solution of (8.61), which is any arbitrary function $H_1(x)$ depending on x only, plus a particular solution. Note however that $H_1(x)$ will have to be compatible with the normalization condition (8.54). The particular solution of (8.61) is easily found to be $-I_1(x,z)$. Hence

$$r_1(x,z) = H_1(x) - I_1(x,z) \tag{8.68}$$

is the general solution of (8.61). So far we seem to have worsened our lot. We now have two functions on our hands, $r_0(x)$ and $H_1(x)$, which still have to be determined. Let us, however, proceed to the next order. At the order ε^0, (8.62) reads,

$$F_1^+ r_2(x,z) = z\partial_x g(x)[H_1(x) - z\partial_x g(x) r_0(x)] + \partial_x f(x) r_0(x) , \tag{8.69}$$

where we have expressed r_1 by (8.68). Fortunately in this order the Fredholm alternative is no longer trivially satisfied and furnishes the means to determine $r_0(x)$. It yields the following equation:

$$-\frac{\sigma^2}{2} \partial_x g(x) \partial_x g(x) r_0(x) + \partial_x f(x) r_0(x) = 0 , \tag{8.70}$$

or the equivalent

$$-\partial_x \left[f(x) + \frac{\sigma^2}{2} g'(x) g(x) \right] r_0(x) + \frac{\sigma^2}{2} \partial_{xx} g^2(x) r_0(x) = 0 . \tag{8.71}$$

We recognize the Fokker-Planck operator corresponding to the white-noise version of (8.39), namely the Stratonovich stochastic differential equation

$$dX_t = f(X_t) dt + \sigma g(X_t) \circ dW_t . \tag{8.72}$$

The result that the SDE (8.72) has to be interpreted in the sense of Stratonovich is of course not unexpected in the light of the theorem of *Wong* and *Zakai* which was discussed in Chap. 5 [5.8]. As established by *Blankenship*, *Papanicolaou* [5.9], the above result, i.e., that the Stratonovich SDE (8.72) is the white-noise limit of (8.39), holds not only for the O-U process but for a very large class of colored-noise processes ζ_t, including even jump processes. The technique they used to derive this generalization of Wong and Zakai's theorem inspired the perturbative method presented in this section [8.6]. Since $r_0(x)$ has to be normalized to one, it coincides with the stationary probability density p_s of the diffusion process X_t given by (8.72): $p_s(x) \equiv r_0(x)$. Now the lowest order of the stationary probability density is completely determined. We see that to determine the zeroth order of the perturbation expansion completely, we have to proceed to the second order and consider the corresponding Fredholm alternative. It turns out to be a general feature that to evaluate the stationary probability density $p_s^\varepsilon(x)$ up to the kth order one must proceed with the perturbation scheme up to the order $k+2$. Using the results obtained so far, we can write $r_1(x,z)$ as

$$r_1(x,z) = H_1(x) - \frac{2}{\sigma^2}\,\frac{f(x)}{g(x)}\,p_s(x)\,z\,. \tag{8.73}$$

Here $H_1(x)$ still remains undetermined. As we just remarked, this correction term of order ε can be completely specified only by proceeding to the order ε^3. This requires the calculation of r_2 which appears in the Fredholm alternative of the order ε^3. The equation for the second correction term reads

$$F_1^+ r_2(x,z) = -z^2\partial_x\left[\frac{2}{\sigma^2}f(x)p_s(x)\right] + z\partial_x[g(x)H_1(x)] + \partial_x f(x)p_s(x)\,. \tag{8.74}$$

The general solution of (8.74) is:

$$r_2(x,z) = \frac{z^2}{\sigma^2}\,\partial_x[f(x)p_s(x)] - z\partial_x[g(x)H_1(x)] + H_2(x)\,, \tag{8.75}$$

where again $H_2(x)$ is an arbitrary function of x only, but has to be compatible with the normalization condition (8.54). We can now proceed to the order ε^3 and apply the Fredholm alternative to the right-hand side of (8.62) for $k=3$. This yields

$$-\frac{\sigma^2}{2}\,\partial_x[g(x)\partial_x g(x)H_1(x)] + \partial_x f(x)H_1(x) = 0\,. \tag{8.76}$$

Putting $H_1(x) = p_s(x)\tilde{H}_1(x)$, it follows that

$$\frac{\sigma^2}{2}\,\partial_x[g^2(x)p_s(x)\,\partial_x\tilde{H}_1(x)] = 0 \tag{8.77}$$

and thus

$$\begin{aligned}\tilde{H}_1(x) &= \frac{2C_1}{\sigma^2}\int^x \exp\left[-\frac{2}{\sigma^2}\int^{x'}\frac{f(u)}{g^2(u)}\,du\right]dx' + C_2\\ &= C_1G(x) + C_2\,,\end{aligned} \tag{8.78}$$

where

$$\tilde{f}(u) = f(u) + \frac{\sigma^2}{2} g'(u) g(u) \tag{8.79}$$

and C_1, C_2 are constants. The normalization condition (8.54) applied to the first correction reads:

$$\int_{b_1}^{b_2} p_s(x) \left[\tilde{H}_1(x) - \frac{2}{\sigma^2} \frac{f(x)}{g(x)} z \right] dx = 0 \,. \tag{8.80}$$

The last integral in (8.80) vanishes for systems having intrinsic inaccessible boundaries b_1, b_2, since

$$\int_{b_1}^{b_2} p_s(x) \frac{2}{\sigma^2} \frac{f(x)}{g(x)} dx = \int_{b_1}^{b_2} \partial_x [g(x) p_s(x)] \, dx = g(x) p_s(x) \Big|_{b_1}^{b_2} \,. \tag{8.81}$$

This implies that

$$\int_{b_1}^{b_2} p_s(x) \tilde{H}_1(x) \, dx = C_1 \int_{b_1}^{b_2} p_s(x) G(x) \, dx + C_2 = 0 \tag{8.82}$$

must hold.

It is generally the case in applications that at least one of the boundaries b_1 and b_2 of the diffusion process describing the system in a white-noise environment is an intrinsic inaccessible boundary. In all the model systems considered in the preceding chapters *both* boundaries are intrinsic and inaccessible, except in the Nitzan-Ross system where the lower boundary is regular and only the upper boundary is natural. Here we restrict ourselves to the most common case as far as applications are concerned. We will formulate the perturbation scheme explicitly only for systems where both boundaries are G-S natural. These considerations on the nature of the boundaries are motivated by the fact that the function $\tilde{H}_1(x)$, as given by (8.78), is connected with the classification criterion L_1. Indeed: $L_1(b_i) = \tilde{H}_1(b_i)$. Since we consider only systems with natural boundaries, the function $\tilde{H}_1(x)$ diverges at b_i. Recall however that

$$p_s^{\varepsilon}(x,z) = p_s(x) p_s(z) \left\{ 1 + \varepsilon \left[\tilde{H}_1(x) - \frac{2}{\sigma^2} \frac{f(x)}{g(x)} z \right] + O(\varepsilon^2) \right\} . \tag{8.83}$$

Consequently, as is already evident from (8.80, 82), it is unfortunately not the behavior of $\tilde{H}_1(x)$ alone that matters, but that of $p_s(x) \tilde{H}_1(x)$. The divergence of $\tilde{H}_1(x)$ near b_i does not necessarily imply the divergence of (8.82). However, a comparison of (8.82) with Feller's classification scheme shows that

$$\int_{b_1}^{b_2} p_s(x) G(x) \, dx = \infty \,, \tag{8.84}$$

if at least one boundary b_i is also natural in Feller's sense. Consider the particularly important case that one of the boundaries, say b_1, is finite. Then this case can always be transformed into the case $b_i = 0$ by simply shifting the variable from x to $x - b_1$. For the often encountered case that the drift and diffusion vanish linearly near this boundary, i.e., $f(x) = O(x)$ and $g(x) = O(x)$, it is easily verified that b_i is natural in Feller's sense, if it is natural in the sense of Gihman and Skorohod. Indeed, with $f(x) = \bar{f}_1 x + O(x^2)$ and $g(x) = x + O(x^2)$, we have for $h_2(x)$ near 0

$$\begin{aligned} h_2(x) &= x^{-2} \exp\left(\frac{2}{\sigma^2} \int^x \frac{f_1 x'}{x'^2} dx'\right) \int^x \exp\left(-\frac{2}{\sigma^2} \int^z \frac{f_1 z'}{z'^2} dz'\right) dz \\ &= x^{([2f_1/\sigma^2]-2)} \int^x z^{-(2f_1/\sigma^2)} dz \\ &= x^{-1}, \quad f_1 = \bar{f}_1 + \frac{\sigma^2}{2} \end{aligned} \tag{8.85}$$

which is *not* integrable near zero.

The important result to retain is that if the boundaries b_i of the diffusion process X_t, given by (8.72), are natural in the Gihman-Skorohod and at least one in the Feller sense, then the integral in (8.82) diverges. This implies that C_1 has to be zero and in turn that $C_2 = 0$. This means that the function $\tilde{H}_1(x)$ vanishes identically for all those systems which have at least one Feller natural boundary. While this covers most of the model systems treated earlier, i.e., all those with $f(x) \sim x$ and $g(x) \sim x$ for small x, the genetic model is a notable exception. In fact as we have seen in Chap. 5, boundaries that are natural in the Gihman-Skorohod scheme can be natural or entrance boundaries in the Feller scheme. Since entrance boundaries are characterized by the fact that any probability that is initially assigned to them flows into the interval (b_1, b_2), we expect that any model where the drift is positive (negative) at a lower (upper) boundary and where the diffusion vanishes at that boundary possesses an entrance boundary. Let us consider the class of models for which $b_1 = 0$ (which as remarked above can be achieved by a simple shift if the state space is finite or semifinite), $f(x) = f_0 + \bar{f}_1 x + O(x^2)$ with $f_0 > 0$ and $g(x) = x + O(x^2)$. Obviously, the genetic model displays this behavior near $b_1 = 0$. Since the genetic model is symmetric with respect to $\tilde{x} = 1/2$ (Sect. 6.8), the nature of the boundary $b_2 = 1$ is the same as that of b_1. It was established earlier that both boundaries are G-S natural. To establish this in general for the above class of models consider $\phi(x)$ near 0:

$$\begin{aligned} \phi(x) &\simeq \exp\left[-\frac{2}{\sigma^2} \int^x dx' (f_0 + f_1 x' + \ldots)/x'^2\right] \\ &= \exp\left\{-\frac{2}{\sigma^2}\left[-\frac{f_0}{x} + f_1 \ln x + O(x)\right]\right\} \\ &= x^{(-2f_1/\sigma^2)} \exp\left[\frac{2f_0}{\sigma^2 x} + O(x)\right], \end{aligned} \tag{8.86}$$

which is obviously not integrable near zero. To show that the boundary $b_1 = 0$ is an entrance boundary we have to determine the integrability of $h_2(x)$ near zero:

$$h_2(x) \simeq x^{-2} \exp\left[\frac{2}{\sigma^2}\left(-\frac{f_0}{x} + f_1 \ln x + f_2 x + \ldots\right)\right]$$
$$\times \int_{x_0}^{x} \exp\left[-\frac{2}{\sigma^2}\left(-\frac{f_0}{z} + f_1 \ln z + \ldots\right)\right] dz\,. \tag{8.87}$$

Let us first investigate the behavior of $\int^x \phi(z)\,dz$ near zero:

$$\int_{x_0}^{x} z^{(-2f_1/\sigma^2)} \exp\left(\frac{2f_0}{\sigma^2}\frac{1}{z}\right) dz = -\int_{1/x_0}^{1/x} u^{([2f_1/\sigma^2]-2)} \exp\left(\frac{2f_0}{\sigma^2}u\right) du \quad \text{with} \quad u = 1/z \tag{8.88}$$

$$= -\int_{x/x_0}^{1} x^{(2-[2f_1/\sigma^2])} v^{([2f_1/\sigma^2]-2)} \exp\left(\frac{2f_0}{\sigma^2}\frac{v}{x}\right)\frac{dv}{x}$$

with $\quad u = v/x\,.$ (8.89)

Since we are interested in the behavior of $\int^x \phi\,(z)\,dz$ near zero, $1/x$ is a large quantity and the above integral can be evaluated by steepest descent techniques. We obtain

$$\int^{x} \phi(z)\,dz \simeq x^{(1-[2f_1/\sigma^2])} \frac{\exp\left(\dfrac{2f_0}{\sigma^2}\dfrac{1}{x}\right)}{\dfrac{2f_0}{\sigma^2 x}}$$
$$= \frac{\sigma^2}{2f_0} x^{(2-[2f_1/\sigma^2])} \exp\left(\frac{2f_0}{\sigma^2}\cdot\frac{1}{x}\right). \tag{8.90}$$

Consequently, we have for the behavior of $h_2(x)$ near zero

$$h_2(x) \simeq x^{([2f_1/\sigma^2]-2)} \exp\left(-\frac{2f_0}{\sigma^2 x}\right)\cdot\frac{\sigma^2}{2f_0} x^{(2-[2f_1/\sigma^2])} \exp\left(\frac{2f_0}{\sigma^2}\frac{1}{x}\right)$$
$$= \frac{\sigma^2}{2f_0}\,. \tag{8.91}$$

Hence $h_2(x)$ is integrable near zero, i.e., $b_1 = 0$ is an entrance boundary for models with $f(x) = f_0 + O(x)$ and $g(x) = x + O(x^2)$. If follows that if none of the boundaries b_1 and b_2 is F natural, i.e., both are entrance boundaries, the integral in (8.82) exists:

$$\int_{b_1}^{b_2} p_s(x)G(x)\,dx = \int_{b_1}^{b_2} h_2(x)\,dx = i < \infty . \tag{8.92}$$

In this case, (8.82) implies

$$C_1 \mathrm{i} + C_2 = 0 . \tag{8.93}$$

There is however a second condition that the stationary probability $p_s^\varepsilon(x,z)$ has to fulfill for any ε, namely $p_s^\varepsilon(x,z) \to 0$ for $x \to 0$, if zero is an entrance boundary. Intuitively this is easily seen. For small x we have

$$dX_t = (f_0 + \bar{f}_1 X_t)\,dt + \frac{\zeta_t}{\varepsilon} X_t dt , \tag{8.94}$$

i.e., the process moves from the point x into the interval (x, b_2) unless ζ_t is sufficiently negative. This reflects only the property of the entrance boundary that any probability assigned to it flows into the interval. Furthermore, (8.94) shows that no probability can accumulate at $b_1 = 0$ for any t, i.e., $p_s^\varepsilon(x,z) \to 0$ for $x \to 0$, if the probability that the noise process has monotonely decreasing unbounded realizations is zero, as it is for the O-U process. It is obvious that only realizations of the noise that become increasingly negative the closer the process X_t comes to the boundary can lead to a nonvanishing value of the stationary probability density $p_s^\varepsilon(x,z)$ at the boundary $b_1 = 0$. These arguments can be made precise by using the qualitative theory of stochastic process as exposed for instance in [8.1], in particular by resorting to controllability arguments (tube method, [6.8]). However, since the result is so intuitively obvious and a precise formulation of the central arguments of the qualitative theory is rather involved, we shall not present the latter here.

The fact that $p_s^\varepsilon(0,z) = 0$ must hold implies in particular that $p_s(0)p_s(z) \cdot [1 + \varepsilon r_1(0,z)] = O(\varepsilon^2)$. Hence $p_s(x)\tilde{H}_1(x)$ must tend to zero for x going to zero because $p_s(x)\ (2/\sigma^2) f(x) g^{-1}(x) \to 0$ for an entrance boundary. Since $h_2(x) \to \sigma^2/2 f_0$ for $x \to 0$ (8.91), we have in addition to (8.93):

$$C_1 \sigma^2/2 f_0 = 0 . \tag{8.95}$$

This implies that $C_1 = 0$ and (8.93) yields that $C_2 = 0$. Therefore we can conclude that also for an entrance boundary the function $\tilde{H}_1(x)$ has to vanish identically. The general result for our perturbation scheme is

$$\tilde{H}_1(x) \equiv 0 . \tag{8.96}$$

The first-order correction is now completely determined and reads

$$r_1(x,z) = -\frac{2}{\sigma^2}\frac{f(x)}{g(x)} p_s(x) z . \tag{8.97}$$

So far we have completely determined the zeroth and first-order terms for the joint probability density $p_s^\varepsilon(x,z)$. However, the quantity we really are interested in is of course the probability density $p_s^\varepsilon(x)$ for the state variable x alone:

$$p_s^\varepsilon(x) = \int_{\mathbb{R}} p_s^\varepsilon(x,z)\,dz = \int_{\mathbb{R}} dz\,p_s(z)\,[p_s(x) + \varepsilon r_1(x,z) + O(\varepsilon^2)]\,. \tag{8.98}$$

As $r_1(x,z)$ is proportional to z, the first-order correction to $p_s^\varepsilon(x)$ vanishes identically. For this reason one must proceed to determine $p_s^\varepsilon(x)$ up to the order ε^2. Let us again remark that this requires one to consider the Fredholm alternative for the order ε^4 to fix the so far unknown part $H_2(x)$. At the order ε^4 the Fredholm alternative necessitates knowing $r_3(x,z)$, which can be determined straightforwardly to be

$$\begin{aligned} r_3(x,z) = &-\frac{z^3}{3\sigma^2}\partial_x[g(x)\,\partial_x f(x)p_s(x)] - z\partial_x\Bigg[g(x)H_2(x) + g(x)\,\partial_x f(x)p_s(x) \\ &- \frac{2}{\sigma^2}\frac{f^2(x)}{g(x)}p_s(x)\Bigg] + H_3(x)\,. \end{aligned} \tag{8.99}$$

The Fredholm alternative for the fourth order reads then

$$\begin{aligned} &-\frac{\sigma^2}{4}\partial_x[g(x)\,\partial_x g(x)\,\partial_x f(x)p_s(x)] \\ &+\frac{\sigma^2}{2}\partial_x g(x)\,\partial_x\left[\frac{2}{\sigma^2}\frac{f^2(x)}{g(x)}p_s(x) - g(x)H_2(x) - g(x)\,\partial_x f(x)p_s(x)\right] \\ &+\tfrac{1}{2}\partial_x f(x)\,\partial_x f(x)p_s(x) + \partial_x f(x)H_2(x) = 0\,. \end{aligned} \tag{8.100}$$

Exploiting the fact that $p_s(x)$ is given by (6.13) and putting $H_2(x) = p_s(x)\tilde{H}_2(x)$ we can write (8.100) as:

$$\begin{aligned} &\frac{1}{2}\left[f(x) - \frac{\sigma^2}{2}g(x)g'(x)\right]\partial_x f(x)p_s(x) - \frac{3}{4}\sigma^2 g^2(x)\,\partial_{xx} f(x)p_s(x) \\ &+ g(x)\,\partial_x\frac{f^2(x)}{g(x)}p_s(x) - \frac{\sigma^2}{2}g^2(x)p_s(x)\tilde{H}_2'(x) \\ &- \frac{\sigma^2}{2}g(x)g'(x)\,\partial_x f(x)p_s(x) = \tilde{C}_1\,. \end{aligned} \tag{8.101}$$

After integration and rearrangement this yields:

$$\tilde{H}_2(x) = -\frac{2}{\sigma^2}\tilde{C}_1\int\frac{dx}{g^2(x)p_s(x)} - \frac{3}{2}\left[\frac{1}{p_s(x)}\partial_x f(x)p_s(x)\right] + \frac{1}{\sigma^2}\frac{f^2(x)}{g^2(x)} + C\,. \tag{8.102}$$

For the same reasons as above, at least the constant $\tilde{C}_1$ has to be equal to zero. Indeed, if at least one boundary b_i were F natural, the normalization condition (8.54) would again be violeted. In the other possible case that none of the boundaries is F natural, but one, say b_1, is an entrance boundary, $\tilde{C}_1 = 0$ since it turns out that all the other terms in $r_2(x,z)$ are such that $r_2(x,z) \to 0$ for $x \to b_1$. Thus we have

$$\tilde{H}_2(x) = \frac{1}{\sigma^2}\frac{f^2(x)}{g^2(x)} - \frac{3}{2}\frac{1}{p_s(x)}\partial_x f(x) p_s(x) + C\,. \tag{8.103}$$

The second correction term now reads

$$r_2(x,z) = \frac{z^2}{\sigma^2}\partial_x[f(x)p_s(x)] + \left[\frac{1}{\sigma^2}\frac{f^2(x)}{g^2(x)} - \frac{3}{2}\frac{1}{p_s(x)}\partial_x f(x)p_s(x) + C\right]p_s(x)\,.$$

The constant C is determined by the normalization condition (8.54)

$$\int_{b_1}^{b_2} r_2(x,z)\,dx = 0\,. \tag{8.104}$$

For natural boundaries

$$\int_{b_1}^{b_2} (\partial_x f(x)p_s(x))\,dx = f(b_2)p_s(b_2) - f(b_1)p_s(b_1) = 0\,.$$

Hence

$$C = -\frac{1}{\sigma^2}\int_{b_1}^{b_2} p_s(x)\frac{f^2(x)}{g^2(x)}\,dx\,. \tag{8.105}$$

Thus we obtain up to the first significant order in ε the following expression for the stationary probability density $p_s^\varepsilon(x)$ of the system:

$$\begin{aligned} p_s^\varepsilon(x) &= p_s(x)\left\{1 + \varepsilon^2\left[C - f'(x) + f(x)\frac{g'(x)}{g(x)} - \frac{1}{\sigma^2}\frac{f^2(x)}{g^2(x)}\right]\right\} \\ &= p_s(x)\{1 + \varepsilon^2[C - u(x)]\}\,. \end{aligned} \tag{8.106}$$

Accordingly the extrema of $p_s^\varepsilon(x)$ can immediately be found to be the roots of

$$\left[f(x_m) - \frac{\sigma^2}{2}g(x_m)g'(x_m)\right]\{1 + \varepsilon^2[C - u(x_m)]\} - \frac{\varepsilon^2\sigma^2}{2}g^2(x_m)u'(x_m) = 0\,. \tag{8.107}$$

As is clear from the structure of (8.107), the position and number of the extrema will mainly be decided by the white-noise contribution, i.e., the left factor in the first term of (8.107). The foregoing systematic perturbation expansion furnishes rigorous answers to the questions raised at the beginning of this chapter. To summarize, we have established the following facts.

i) The solution of a SDE with colored noise converges weakly, i.e., in distribution, to the solution of the corresponding SDE with white noise, where the latter has to be interpreted as a Stratonovich equation. In the preceding, this result was proven for the concrete case of external O-U noise. However, it is obvious that the perturbation scheme can be formulated for any colored noise. The only modification that has to be made is to replace the FP operator governing the evolution of the O-U process by the appropriate operator describing the temporal change in the transition probability (density) of the ergodic Markov noise process considered. Let us further remark that the perturbations expansion is also easily extended to cover the time-dependent case by replacing F_3 by $F_3 - \partial_t$. It is rather straightforward with these modifications to establish that in general, i.e., also for time-dependent situations and any colored external noise, the diffusion process as given by (8.72) is obtained in the white-noise limit $\varepsilon \to 0$, [5.9]. An explicit evaluation of the correction terms, however, is likely to encounter two principal technical difficulties. First, if the colored noise is not given by the O-U process whose FP operator has some nice properties (e.g., linear drift and constant diffusion), it is rarely feasible to solve explicitly (8.56, 57). Second, in the time-dependent case we encounter the problem that in general the time-dependent solution of the white-noise problem is not known. Nevertheless, the important upshot of the above analysis is that *the predictions of the white-noise idealization are robust* in the sense that they are recovered for *any* colored noise in the limit of short correlation times. In other words, qualitatively the same noise-induced transition phenomena occur in a neighborhood of white noise. Noise-induced transitions are *not* artefacts due to the white-noise idealization but are generic phenomena for rapid external noises.

ii) At the same time that it establishes the robustness of the white-noise analysis, our *perturbation scheme furnishes* also *explicit expressions*, at least in the important case of O-U noise, *for the quantitative modifications due to nonvanishing correlations in the noise.* Note that the perturbation scheme is a *systematic* expansion in ε, i.e., roughly speaking in the correlation time or in the inverse of the bandwidth, and thus, if it is desired, corrections to the white-noise analysis can be determined up to an arbitrarily high order. To conclude, let us comment here on a technical point of the perturbation expansion: the convergence of $p_s^\varepsilon(x)$ towards $p_s(x)$ is in general not uniform. In fact, commonly $\tilde{H}_2(x)$ diverges at one or both boundaries. This implies that one should be careful with the application of the results in an ε neighborhood of such a boundary. It does not, however, invalidate our reasoning that $H_1(x)$ has to vanish identically even for an F entrance boundary. The argument does not imply in any way uniform convergence but simply convergence; in the n-th order approximation $p_s^\varepsilon(0)$ has to be zero with a possible error of order ε^{n+1}. If $H_1(x) \neq 0$, the error term would be of order ε. Furthermore, due to the nonuniformity of convergence and due to the fact that the normalization was imposed to hold in all orders of ε, positivity does not hold in all orders of ε. In other words, in an ε neighborhood of the boundaries the n-th order approximation for $p_s^\varepsilon(x)$ can be slightly negative in such a way that the total error is of the order of $O(\varepsilon^{n+1})$.

8.4.1 Verhulst Model

Assuming that the fluctuations of the growth parameter in the Verhulst model (6.37) are given by an Ornstein-Uhlenbeck process of the form

$$\lambda_t = \lambda + \zeta_t/\varepsilon\,, \tag{8.108}$$

we find that the stationary probability density is given by

$$p^\varepsilon(x) = \left(\frac{2}{\sigma^2}\right)^{(2\lambda/\sigma^2)} \frac{x^{([2\lambda/\sigma^2]-1)}}{\Gamma\left(\frac{2\lambda}{\sigma^2}\right)} \exp\left(-\frac{2x}{\sigma^2}\right)\left[1+\varepsilon\left(x-\frac{\lambda}{2}-\frac{(\lambda-x)^2}{\sigma^2}\right)\right]. \tag{8.109}$$

The equation for the extrema is

$$\left[\left(\lambda-\frac{\sigma^2}{2}\right)x-x^2\right]\left[1+\varepsilon^2\left(x-\frac{\lambda}{2}-\frac{(\lambda-x)^2}{\sigma^2}\right)\right]$$
$$+\varepsilon^2\frac{\sigma^2}{2}x^2\left(1+\frac{2(\lambda-x)}{\sigma^2}\right)=0\,. \tag{8.110}$$

Obviously, $x=0$ is a root of this equation independently of the values of the parameters. It is a double root for $\lambda=\sigma^2/2$. The noise-induced transition point is therefore not affected by the existence of correlation in the external noise. As follows from the fact that in (8.106) $p_s(x)$ appears as a factor, this result is a general feature of all models where the white-noise-induced transition corresponds to a change of the probability density near one of the boundaries, namely a change from divergent to nondivergent behavior.

8.4.2 The Genetic Model

When the selection coefficient λ in (6.51) fluctuates like

$$\lambda_t = \zeta_t/\varepsilon\,, \tag{8.111}$$

we find that $p_s^\varepsilon(x)$ is given by

$$p_s^\varepsilon(x) = \frac{1}{2}\,\frac{\exp(2/\sigma^2)}{\mathrm{K}_0(2/\sigma^2)}\exp\left(-\frac{1}{\sigma^2x(1-x)}\right)\cdot\left\{1+\varepsilon^2\left[1-\frac{\mathrm{K}_1(2/\sigma^2)}{\mathrm{K}_0(2/\sigma^2)}\right.\right.$$
$$\left.\left.+\frac{1}{2}\,\frac{(1-2x^2)}{x(1-x)}\left(1-\frac{1}{2\sigma^2x(1-x)}\right)\right]\right\}, \tag{8.112}$$

where K_0 and K_1 are the modified Bessel functions. Note that $p_s^\varepsilon(x)$ is symmetric around $x = 1/2$. This implies trivially that $x = 1/2$ is always an extremum. As we saw in Chap. 6, in the white-noise limit the fluctuations of λ induce a cusp type of transition with a critical point at $\sigma_c^2 = 4$, $\lambda = 0$. This critical point corresponds to a triple root at $x = 1/2$ of the equation for the extrema. The influence of the correlations of the noise shifts this critical point at $\lambda = 0$, $x = 1/2$, to the value

$$\sigma^2/4 = [1+\varepsilon^2(C+2)]/[1+\varepsilon^2(C+3)]\,. \tag{8.113}$$

Up to the first order in ε^2 this yields

$$\sigma^2 = 4 - 4\varepsilon^2\,. \tag{8.114}$$

8.5 Switching-Curve Approximation

In this section we consider the opposite limit, namely that the correlation time of the external noise is long compared to the characteristic evolution time of the state variable x. Of course the appropriate scaling to take this limit is different from the one considered when the correlations are vanishingly small. The following set of stochastic differential equations for the pair process (X_t, ζ_t) has to be used to explore the effects of slow noise:

$$\begin{aligned} dX_t^\varepsilon &= [h(X_t^\varepsilon) + \lambda g(X_t^\varepsilon)]\,dt + \zeta_t g(X_t^\varepsilon)\,dt \\ &= f(X_t^\varepsilon)\,dt + \zeta_t g(X_t^\varepsilon)\,dt\,, \end{aligned} \tag{8.115}$$

$$d\zeta_t = -\varepsilon^2\zeta_t dt + \varepsilon\sigma\,dW_t\,. \tag{8.116}$$

The correlation function of ζ_t is

$$C(t) = \frac{\sigma^2}{2}\exp(-\varepsilon^2 t)\,, \tag{8.117}$$

and the correlation time is now given by

$$\tau_{\text{cor}} = 1/\varepsilon^2\,. \tag{8.118}$$

The spectral density of the noise process ζ_t is the Lorenzian

$$S^\varepsilon(\nu) = \sigma^2\frac{\varepsilon^2}{2\pi(\varepsilon^4+\nu^2)} \tag{8.119}$$

and

$$\lim_{\varepsilon\to 0} S^\varepsilon(\nu) = \frac{\sigma^2}{2}\delta(\nu)\,. \tag{8.120}$$

The stationary form of the Fokker-Planck equation for the pair process (X_t, ζ_t) is

$$\varepsilon^2\left(\partial_z z + \frac{\sigma^2}{2}\partial_{zz}\right) - \partial_x[f(x,\lambda) + zg(x)]p(x,z) = 0\,. \tag{8.121}$$

Expanding the stationary probability density $p_s(x,z)$ in the form (8.50) yields to the lowest order in ε:

$$\partial_x[f(x,\lambda) + zg(x)]p_0(x,z) = 0\,. \tag{8.122}$$

As is easily verified, this implies that

$$p_0(x,z) = p_s(z)\,\delta(x-u(z))\,, \tag{8.123}$$

where $u(z)$ is defined by the condition

$$f(u(z)) + zg(u(z)) = 0\,. \tag{8.124}$$

Since

$$\delta(u(z)-x) = |h'(u^{-1}(x))|^{-1}\delta(z-u^{-1}(x))\,, \tag{8.125}$$

integrating out the noise variable, we obtain at the lowest order for the probability density of the system

$$p_0(x) = p_s(u^{-1}(x))\,|u^{-1'}(x)|. \tag{8.126}$$

Here $p_s(.)$ is the stationary probability density of the Ornstein-Uhlenbeck process.

This approximation is known as the switching-curve approximation and has been introduced following more intuitive arguments in [6.8]. The above procedure obviously corresponds to the adiabatic elimination of the state variable x, i.e., the system is always in a quasi-stationary state with respect to the instantaneous value of the fluctuating parameters. The physical picture underlying this is that since the noise is much slower than the evolution of the system, the latter always stays in the immediate vicinity of the switching curve. This implies that the probability density of the system should be given by the probability density of the noise transformed via the switching curve $f(x) + zg(x) = 0$, as is indeed confirmed by (8.126).

Since the procedure is completely analogous to the perturbation expansion presented in the preceding section, only a different scaling is used here, in principle the higher-order correction terms to the probability density can be calculated in a systematic manner. However, practically the evaluation meets the following difficulties: the lowest-order operator in the perturbation expansion is the operator which appears in the continuity equation describing the deterministic motion

$$\partial_x [f(x) + zg(x)] . \tag{8.127}$$

It does not have the nice properties of the Fokker-Planck operator for the Ornstein-Uhlenbeck process which played the main role in the white-noise limit. Indeed, (8.125) being a function of both variables, the joined probability density $p_0(x,z)$ does not factorize to the lowest order. Furthermore, not being the evolution operator of the diffusion process but of the deterministic motion, generalized functions as the Dirac δ function in (8.123) have to be used. This makes the explicit determination of the higher-order correction terms a rather intractable problem.

8.6 An Approximate Evolution Operator for Systems Coupled to Colored Noise

In Sect. 8.4, we established the robustness of the white-noise analysis and determined the quantitative modifications to which nonvanishing correlations in the external noise give rise by studying the behavior of the composite process $(X_t^\varepsilon, \zeta_t)$ in the neighborhood of white noise via a perturbation expansion for the probability density. This is a systematic approximation scheme in the correlation time and is formulated within the framework of Markov theory. As explained at the beginning of this chapter, the price that has to be paid in order to be able to describe systems coupled to a colored-noise environment with the help of Markov processes is the enlarging of the state space. Only the pair process, consisting of the state variable of the system X_t and the noise process ζ_t, is Markovian. Due to the structure of the Fokker-Planck operator for the evolution of the transition probability density of the pair process, in this context the bandwidth perturbation scheme is obviously a convenient choice to investigate the behavior of the system. However, an alternative route exists to gain insight into the effect of colored noise on nonlinear systems. It consists of the determination of an approximate evolution operator for the one-time probability density of the non-Markovian process X_t describing the system.

This way has been chosen and extensively used by *Sancho* and *San Miguel* [8.7 – 10] to study the behavior of systems subjected to external noise with nonvanishing correlations. Note that while this approach avoids the enlarging of the space of variables, it implies that one has to work outside the framework of the Markov theory. Consequently some heavy mathematical arsenal has to be used to overcome the obstacles which any treatment of non-Markovian processes faces. Sancho and San Miguel indeed have to resort to rather involved nonprobabilistic techniques to determine the form of the approximate evolution operator.

Let us now outline the basic features of this non-Markovian approximation scheme. To evaluate explicitly at least the stationary solution of the approximate evolution operator, the latter should of course possess some suitable properties. It would be most convenient if it were Fokker-Planck type, i.e., containing only first- and second-order derivatives with a nonnegative coefficient for the ∂_{xx}-

term. This is advantageous for two principal reasons. First, the steady-state solution is guaranteed to be positive and can thus be interpreted as a probability density. Second, its form is explicitly known, since it would also be given by (6.15) under suitable boundary conditions. Both features are in general lost if the operator involves third- or higher-order derivatives. In such a case, neither the positivity of the solution can be ensured nor can an explicit expression for it be obtained. Note that an evolution operator of the Fokker-Planck type is not incompatible with the non-Markovian character of the process X_t as already pointed by *Hänggi* et al. [4.4]. This operator describes only the temporal evolution of the one-time probability density $p(x,t)$, but not that of a transition probability density. It was emphasized in Chap. 4 that this property, i.e., $p(x,t)$ obeys a Fokker-Planck type equation, does *not* imply any Markov property for the process X_t. In the following, the name Fokker-Planck operator and Fokker-Planck equation will be reserved strictly for diffusion processes. The word "type" will be added for an evolution operator or equation for the one-time probability density of a non-Markovian process.

As Sancho and San Miguel showed, it is indeed possible to obtain an FP type operator up to first order in the correlation time of the external noise for one-variable systems subjected to an O-U process. To establish this result, consider the SDE

$$dX_t = [h(X_t) + \lambda g(X_t)]\,dt + \zeta_t g(X_t)\,dt\,, \tag{8.128}$$

where ζ_t is a stationary O-U process given by

$$d\zeta_t = -\gamma\zeta_t dt + \gamma\sigma dW_t\,, \qquad \zeta_0 N\left(0, \frac{\gamma\sigma^2}{2}\right). \tag{8.129}$$

Let the (deterministic) functions $h(x)$ and $g(x)$ be such that a unique R solution of (8.128) exists [8.5, 11]. In other words, there is a stochastic process X_t such that almost surely

$$dX_t(\omega) = [h(X_t(\omega)) + \lambda g(X_t(\omega))]\,dt + \zeta_t(\omega)\,g(X_t(\omega))\,dt$$

with initial condition $X_0(\omega)$. Recall that

$$P(X_t \in B) = E\{I_B(X_t)\}\,.$$

Choosing $B = (-\infty, x]$, we have

$$\begin{aligned} F(x,t) &= \int_\Omega I_B(X_t(\omega))\,dP(\omega) \\ &= \int_\Omega \left[\int_{-\infty}^{x} \delta(X_t(\omega) - x')\,dx'\right] dP(\omega)\,. \end{aligned}$$

Assuming a density exists

$$p(x,t) = \partial_x \int_\Omega \left[\int_{-\infty}^{x} \delta(X_t(\omega)-x')\,dx' \right] dP(\omega)$$

and at least formally,

$$\begin{aligned} p(x,t) &= \int_\Omega \delta(X_t(\omega)-x)\,dP(\omega) \\ &= E\{\delta(X_t-x)\}\,. \end{aligned}$$

For the time derivative of the distribution function, we have

$$\begin{aligned} \partial_t F(x,t) &= \partial_t \int_\Omega I_{(-\infty,x]}(X_t(\omega))\,dP(\omega) \\ &= \int_\Omega I'_{(-\infty,x]}(X_t(\omega))\,\dot{X}_t(\omega)\,dP(\omega) \\ &= \int_\Omega I'_{(-\infty,x]}(X_t(\omega))\,[f(X_t(\omega)) + \zeta_t(\omega)\,g(X_t(\omega))]\,dP(\omega) \\ &= \int_\Omega \int_{-\infty}^{x} -\partial_{x'}\delta(X_t(\omega)-x')\,[f(X_t(\omega)) + \zeta_t(\omega)\,g(X_t(\omega))]\,dx'\,dP(\omega) \end{aligned}$$

and at least formally,

$$\partial_t p(x,t) = -\partial_x f(x)\,p(x,t) - \partial_x g(x)\,E\{\zeta_t \delta(X_t-x)\}\,. \qquad (8.130)$$

The same result is also obtained by the use of the so-called stochastic Liouville equation, based on *Van Kampen's* lemma [8.12–15]. The above derivation is rather formal and certainly needs further justification. We shall return to this point later. Note that (8.130) is not a closed equation for $p(x,t)$ due to the presence of the term $E\{\zeta_t\delta(X_t-x)\}$. To make further progress, we have to exploit the Gaussian character of the colored external noise. *Novikov* [8.16] (see also [4.7]) has proven the following theorem for Gaussian random processes: Let Z_t be a Gaussian process and $\varphi(Z_t)$ be a function of Z_t. Then

$$E\{Z_t \varphi(Z_t)\} = \int_0^t dt'\,C_Z(t,t')\,E\left\{\frac{\delta\varphi(Z_t)}{\delta Z_{t'}}\right\}, \qquad (8.131)$$

where $\delta\varphi(Z_t)/\delta Z_{t'}$ denotes the functional derivative. Obviously X_t is a function of the noise process ζ_t. Applying Novikov's theorem to the expectation of the product of ζ_t and $\delta(X_t-x)$ we obtain:

$$\partial_t p(x,t) = -\partial_x f(x) p(x,t) - \partial_x g(x) \int_0^t dt' C_\zeta(t,t') E\left\{\frac{\delta[\delta(X_t - x)]}{\delta\zeta_{t'}}\right\}$$
$$= -\partial_x f(x) p(x,t) + \partial_x g(x)\, \partial_x \int_0^t dt' C_\zeta(t,t') E\left\{\delta(X_t - x)\frac{\delta X_t}{\delta\zeta_{t'}}\right\}. \tag{8.132}$$

To make further progress, the so-called response function $\delta X_t / \delta\zeta_{t'}$ has to be evaluated. Let us first consider the linear case, i.e., $h(x) = \alpha x$ and $g(x) = 1$. As we have seen in Sect. 8.3, the R solution is then given by

$$X_t(\omega) = X_0(\omega) \mathrm{e}^{\alpha t} + \int_0^t [\lambda + \zeta_s(\omega)] \mathrm{e}^{\alpha(t-s)} ds. \tag{8.133}$$

It follows straightforwardly that the response function is given by the following expression

$$\frac{\delta X_t}{\delta \zeta_s} = \mathrm{e}^{\alpha(t-s)}. \tag{8.134}$$

Hence we have in this case

$$\partial_t p(x,t) = -\partial_x f(x) p(x,t) + \partial_{xx} \int_0^t dt' C_\zeta(t,t') \mathrm{e}^{\alpha(t-t')} E\{\delta(X_t - x)\}$$
$$= -\partial_x f(x) p(x,t) + \partial_{xx} D(t) p(x,t) \tag{8.135}$$

with the obvious definition of $D(t)$. If ζ_t is the O-U process then

$$D(t) = (1/2) \int_0^t dt' \gamma\sigma^2 \exp[(\alpha - \gamma)(t - t')]$$
$$= \frac{1}{2}\frac{\gamma\sigma^2}{\gamma - \alpha}[1 - \exp(\alpha - \gamma) t] \qquad \text{for} \quad \alpha \neq \gamma$$
$$= \frac{1}{2}\gamma\sigma^2 t \qquad \text{for} \quad \alpha = \gamma. \tag{8.136}$$

Thus for the linear problem, i.e., $f(x) = \alpha x + \lambda$ and $g(x) = 1$, an exact Fokker-Planck type equation for the one-time probability density exists, if the external colored noise is Gaussian. Note that this result holds true for any real Gaussian noise, since the Markov property does not enter into the derivation of (8.135). It is valid for any form of the correlation function, if the noise is Gaussian. This result establishes furthermore that the one-time probability density of the models belonging to the exactly soluble class described in Sect. 8.3 obeys an exact

Fokker-Planck type equation. Indeed, the particularity of those models was that via the transformation $u = H(x)$ a linear SDE is obtained. In this new variable an exact FP type equation exists according to the above result, and transforming back to the original variable we have

$$\partial_t p(x,t) = -\alpha \partial_x g(x) H(x) p(x,t) + D(t) \partial_x g(x) \partial_x g(x) p(x,t) . \tag{8.137}$$

This shows once again that in these models the only effect of colored noise, as compared to the white-noise situation, is a renormalization of the diffusion coefficient. In the case of O-U noise, σ^2 is replaced by $\sigma^2(1-\alpha/\gamma)^{-1} \cdot [1-\exp(\alpha-\gamma)t]$, which in the white-noise limit, corresponding here to $\gamma \to \infty$ (8.129) converges obviously to σ^2.

Outside the linear case it is generally not possible to evaluate the response function $\delta X_t/\delta\zeta_{t'}$, in an exact explicit manner. In this method too it is necessary to resort to approximation techniques in order to discuss the general case of a nonlinear system in a colored-noise environment.

As in Sect. 8.4, let us first consider the neighborhood of white noise. The appropriate smallness parameter is then the correlation time $\tau_{\text{cor}} = \gamma^{-1}$. If the correlations in the external noise decay very rapidly, then it is intuitively clear that the main contribution to the response function arises from the vicinity of $t = t'$. Expanding in powers of $(t-t')$, we have

$$\frac{\delta X_t}{\delta\zeta_{t'}} = \left.\frac{\delta X_t}{\delta\zeta_{t'}}\right|_{t'=t} + \left.\frac{d}{dt'}\frac{\delta X_t}{\delta\zeta_{t'}}\right|_{t'=t} (t-t') + \dots , \tag{8.138}$$

and the main problem is now to evaluate the response function and its derivatives at equal times. While the first term is easy to guess

$$\left.\frac{\delta X_t}{\delta\zeta_{t'}}\right|_{t=t'} = g(X_t) , \tag{8.139}$$

the second term is less obvious

$$\begin{aligned} \left.\frac{d}{dt'}\frac{\delta X_t}{\delta\zeta_{t'}}\right|_{t=t'} &= f(X_t)g'(X_t) - g(X_t)f'(X_t) \\ &= -g^2(X_t)[f(X_t)/g(X_t)]' . \end{aligned} \tag{8.140}$$

Since the evaluation of these expressions uses techniques, namely an extension of the Martin-Siggia-Rose formalism to Fokker-Planck type dynamics [8.17, 18], with which we do not assume the reader to be familiar and which are outside the scope of this monograph, we shall skip the explicit derivation of (8.139, 140). It can be found in [8.7]. For the second derivative, Sancho and San Miguel obtained

$$\left.\frac{d^2}{dt'^2}\frac{\delta X_t}{\delta\zeta_{t'}}\right|_{t'=t} = f(X_t)^2\left[f(X_t)\left(\frac{g(X_t)}{f(X_t)}\right)'\right]' - g^2(X_t)\left[g(X_t)\left(\frac{f(X_t)}{g(X_t)}\right)'\right]'\zeta_t . \tag{8.141}$$

In contradistinction to the first two terms in the expansion, the third one depends on the noise process. This implies that any inclusion of terms of order $(t-t')^2$ or higher necessitates a repeated application of Novikov's theorem. Let us therefore first consider only terms up to $(t-t')$, which corresponds to an expansion in τ_{cor} up to first order as will soon become clear. Substituting expansion (8.138) in (8.132), we obtain the following evolution equation:

$$\begin{aligned}\partial_t p(x,t) = &-\partial_x f(x)p(x,t) + \partial_x g(x)\,\partial_x \int_0^t dt' \frac{\gamma\sigma^2}{2}\exp[-\gamma|t-t'|]\\ &\times E\{\delta(X_t-x)[g(X_t) + (fg'-gf')(t-t')]\}\\ = &-\partial_x f(x)p(x,t) + \partial_x g(x)\,\partial_x \frac{\sigma^2}{2}(1-\mathrm{e}^{-\gamma t})g(x)p(x,t)\\ &+\partial_x g(x)\,\partial_x \frac{\gamma\sigma^2}{2}\left(\frac{1}{\gamma^2} - \frac{\mathrm{e}^{-\gamma t}}{\gamma^2} - \frac{t\mathrm{e}^{-\gamma t}}{\gamma}\right)(fg'-gf')p(x,t)\,.\end{aligned} \tag{8.142}$$

Since this approximation applies to the white-noise neigborhood, i.e., $\tau_{\mathrm{cor}} \ll 1$ or $\gamma \gg 1$, the transient terms in (8.142) decay rapidly and after a short initial layer we find the following FP type equation for the one-time probability density

$$\begin{aligned}\partial_t p(x,t) = &-\partial_x\left[f(x) + \frac{\sigma^2}{2}g'(x)v(x)\right]p(x,t)\\ &+\frac{\sigma^2}{2}\partial_{xx}g(x)v(x)p(x,t)\,,\end{aligned} \tag{8.143}$$

where

$$v(x) = g(x) + \gamma^{-1}[f'(x)g(x) - g'(x)f(x)]\,. \tag{8.144}$$

Note that this FP type operator creates an artificial restriction of the state space by inducing unphysical boundaries. Indeed, it is defined, in the sense that its solution is guaranteed to be nonnegative, only for values of x such that $g(x)v(x) > 0$, i.e.,

$$g^2(x)[1 + \gamma^{-1}g(x)(f(x)/g(x))'] > 0\,. \tag{8.145}$$

Thus, if the second factor takes negative values in $[b_1, b_2]$, then the state space has to be restricted to the interval between the zeros of this factor. Obviously these artificial boundaries approach the physical ones in the white-noise limit. These unphysical boundaries arise here for the same reason that the probability density in the bandwidth perturbation expansion takes slightly negative values in the vicinity of the physical boundaries, namely the convergence is not uniform on $[b_1, b_2]$. In contradistinction to the bandwidth perturbation expansion, the range of validity of the approximate evolution operator (8.143) is still an open question at this point. It is rather difficult to establish it in general within the framework of this approach for the following reason: the term involving $d^2(\delta X_t/\delta\zeta_{t'})/dt'^2$ in the expansion of the response function contains a term multiplied by the noise, namely $-g^2[g(f/g)']'\zeta_t$, cf. (8.141). Thus, if the approximation is continued to second order in τ_{cor}, in general no FP type equation exists. This is due to the fact that a repeated application of Novikov's theorem contributes a further derivative with respect to x. The term involving ζ_t vanishes identically and then an exact FP type equation exists, if and only if the system belongs to the class of soluble models in Sect. 8.3. Indeed, the second derivative of the response function at equal times contains no dependence on ζ_t, if

$$g(x)[f(x)/g(x)]' = \text{const}\,. \tag{8.146}$$

This implies that

$$f(x) \sim g(x)\int^x \frac{dz}{g(z)}\,,$$

which is the characterizing feature of (8.11). In general, if one proceeds to the order τ_{cor}^2 in the approximation scheme, the evolution operator contains a third-order derivative. This implies that a truncation after the first-order terms in τ_{cor} amounts not only to neglecting terms multiplied by higher powers of τ_{cor}, but also terms containing higher-order derivatives. This is known as a singular perturbation problem. The danger lies in the fact that the $\tau_{\text{cor}}^2\partial_{xxx}$ terms could be of the same order of magnitude as the $\tau_{\text{cor}}\partial_{xx}$ terms since the one-time probability density $p(x,t)$ is obviously τ_{cor} dependent. However, if the approximate FP operator describes the non-Markovian process X_t correctly up to first order in τ_{cor}, then the probability density as determined by (8.143) has to coincide with (8.49), i.e., that given by the bandwidth perturbation expansion. The latter is obtained by a systematic expansion in τ_{cor} and therefore the range of its validity as well as the magnitude of the error terms are precisely known. The stationary solution of the FP type equation (8.143) is given by

$$p_s^\tau(x) = N\frac{1}{v(x)}\exp\left(\frac{2}{\sigma^2}\int dx\frac{f(x)}{g(x)\,v(x)}\right),$$

if there is no probability flux across the boundaries. Though the FP type operator (8.143) contains only terms up to first order in τ_{cor}, the steady-state solution involves arbitrarily high orders in τ_{cor}. It is obvious that these higher-order terms should not be taken seriously. Let us therefore develop $p_s^\tau(x)$ in powers of τ_{cor} and truncate after the first term. Writing for simplicity of notation $\tau = \tau_{\text{cor}}$ we have:

$$\begin{aligned} p_s^\tau &= \frac{N}{g(x)+\tau(f'g-g'f)}\exp\left(\frac{2}{\sigma^2}\int\frac{f}{g^2+\tau g(f'g-g'f)}\right) \\ &= (N_0+\tau N_1+\ldots)\frac{1}{g}\left(1-\tau\frac{f'g-g'f}{g}+\ldots\right) \end{aligned}$$

$$\times \exp\left(\frac{2}{\sigma^2}\int\frac{f}{g^2}\right)\left[1-\frac{2}{\sigma^2}\tau\int\frac{f}{g}\left(\frac{f}{g}\right)'+\dots\right]$$

$$= N_0\frac{1}{g}\exp\left(\frac{2}{\sigma^2}\int\frac{f}{g^2}\right)\left\{1+\tau\left[N_1-f'+\frac{fg'}{g}-\frac{1}{\sigma^2}\left(\frac{f}{g}\right)^2\right]+\dots\right\}. \quad (8.147)$$

This expression does indeed coincide with the one obtained by the bandwidth perturbation expansion (8.106), i.e., up to terms of order τ_{cor} the bandwidth perturbation expansion and the non-Markovian approximation scheme of Sancho and San Miguel are equivalent. This shows that the approximate FP type operator yields a generally valid approximation, up to first order in τ_{cor}, of the stationary probability density of the non-Markovian system. This also justifies a posteriori the formal steps entering the derivation of the FP type operator. A comparison of the higher-order correction terms is not possible. The evolution operator is then no longer FP type and $p_s^\tau(x)$ cannot be explicitly evaluated anymore.

8.7 Nonlinear External Noise

So far we have dealt only with systems where the external parameter appears in a linear way in the phenomenological equation. As the preceding chapters illustrate, this case covers a broad class of applications. On the other hand, of course situations arise where the state of the environment influences in a nonlinear fashion the evolution of the system. In this and the following chapter, we shall discuss two such applications which are of particular interest, namely nematic liquid crystals and nerve membranes. Before analyzing any particular model system, we shall now elaborate on the theoretical approach of systems subjected to nonlinear external noise.

8.7.1 Theoretical Aspects

Let us consider a phenomenological equation of the type

$$\dot{X}(t) = h(X(t)) + \mu(\lambda)g(X(t)), \quad (8.148)$$

where μ is a nonlinear function of the external parameter λ. To take into account the environmental fluctuations, in the spirit of our general approach we replace the constant external parameter by a stationary stochastic process $\lambda_t = \lambda + \zeta_t$ and obtain the SDE:

$$dX_t = [h(X_t) + \mu(\lambda + \zeta_t)g(X_t)]\,dt. \quad (8.149)$$

As already remarked earlier, principle reasons render impossible any attempt to give a well-defined meaning to nonlinear operations on the Dirac delta function or any other generalized function. This implies that it is meaningless to pass to the white-noise idealization in the above situation by simply setting $\lambda_t = \lambda + \sigma\xi_t$. The concept of a SDE like

$$\dot{X}_t = [h(X_t) + \mu(\lambda + \sigma\xi_t)g(X_t)] \tag{8.150}$$

and a solution to it, can be defined in *no* way. This is not a question of replacing the ordinary calculus by an appropriate version of Ito or Stratonovich calculus. For principle mathematical reasons, any attempts to handle an SDE like (8.150) are doomed to failure. On the other hand, nonlinear operations on ordinary stochastic processes, as for instance on diffusion processes, are mathematically well defined and the SDE (8.149) is meaningful as an ordinary SDE in the sense of Sect. 8.2. In keeping with the spirit of this chapter, we shall assume in the following that ζ_t is colored noise. Independently of the mathematical impossibility to define a nonlinear function of white noise, it remains of course true that in most applications the environment varies on a much faster time scale than the system. In other words, ζ_t is a process with a short correlation time. The validity of the inequality (3.8) is of course in no way influenced by the fact that the external noise affects the system in a nonlinear way. It is therefore natural to look for a possibility to define an appropriate white-noise limit of the SDE (8.149). The obvious short cut is to write $\mu(\lambda) + \tilde{\zeta}_t$ instead of $\mu(\lambda + \zeta_t)$ and then pass to the white-noise idealization by setting $\tilde{\zeta}_t = \sigma\zeta_t$. This approach, employed to describe nematic liquid crystals [8.19] then ends up with a S SDE

$$dX_t = [h(X_t) + \mu(\lambda)g(X_t)]\,dt + \sigma g(X_t) \circ dW_t\,. \tag{8.151}$$

This is not, however, a legitimate procedure. If the environmental fluctuations are, for instance, Gaussian distributed, the common situation as explained in Chap. 1, then $\mu(X_t)$ is a *non-Gaussian* process. To replace $\mu(\lambda_t)$ by $\mu(\lambda) + \sigma\zeta_t$ is an arbitrary procedure, which neglects the physics specific to the problem at hand. There is no solid connection between the results obtained on the basis of the SDE (8.151) and the physical reality they purport to describe. In fact, such results may even be misleading [8.10, 20].

Following the discussion in Sect. 8.1 and for the sake of concreteness, we shall again consider that the external noise ζ_t is given by the O-U process. To explore the behavior of a system subjected to rapid external noise, the bandwidth perturbation expansion (Sect. 8.4) as well as the approximative evolution operator technique (Sect. 8.6) are available. The latter was recently used by *San Miguel, Sancho* [8.9, 10] to address the problem of quadratic external Gaussian noise. In the following, we shall only present the bandwidth perturbation expansion, which in our opinion is more advantageous in dealing with SDE's like (8.149). It is a systematic perturbation expansion in the correlation time; it is also very transparent and easy to employ in the present situation. Defining the external noise process η_t by

$$\begin{aligned}\eta_t &= \mu(\lambda + \zeta_t) - E\{\mu(\lambda + \zeta_t)\} \\ &= \mu(\lambda + \zeta_t) - m(\lambda, \sigma^2)\,,\end{aligned} \tag{8.152}$$

we can write (8.149) as

$$dX_t = [h(X_t) + m(\lambda, \sigma^2) g(X_t)]\, dt + g(X_t)\, \eta_t dt\,. \tag{8.153}$$

Speeding up the O-U process according to the general procedure discussed in Sect. 8.4, i.e., $\eta_t^\varepsilon = \eta_{t/\varepsilon^2} = \mu(\lambda + \zeta_{t/\varepsilon^2}) - m(\lambda, \sigma^2)$, to take account of the fact that we are particularly interested in the white-noise neighborhood, we obtain completely similar to (8.39),

$$dX_t^\varepsilon = [h(X_t^\varepsilon) + m(\lambda, \sigma^2) g(X_t^\varepsilon)]\, dt + \frac{1}{\varepsilon} g(X_t)\, \eta_t^\varepsilon dt\,, \tag{8.154}$$

$$d\zeta_t = -\frac{1}{\varepsilon^2} \zeta_t dt + \frac{\sigma}{\varepsilon} dW_t\,. \tag{8.155}$$

Recall that for the convenience of calculations time is rescaled, $t \to t/\tau_{\text{cor}}$, such that the original bandwidth of the colored noise is $\gamma = 1$. It goes without saying that the pair process $(X_t^\varepsilon, \zeta_t)$ is a diffusion process and its transition probability density is governed by a FPE like (8.45), where now

$$F_1 = \partial_z z + \frac{\sigma^2}{2} \partial_{zz}\,, \tag{8.156}$$

$$F_2 = -[\mu(\lambda + z) - m(\lambda, \sigma^2)]\, \partial_x g(x)\,, \tag{8.157}$$

$$\begin{aligned} F_3 &= -\partial_x [h(x) + m(\lambda, \sigma^2) g(x)]\,, \\ &= -\partial_x f(x)\,. \end{aligned} \tag{8.158}$$

The procedure is completely analogous to that of the linear noise case. Writing again

$$p^\varepsilon(x, z, t) = p^0(x, z, t)[1 + r_1(x, z, t) + \varepsilon^2 r_2(x, z, t) + \ldots] \tag{8.159}$$

and following the same steps as in Sect. 8.4, we obtain

$$F_1 p_0(x, z, t) = 0\,, \tag{8.160}$$

which similar to (8.55), implies

$$p_0(x, z, t) = p_0(x, t) p_s(z)\,. \tag{8.161}$$

Here $p_s(z)$ is given by (8.41) and $p_0(x, t)$ has to be determined by the Fredholm alternative for the second order in ε. For the first order in the expansion parameter we have

$$F_1^+ r_1(x, z, t) = [\mu(\lambda + z) - m(\lambda, \sigma^2)]\, \partial_x g(x) p_0(x, t)\,. \tag{8.162}$$

In this order the Fredholm alternative is, as in the linear noise case, trivially satisfied, since

$$E\{\mu(\lambda+z) - m(\lambda,\sigma^2)\} = 0\,. \tag{8.163}$$

The solution of (8.162) is given by

$$r_1(x,z,t) = H_1(x,t) - \tilde{\mu}(\lambda+z)\,\partial_x g(x) p_0(x,t)\,,$$

where $\tilde{\mu}(\lambda+z)$ is a particular solution of

$$F_1^+\tilde{\mu} = -\,[\mu(\lambda+z) - m(\lambda,\sigma^2)]\,. \tag{8.164}$$

As shown below, $\tilde{\mu}$ is easily calculated in applications where, for instance, μ is a polynomial in z. Passing on to the second order in the perturbation scheme, we have

$$F_1^+ r_2(x,z,t) = [\mu(\lambda+z) - m(\lambda,\sigma^2)]\,\partial_x g(x)\,. \\ [H_1(x,t) - \tilde{\mu}\partial_x g(x)\,p_0(x,t)] + (\partial_x f(x) + \partial_t)p_0(x,t)\,. \tag{8.165}$$

The Fredholm alternative applied to the rhs yields the following FPE for $p_0(x,t)$:

$$-\frac{\tilde{\sigma}^2}{2}\partial_x g(x)\,\partial_x g(x) p_0(x,t) + [\partial_x f(x) + \partial_t]\,p_0(x,t) = 0\,, \tag{8.166}$$

where

$$\tilde{\sigma}^2 = 2\int_{\mathbb{R}} dz p_s(z)\,[\mu(\lambda+z) - m(\lambda,\sigma^2)]\,\tilde{\mu}(\lambda+z)\,. \tag{8.167}$$

Writing the FPE (8.166) in the standard form

$$\partial_t p_0(x,t) = -\,\partial_x\left[f(x) + \frac{\tilde{\sigma}^2}{2}g'(x)\,g(x)\right]p_0(x,t) \\ + \frac{\tilde{\sigma}^2}{2}\partial_{xx}g^2(x)p_0(x,t)\,, \tag{8.168}$$

we conclude that the following SDE is the appropriate white-noise limit in the case of nonlinear external noise:

$$dX_t = [h(X_t) + m(\lambda,\sigma^2)g(X_t)]\,dt + \tilde{\sigma}g(X_t)\circ dW_t\,. \tag{8.169}$$

Note that the nonlinearity of the external noise has two consequences: (i) In addition to the noise-induced drift $\tilde{\sigma}^2 g'(x)\,g(x)/2$, which is also present in the linear case, the drift is further modified due to the fact that for $\sigma^2 \neq 0$, $m(\lambda,\sigma^2) \neq \mu(\lambda)$. In general it is only in the limit σ^2 tending to zero that the two expressions are equal: $m(\lambda, 0) = \mu(\lambda)$. (ii) The intensity of the white noise depends on the mean value of the external parameter, as is clear from (8.167). Obviously, in a way completely similar to that described in Sect. 8.4, higher-order corrections to the white-noise limit (8.168) can be obtained in a systematic manner. For the sake of concreteness, we shall now apply these general results to the case where $\mu(\lambda) = \lambda^2$, which is of interest in applications. Evaluating the various expressions, we obtain

$$\eta_t = \lambda^2 + 2\lambda\zeta_t + \zeta_t^2 - \lambda^2 - \sigma^2/2 \tag{8.170}$$

$$= \zeta_t^2 + 2\lambda\zeta_t - \sigma^2/2\,, \tag{8.171}$$

$$m(\lambda, \sigma^2) = \lambda^2 + \sigma^2/2 \tag{8.172}$$

$$F_1^+\,\tilde{\mu}(\lambda + z) = -(z^2 + 2\lambda z - \sigma^2/2)\,, \tag{8.173}$$

which implies

$$\tilde{\mu}(\lambda + z) = \tfrac{1}{2}(z^2 + 4\lambda z - \sigma^2/2)\,, \tag{8.174}$$

and

$$\tilde{\sigma}^2 = \frac{\sigma^4}{2} + 4\lambda^2\sigma^2\,. \tag{8.175}$$

Thus we can conclude that the system

$$\dot{X}(t) = h(X(t)) + \lambda^2 g(X(t)) \tag{8.176}$$

is described in a rapidly fluctuating environment by the diffusion process which obeys the SDE

$$dX_t = \left[h(X_t) + \left(\lambda^2 + \frac{\sigma^2}{2}\right) g(X_t)\right] dt + \left(\frac{\sigma^4}{2} + 4\lambda^2\sigma^2\right)^{1/2} g(X_t) \circ dW_t\,. \tag{8.177}$$

The same result is obtained by using the approximate evolution operator technique [8.10]. The particular features of nonlinear noise are the contribution

$$\frac{\sigma^2}{2} g(x) = E\{\zeta_t^2\} g(x)$$

to the drift and the modification of the white-noise intensity σ by the factor $(\sigma^2/2 + 4\lambda^2)^{1/2}$ which depends on the deterministic value of the external parameter, i.e., λ, as well as on the nonzero mean value of the fluctuations, i.e., $\sigma^2/2 = E\{\zeta_t^2\}$.

8.7.2 The Freedericksz Transition in Nematic Liquid Crystals

A nematic liquid crystal is a liquid consisting of elongated molecules, i.e., little rods, which on the average align parallel to a prefered direction. This long-range orientational order is responsible for the crystalline properties of the nematic state. However, the position of the molecules is not fixed in contrast to a crystal. This absence of long-range translational order is at the origin of the fluidity of the nematic phase. If the nematic liquid is enclosed between two parallel plates, the preferred direction can be prescribed by treating the surface of the plates. Rubbing the surface along a fixed direction will create grooves and the nematic molecules will lie parallel to the rubbed direction. As far as the elastic energy of the material near the surface is concerned, this is the most favorable situation. The extra energy required to leave the alignment parallel to the grooves is quite appreciable and leads to a firm anchoring of the director near the surface [8.21]. The director is a vector $\boldsymbol{n}$ chosen to represent the direction of preferred orientation of the molecules in the neighborhood of any point. Due to the elongated structure of the molecules, the influence of a magnetic field $\boldsymbol{H}$ or electric field $\boldsymbol{E}$ depends on the direction of the fields relative to $\boldsymbol{n}$. For instance, the susceptibility $\chi_{\parallel}$ for $\boldsymbol{H}$ parallel to $\boldsymbol{n}$ is larger than the susceptibility $\chi_{\perp}$ for $\boldsymbol{H}$ perpendicular to $\boldsymbol{n}$. This anisotropy,

$$\chi_{\mathrm{a}} = \chi_{\parallel} - \chi_{\perp} \,,$$

gives rise to an excess magnetization in the direction of $\boldsymbol{n}$, which in turn produces a magnetic torque [8.22]. Thus an externally applied magnetic or electric field tries to distort the nematic liquid, or in other words change the direction of $\boldsymbol{n}$.

Let us now consider more specifically the following setup, illustrated in Fig. 8.1. We choose the x axis to coincide with the direction along which the two parallel confining plates have been rubbed. The z axis is the axis perpendicular to the plates and thus the y axis is parallel to the plates and perpendicular to the x axis. In the absence of any external influence, the director $\boldsymbol{n}$ in the whole nematic layer is parallel to the x axis. In the following, it will be convenient to represent the preferred direction of the molecules by the angle θ which $\boldsymbol{n}$ forms with the x axis. If a magnetic field H is applied parallel to the y axis, it will create a torque on the rodlike molecule, which will tend to align the molecules parallel to H.

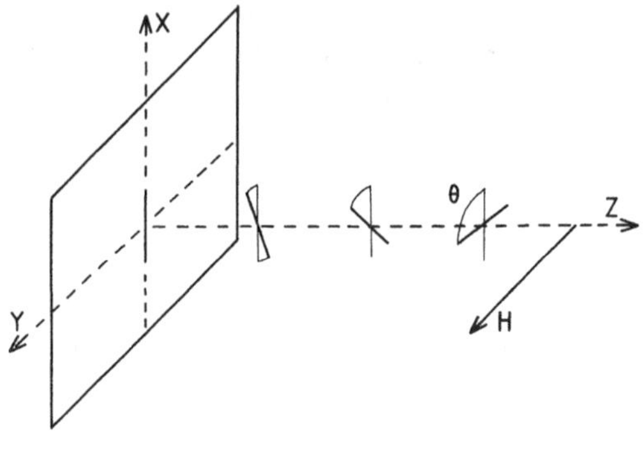

Fig. 8.1. Twist mode: on the wall ($z = 0$) the molecules are aligned parallel to the x axis; in the bulk, due to the magnetic field parallel to y, their director is twisted to form an angle θ with respect to x. Angle θ increases with the distance z from the wall and the intensity H of the magnetic field

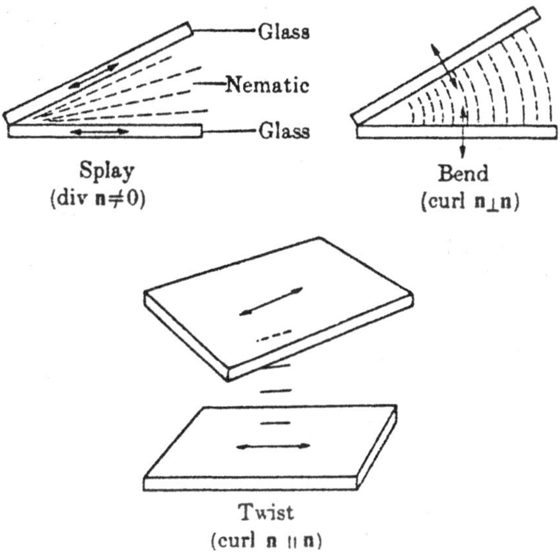

Fig. 8.2. Types of deformations observed in nematics. The figure shows how each type may be obtained separately by suitable glass walls [8.24]

However, if the surfaces of the end plates have been properly treated, the anchoring of the molecule is quite firm and almost impossible to overcome by a magnetic field [8.21]. Therefore, the director will remain parallel to the x axis near the lower and upper plate. Calling the position of the lower plate $z = -d/2$ and that of the upper plate $z = d/2$, we have

$$\theta(\pm d/2) = 0 \tag{8.178}$$

for all values of H. Consequently, in the layer the magnetic torque is counteracted by an elastic torque stemming from the preferred parallel alignment of the molecules and a viscous torque opposing any rapid rotation of the director. Balancing the torques, one obtains [8.23]

$$\lambda_1 \partial_t \theta = k_{22} \partial_{zz} \theta + \chi_a H^2 \sin\theta \cos\theta\,, \tag{8.179}$$

where λ_1 is the twist viscosity (nematics are anisotropic liquids) and k_{22} is an elastic constant, namely the twist modulus. The simplicity of the theoretical treatment of the above situation is due to the fact that we deal with a twist deformation. In contrast to the splay or bend deformation (Fig. 8.2), the center of gravity of the molecules does not undergo any translational motion; no flow effects have to be taken into account. Furthermore, the magnetic field remains constant across the nematic slab even if the latter is distorted, since the anisotropic susceptibility is small. This allows us to neglect any modification of the magnetic field due to the spatial structure in the nematic that develops eventually.

Before we investigate the influence of a fluctuating magnetic field on a nematic layer, it is of course necessary to get acquainted with the results of the deterministic case. For the steady-state

$$0 = k_{22}\partial_{zz}\theta + \chi_a H^2 \sin\theta \cos\theta \,. \tag{8.180}$$

Obviously, $\theta(z) \equiv 0$ is a solution of (8.180) and it prevails for $H = 0$. Our task is to determine the stability of the trivial homogeneous steady-state $\theta(z) = 0$, once the magnetic field H is applied. As mentioned before, $\theta(\pm d/2) = 0$ for all values of H. Thus the admissible class of spatically inhomogeneous perturbations $\hat{\theta}(z,t)$ is given by:

$$\hat{\theta}(z,t) = \hat{\theta}_0 \exp(\omega_\mu t) \cos[(2\mu+1)\pi z/d] \,, \qquad \mu = 0,1,2,\ldots \tag{8.181}$$

with

$$|\hat{\theta}_0| \ll 1 \,.$$

Inserting (8.181) into (8.179) and taking into account that

$$\sin\theta \cos\theta \approx \theta$$

for small θ, we obtain

$$\omega_\mu = (1/\lambda_1)[\chi_a H^2 - k_{22}(2\mu+1)^2(\pi^2/d^2)] \,.$$

The μ-th mode becomes unstable for $\omega_\mu \geqslant 0$, i.e.,

$$H^2 \geqslant (k_{22}/\chi_a)(\pi^2/d^2)(2\mu+1)^2 \,. \tag{8.182}$$

This implies that at

$$H_c^2 = (k_{22}/\chi_a)(\pi^2/d^2) \tag{8.183}$$

the homogeneous steady state $\theta(z) \equiv 0$ becomes unstable against perturbations of the form $\cos \pi z/d$. This transition phenomenon is known as the *Freedericksz transition.* Only if the value of the applied magnetic field is larger than the critical value H_c can the external field twist the director in the nematic layer away from the preferred direction imposed by the plates. The state of the system becomes spatially inhomogeneous. For $H_c < H < 3H_c$ only the lowest spatial mode is unstable and will dominate the spatial behavior in the neighborhood of the threshold condition (8.183). To describe the state of the nematic, we can neglect higher spatial modes and write

$$\theta(z,t) = \theta_m(t)\cos(\pi z/d) \,. \tag{8.184}$$

The maximum deformation $\theta_m(t)$ occurs for symmetry reasons at $z = 0$. For not too large values of θ_m, i.e., in the vicinity of the critical point, (8.179) then yields:

$$\lambda_1 \dot{\theta}_m(t) = -k_{22}(\pi^2/d^2)\theta_m(t) + \chi_a H^2(\theta_m - \tfrac{1}{2}\theta_m^3) \tag{8.185}$$

with

$$\sin\theta\cos\theta \simeq \theta - \tfrac{1}{2}\theta^3 + \ldots \,. \tag{8.186}$$

Note that (8.185) resembles the (generalized) Verhulst equation. The essential difference is that the external parameter, i.e., the magnetic field H, appears in a nonlinear way in the phenomenological equation.

For the stationary state of (8.185) we have:

$$\theta_{\mathrm{m}} = 0 \tag{8.187}$$

and

$$\theta_{\mathrm{m}} = \pm [-k_{22}(\pi^2/d^2) + \chi_{\mathrm{a}} H^2]^{1/2} \quad \text{for} \quad H \geqslant H_{\mathrm{c}} .$$

Note that the size of the system has a direct influence on the threshold value of the Freedericksz transition, namely,

$$H_{\mathrm{c}}^2 \sim d^{-2} .$$

The lowering of the critical field strength by increasing the sample thickness is qualitatively easily understood: the magnetic torque induced by the external field is counter-balanced by an elastic torque, which has its ultimate origin in the anchoring of the director at the two end plates.

Since the dielectric constant of a nematic is strongly anisotropic, the Freedericksz transition can also be induced by applying an electric field across the slab. However, compared to the magnetic case two complicating factors enter the picture. First, the electric field is in general not uniform across the sample. The distortion of the nematic layer gives rise to a z-dependent electric field $E(z)$. Only in the limit of small distortions, i.e., near the critical threshold, is $E(z)$ nearly constant. This implies that for electric field strengths below and slightly above the critical strength the problem completely resembles the Freedericksz transition in a magnetic field. It can be described by an equation similar to (8.185). To go beyond the vicinity of the critical point requires not only that the nonuniformity of E is taken into account but also, secondly, that the distinction between a conducting nematic and an insulating nematic is made. In the former case, the Freedericksz transition can become a first-order transition with hysteresis effects [8.22]. The situation is further complicated by the fact that in this case conduction-induced instabilities can occur [8.21].

After this digression on electric fields, let us now study the influence of a fluctuating magnetic field on a nematic slab and determine the way in which the Freedericksz transition is modified. As already stressed, the external parameter enters nonlinearly in this problem, namely the square of the magnetic field appears in (8.179). We are thus faced with a typical application of the theoretical methods developed in the preceding section. We shall restrict considerations to external fluctuations whose strength is not too large, returning to this point later on. In this case, the assumptions leading to the phenomenological equation (8.185) remain valid. Thus (8.185) can serve as a legitimate starting point to study the influence of external noise on the Freedericksz transition. For reasons expounded earlier, we shall again represent the external noise by an O-U process and we have, rescaling the time $t \to \gamma t$ and introducing $\alpha = \tau_{\mathrm{cor}} k_{22} \pi^2/(\lambda_1 d^2)$ and $\beta = \tau_{\mathrm{cor}} \chi_{\mathrm{a}}/\lambda_1$,

$$\begin{aligned}\dot{\theta}_{\mathrm{m}} &= -\alpha\theta_{\mathrm{m}} + \beta(H+\zeta_t)^2(\theta_{\mathrm{m}} - \tfrac{1}{2}\theta_{\mathrm{m}}^3) \\ &= [-\alpha + \beta(H+\zeta_t)^2]\,\theta_{\mathrm{m}} - \tfrac{1}{2}\beta(H+\zeta_t)^2\theta_{\mathrm{m}}^3 .\end{aligned} \tag{8.188}$$

According to Sect. 8.7.1, in particular (8.176, 177), the white-noise limit of (8.188) is given by

$$\begin{aligned}d\theta_{\mathrm{m}} &= \left[-\alpha\theta_{\mathrm{m}} + \beta\left(H^2 + \frac{\sigma^2}{2}\right)\left(\theta_{\mathrm{m}} - \frac{1}{2}\theta_{\mathrm{m}}^3\right)\right] dt \\ &\quad + \beta\left(\frac{\sigma^4}{2} + 4H^2\sigma^2\right)^{1/2}\left(\theta_{\mathrm{m}} - \frac{1}{2}\theta_{\mathrm{m}}^3\right) \circ dW_t .\end{aligned} \tag{8.189}$$

As in the Verhulst equation, $\theta_{\mathrm{m}} = 0$ is a boundary of the state space and a stationary point. To be precise, the state space $(-\pi/2, +\pi/2)$ is split into two noncommunicating halves by the intrinsic boundary $\theta_{\mathrm{m}} = 0$. The stationary point loses its stability at the same time when the nature of the boundary changes, namely when 0 becomes a natural boundary. Using the analytical condition (5.89),

$$L_1(0) = \infty ,$$

we find that $\theta_{\mathrm{m}} = 0$ loses stability at

$$H^2 = H_{\mathrm{c}}^2 - \sigma^2/2 . \tag{8.190}$$

This is an expected result. Nonlinear noise changes the mean value of the external parameter. Here it increases it from H^2 to $H^2 + \sigma^2/2$. This effect is trivially associated with any *nonlinear* noise. In our example, the noise situation with H and $E\{\zeta_t^2\} = \sigma^2/2$ corresponds to the deterministic case with H replaced by $(H^2 + \sigma^2/2)^{1/2}$. This explains the lowering of the threshold as given by (8.190). Note that no size effect appears. Independent of the thickness of the layer the fluctuations of the magnetic field decrease the threshold by their variance. Equation (8.190) seems to predict that the stationary point $\theta_{\mathrm{m}} = 0$, corresponding to the homogeneous steady-state, can be destabilized with an arbitrarily small external field if the fluctuations are strong enough. Remember, however, that according to the deterministic stability analysis the lowest spatial mode is expected to be dominant only for $H_{\mathrm{c}} < H < 3H_{\mathrm{c}}$. Therefore, (8.185), upon which our analysis is based, ceases to be valid when the external noise becomes strong enough to excite higher spatial modes. The noise intensity at which this occurs should ideally be determined by an analysis of the partial differential equation (8.179) or, equivalently, by an analysis of the hierarchy of equations of the mode amplitudes $\theta_{\mathrm{m},\mu}(t)$. Both problems, even if the second one is truncated to only two modes, are analytically intractable at this stage in the development of the theory of SDE's. However, an estimate on the upper bound of the noise intensity can be obtained in the following way. If the deterministic magnetic field strength is in the neighborhood of H_{c}, the probability that $H + \zeta_t$ exceeds $3H_{\mathrm{c}}$ should be negligible. This implies that $H+3\sigma<3H_c$ for $\text{Prob}[(H+\zeta_t)\leqslant 3H_c]=0.99865$. We obtain as an order of magnitude estimate that roughly

$$3\sigma<H_c \quad \text{or} \quad \sigma^2<\sigma_1^2\simeq 0.1\,H_c^2\,, \tag{8.191}$$

as the condition that no higher modes are excited by the noise.

Let us now turn our attention to the second noise-induced transition point, where the most probable value for θ_m becomes nonzero. We have the following equation for the extrema of the stationary probability density:

$$\begin{aligned}&-\alpha\theta_m+\beta\left(H^2+\frac{\sigma^2}{2}\right)\left(\theta_m-\frac{1}{2}\theta_m^3\right)\\&-\frac{1}{2}\beta^2\left(\frac{\sigma^2}{2}+4H^2\sigma^2\right)\left(1-\frac{3}{2}\theta_m^2\right)\left(\theta_m-\frac{1}{2}\theta_m^3\right)=0\,,\end{aligned} \tag{8.192}$$

or

$$\theta_m=0 \tag{8.193}$$

and

$$\begin{aligned}&-\alpha+\beta\left(H^2+\frac{\sigma^2}{2}\right)\left(1-\frac{1}{2}\theta_m^2\right)\\&-\frac{1}{2}\beta^2\left(\frac{\sigma^4}{2}+4H^2\sigma^2\right)\left(1-\frac{3}{2}\theta_m^2\right)\left(1-\frac{1}{2}\theta_m^2\right)=0\,.\end{aligned} \tag{8.194}$$

The Freedericksz transition is characterized by the fact that $\theta_m=0$ is a critical point, i.e., a triple root of (8.192). This occurs for

$$-\alpha+\beta\left(H^2+\frac{\sigma^2}{2}\right)-\frac{1}{2}\beta^2\left(\frac{\sigma^4}{2}+4H^2\sigma^2\right)=0\,, \tag{8.195}$$

implying

$$\begin{aligned}H^2&=\frac{1}{(1-2\beta\sigma^2)}\left(\frac{\alpha}{\beta}+\frac{\beta\sigma^4}{4}-\frac{\sigma^2}{2}\right)\\&=\left(H_c^2+\frac{\beta\sigma^4}{4}-\frac{\sigma^2}{2}\right)\Big/(1-2\beta\sigma^2)\,.\end{aligned} \tag{8.196}$$

This predicts that the transition will disappear if the intensity of the noise becomes too large, namely at

$$2\beta\sigma_*^2 = 1\,,$$

or

$$\sigma_*^2 = \frac{1}{2\beta} = \frac{H_c^2}{2\alpha}$$
$$= \frac{1}{2}\,\frac{\lambda_1 d^2}{k_{22}\pi^2\tau_{cor}} H_c^2\,. \tag{8.197}$$

Remember that σ^2 has to fulfill condition (8.191). Otherwise our approach breaks down, since higher modes would be excited by the noise. Comparing the order of magnitude estimate (8.191) with (8.197), we notice that σ_*^2 is proportional to H_c^2. In contrast to (8.191), however, the proportionality factor depends on the thickness of the sample, to be precise on d^2. This proportionality factor plays the role of a decay rate for the deterministic situation. If the field is switched off from $H > H_c$ to zero, the distortion of the nematic layer decays with the characteristic time

$$\tau_{macro} = (\lambda_1 d^2/k_{22}\pi^2)\cdot(1/\tau_{cor})$$
$$= \tau(0)/\tau_{cor}\,. \tag{8.198}$$

The typical value of $\tau(0)$ for a nematic layer of 25 μm is about 0.1 s according to *Chandrasekhar* [Ref. 8.21, p. 151]. Since the thickness enters in a quadratic way in $\tau(0)$, the decay time increases rapidly with the thickness of the sample. For the white noise limit to be physically meaningful, the bandwidth of the original colored noise must be much larger than the "characteristic frequency" of the system τ_{macro}^{-1}, i.e. $\gamma \gg \tau(0)^{-1}$ or $\tau_{cor} \ll \tau(0)$. Only if this inequality is fulfilled, can we replace the original colored noise problem by a white noise idealization without qualitatively changing the physics. This implies that in the domain of validity of the white noise approximation $\sigma_*^2 = \tau(0)H_c^2/(2\tau_{cor})$ is orders of magnitude larger than $\sigma_1^2 \simeq 0.1\,H_c^2$. The disappearance of the Freedericksz transition at $\sigma^2 = \sigma_*^2$ is thus outside the validity of our theoretical description (8.188). For $\sigma^2 > \sigma_1^2$, we expect that higher spatial modes are excited. This would invalidate (8.188) and consequently (8.196). As is clear from the above remarks, the validity of the inequality $\tau_{cor} \ll \tau(0)$, and thus the quality of the white noise approximation depends on the bandwidth of the noise used and on the sample size. For $\sigma^2 < \sigma_1^2 \ll \sigma_*^2$ we have

$$H^2 = \left(H_c^2 + \beta\frac{\sigma^4}{4} - \frac{\sigma^2}{2}\right)[1 + 2\beta\sigma^2 + O(\sigma^4)]$$
$$\simeq H_c^2 + \frac{\sigma^2}{2}(4\beta H_c^2 - 1) \tag{8.199}$$

$$= H_c^2 + \frac{\sigma^2}{2}\left(4\,\frac{k_{22}\pi^2\tau_{cor}}{\lambda_1 d^2} - 1\right) = H_c^2 + \frac{\sigma^2}{2}\left(4\,\frac{\tau_{cor}}{\tau(0)} - 1\right) \tag{8.200}$$

$$H^2 = H_c^2 - \frac{\sigma^2}{2} + 2\sigma^2 \frac{\tau_{\text{cor}}}{\tau(0)} + O(\sigma^4) . \tag{8.201}$$

Note that this is the white-noise idealization to the colored noise problem despite the explicit appearance of the correlation time or bandwidth of the colored noise. Comparing the result (8.201) with (8.190) we conclude that in the white-noise idealization of a nonlinear colored noise with a large bandwidth γ, the two transitions of the Verhulst model, namely where the stationary point $\theta = 0$ becomes unstable and where the most probable value becomes non-zero, almost coincide. This is due to the fact that the modification of the drift by the term $E\{\zeta^2\}g(X_t) = (\sigma^2/2)g(X_t)$ dominates the effects which arise from the multiplicative nature of the noise. This feature of nonlinear noise was first found by *San Miguel* and *Sancho* [8.10]. This property is also responsible for the fact that any size effects are negligible for the noise-induced transition. Only the small correction term containing $\tau(0)$ depends on the thickness of the sample.

These theoretical predictions clearly call for an experimental investigation of the Freedericksz transition in a fluctuating magnetic field. In particular, an experimental test of (8.196) and the absence of a size effect, (8.201), would be very welcome. Furthermore, experimental data on the possibility of random excitation of higher spatial modes are necessary for further progress in the modelling of this system. The Freedericksz transition is in our opinion a very promising system, experimentally as well as theoretically, to gain a first insight into the interplay of spatial inhomogeneity and external noise. Experimental studies on the effect of noise on nematic layers have been carried out, however, for historical reasons on transitions of a more complicated nature than the Freedericksz transition. These experiments will be discussed in the next section.

8.7.3 Electrohydrodynamic Instabilities and External Noise

Two Japanese groups [8.25, 26] have performed experiments on so-called electrohydrodynamic transitions in nematics. These transitions, though they share certain features with the Freedericksz transition, are of a more complicated nature since they involve flow effects. They are similar to the well-known Rayleigh-Bénard problem of a thin fluid layer heated from below [8.27 – 34].

If an ac voltage is applied to a nematic sample, the following transition phenomena are observed. If for a fixed frequency, which is sufficiently low to be in the so-called conduction regime [8.35], the amplitude of the ac voltage is increased, the rest state of the liquid crystal becomes unstable above a critical voltage V_c. Convection sets in and a roll pattern, a two-dimensional flow, is observed, as illustrated in Fig. 8.3. In electrohydrodynamics this regime is known as the Williams domain [8.36]. Increasing the electric field still further, a succession of different flow patterns is observed, till a transition to a turbulent flow, the "dynamic scattering mode", takes place [8.35]. The different flow patterns can easily be visualized due to the anisotropy of the refractive index of nematics. In the Williams domain, for instance, dark and light bands corresponding to the rolls alternate.

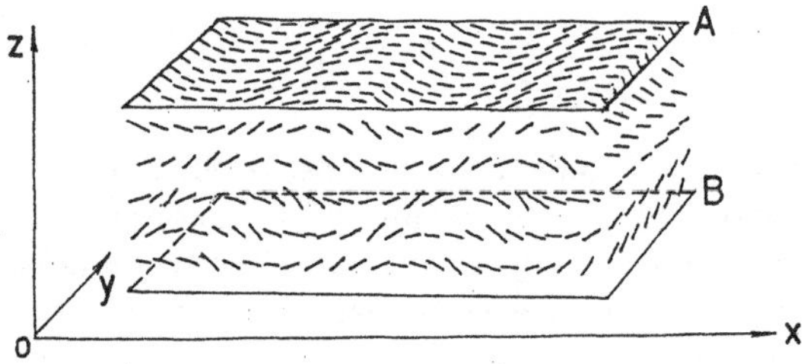

Fig. 8.3. Sketch of the type of patterns obtained in the Williams domain with a nematic layer enclosed between two parallel plates A and B, and submitted to an ac voltage parallel to the z axis

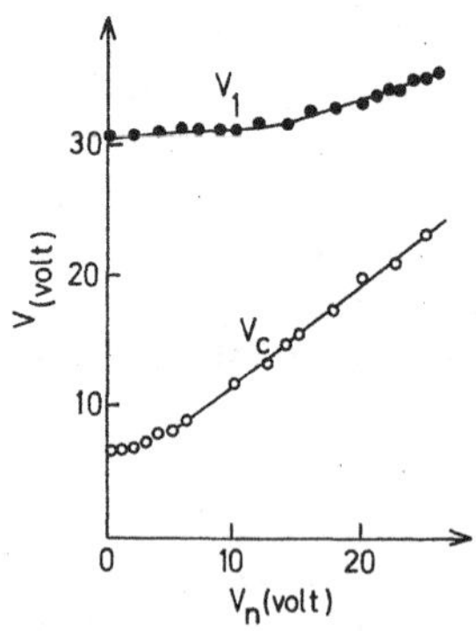

Fig. 8.4. Shift of the critical points corresponding to the onset of Williams domain V_c and turbulence V_1 as a function of the voltage noise intensity [8.25]

In the particular experiments of *Kai* et al. [8.25] and *Kawakubo* et al. [8.26], the effect of externally applied voltage noise on the transition to the Williams domain in the nematic MBBA was investigated. The frequency of the ac voltage is 25 Hz (Kai) and 50 Hz (Kawakubo). The external quasi-white noise V_t^n has a flat spectrum from 20 to 20 kHz (Kai) and up to 5 kHz (Kawakubo), respectively. For further details on the experimental procedures the reader is invited to consult the original references. The experimental results of both groups agree and are represented in Fig. 8.4.

The theoretical models adopted by the two groups are quite different and will be discussed separately in the following. The electrohydrodynamic description of transition phenomena in nematics is rather complicated; liquid crystals are anisotropic fluids. Kai and co-workers argue that in the immediate vicinity of the critical point flow effects can be neglected. The situation then resembles the Freedericksz transition induced by an electric field. Retaining only the lowest spatial mode

$$\theta = \theta_{\mathrm{m}} \cos(\pi z/d) \cos(2\pi k_x x) \tag{8.202}$$

and balancing the viscous, elastic and electric torques, Kai and co-workers obtain the following equation:

$$\dot{\theta}_{\mathrm{m}} = \frac{\varepsilon_{\mathrm{a}}}{4\pi\gamma_1 d^2}(V^2 - V_{\mathrm{c}}^2)\,\theta_{\mathrm{m}} - \frac{\varepsilon_{\mathrm{a}} V^2}{6\pi\gamma_1 d^2}\,\theta_{\mathrm{m}}^3\,. \tag{8.203}$$

Here θ is the director angle with respect to the x axis, k_x the wave number in the x direction, ε_a the anisotropic dielectric constant, γ_1 the anisotropic viscosity and

$$V_c^2 = (4\pi^3/\varepsilon_a)(k_{11} + 4k_{33}k_x^2 d^2)\,. \tag{8.204}$$

As was expected, (8.203) is completely similar to (8.185). *Kai* et al. [8.25] argue that the increase in the threshold value with increasing noise intensity can be understood by the fact that the wave number k_x increases with the applied frequency as well as the voltage. The external noise randomly excites many different rolls with higher wave number, leading to an enhanced small-scale convection and to a higher threshold value for V^2. These arguments seem to be partially supported by the experimental results of *Kawakubo* et al. [8.26], who determined the wave number spectral density. The strong peak, reflecting the onset of the Williams domain, is *not* appreciably broadened by the external noise, but is shifted from $k_x = 415\ \text{cm}^{-1}$ for $(E\{V_t^{n2}\})^{1/2} = 0$ to $k_x = 553\ \text{cm}^{-1}$ for $(E\{V_t^{n2}\})^{1/2} = 9.92$ V. However, an analysis of the data[2] shows that the increase in k_x might not totally account for the increase in the threshold value. This fact gains in importance in light of our analysis of the Freedericksz transition. We found that a second mechanism which *decreases* the threshold plays a role, namely a noise induced transition due to the interplay of the nonlinear dynamics of the system and the external noise. This mechanism operates independently from the change in wave numbers of the rolls. Transcribing the results for the Freedericksz transition to this case, we have, rescaling again the time, with

$$a = \tau_{\text{cor}}\,\varepsilon_a/(2\pi\gamma_1 d^2)$$

$$\dot{\theta}_m = \frac{a}{2}[(V+\zeta_t)^2 - V_c^2]\,\theta_m - \frac{a}{3}(V+\zeta_t)^2\theta_m^3\,, \tag{8.205}$$

and the white-noise limit is given by:

$$d\theta_m = a\left[\frac{1}{2}\left(V^2 + \frac{\sigma^2}{2} - V_c^2\right)\theta_m - \frac{1}{3}\left(V^2 + \frac{\sigma^2}{2}\right)\theta_m^3\right]dt$$
$$+\, a\left(\frac{\sigma^4}{2} + 4V^2\sigma^2\right)^{1/2}\left(\frac{1}{2}\theta_m - \frac{1}{3}\theta_m^3\right)\circ dW_t\,. \tag{8.206}$$

The stationary point $\theta_m = 0$ becomes unstable at

$$V^2 = V_c^2 - (\sigma^2/2)\,. \tag{8.207}$$

This is a trivial noise effect, already discussed for the Freedericksz transition, and again σ^2 has to be sufficiently small so that the excitation of higher modes is

[2] Unfortunately only three values of noise intensity are reported.

negligible. The experimental results of Kawakubo show that this is indeed the case in the experiments. Here k_x varies with σ^2, but the wave number spectral density clearly reveals that essentially only one spatial mode is excited; the peak is *not* broadened by external noise and no new peaks appear near the transition. The equation for the extrema of the stationary probability density reads:

$$\frac{1}{2}\left(V^2+\frac{\sigma^2}{2}-V_c^2\right)\theta_m-\frac{1}{3}\left(V^2+\frac{\sigma^2}{2}\right)\theta_m^3 - a\left(\frac{\sigma^4}{4}+2V^2\sigma^2\right)\left(\frac{1}{2}-\theta_m^2\right)\left(\frac{1}{2}\theta_m-\frac{1}{3}\theta_m^3\right)=0\,. \tag{8.208}$$

The threshold condition is obtained by the fact that $\theta_m = 0$ has to be a triple root:

$$\begin{aligned}&\frac{1}{2}\left(V^2+\frac{\sigma^2}{2}-V_c^2\right)-\frac{a}{4}\left(\frac{\sigma^2}{4}+2V^2\sigma^2\right)=0\,,\\ &V^2=\frac{V_c^2}{1-a\sigma^2}+\left(\frac{a\sigma^4}{8}-\frac{\sigma^2}{2}\right)\Big/(1-a\sigma^2)\,.\end{aligned} \tag{8.209}$$

As for the Freedericksz transition, we find that the noise intensity has to be smaller than σ_*^2 for a critical point to occur in this description;

$$\sigma_*^2 = 2\pi\gamma_1 d^2/\varepsilon_a\,. \tag{8.210}$$

Inspection of (8.203) shows that the zero field decay time is given by $\tau(0) = (4\pi\gamma_1 d^2)/(\varepsilon_a V_c^2)$ and thus

$$a = (2\tau_{cor})/(\tau(0)\,V_c^2)\,. \tag{8.211}$$

For $\sigma^2 \ll \sigma_*^2$ we obtain

$$V^2 \simeq V_c^2 - \sigma^2/2 + 2(\tau_{cor}/\tau(0))\,\sigma^2\,. \tag{8.212}$$

As is clear from (8.204), deterministically this transition is also characterized by a size effect. However, for the external noise case we find a phenomenon similar to the Freedericksz transition: The effect due to the nonlinearity of the noise dominates the effect due to the multiplicative nature of the noise. Again the two transition points almost coincide and size effects are negligible.

Kawakubo and co-workers base their theoretical explanation on the theory by *Dubois-Violette* et al. [8.37] (see also [8.21]) for electrohydrodynamic instabilities. The latter takes into account the nonuniformity of the electric field and the formation of space charges due to the distortion of the layer. Dubois-Violette et al. considered situations close to the instability and carried out a linear stability analysis. The state of the nematic layer is described in this theory by two variables, the space charge density q, and the local curvature of the molecular alignment, i.e., $\psi = \partial\theta/\partial x$. Neglecting all nonlinear terms, the following equations are obtained:

$$\dot{q} = -q(t)/\tau - \sigma_{\rm H} E\cos(\omega t)\cdot\psi(t)\,, \tag{8.213}$$

$$\dot{\psi} = -E\cos(\omega t)(q(t)/\eta) - \lambda[A + E^2\cos^2(\omega t)]\,\psi(t)\,. \tag{8.214}$$

Here τ is the dielectric relaxation time for the space charge, $\sigma_H = \sigma_{||}(\varepsilon_\perp/\varepsilon_{||} - \sigma_\perp/\sigma_{||})$ the Helfrich conductivity, η the effective viscosity, and λ and A are positive definite constants, determined by the dielectric, conducting, elastic and viscous properties of the nematic. The externally imposed ac field is $E\cos\omega_\tau$. Kawakubo et al. take (8.213, 214) as the starting point for their analysis of the influence of quasi-white voltage noise. Representing also the external noise by an O-U process, they replace $E\cos(\omega t)$ by $E\cos(\omega t) + \zeta_t$ in the above equations. The curvature ψ varies on a much larger time scale than the space charge q. For rapid external noise, namely $\tau_{\rm cor} \ll \tau$, ζ_t is averaged out in (8.213) and ψ is taken constant. This yields:

$$q(t) = -\frac{\sigma_{\rm H} E\tau\psi}{1+\omega^2\tau^2}(\cos\omega t + \omega\tau\sin\omega t)\,. \tag{8.215}$$

Injecting this expression in (8.214) and averaging over one period, Kawakubo et al. obtain:

$$\dot{\psi} = \left[\left(\frac{\sigma_{\rm H}\tau}{\eta(1+\omega^2\tau^2)} - \lambda\right)\frac{E^2}{2} - \lambda A - \lambda E\{\zeta_t^2\}\right]\psi\,, \tag{8.216}$$

furnishing the following condition for an instability to occur:

$$\frac{E^2}{2} > \frac{\lambda\eta(1+\omega^2\tau^2)}{\sigma_{\rm H}\tau - \lambda\eta(1+\omega^2\tau^2)}(A + E\{\zeta_t^2\})\,,$$

or of the voltage

$$V^2 > V_0^2 + \alpha E\{V_t^{n2}\} \tag{8.217}$$

with

$$V_0^2 = \frac{\lambda\eta(1+\omega^2\tau^2)A}{\sigma_{\rm H}\tau - \lambda\eta(1+\omega^2\tau^2)}d^2$$

$$V_t^n = \zeta_t d\,, \qquad \alpha = \frac{\lambda\eta(1+\omega^2\tau^2)}{\sigma_{\rm H}\tau - \lambda\eta(1+\omega^2\tau^2)}\,.$$

Note that this result is based on a linearized description of the system. As already emphasized for the Verhulst model in Sect. 6.4, a linear analysis can describe a transition due to a change in the nature of the boundary only. Thus the above linearized treatment of conduction instabilities can determine the stability of the stationary point $\psi = 0$, but it cannot determine the noise induced transition point, where the most probable value of ψ becomes nonzero. Equation (8.217) must therefore be compared with (8.207) and not with (8.209) or (8.212). As in the Freedericksz transition, *Kawakubo* et al. [8.26] found no size effect for the destabilization of the stationary point. However, their two variable model predicts an increase in the threshold with the variance of the fluctuations. This is obviously in agreement with the experimental results. In order to study theoretically the second noise-induced transition point, the nonlinear version of the *Dubois-Violette* and co-workers theory has to be used. Due to the fact that we have to deal then with a two-variable nonlinear diffusion process, it seems at present not feasible to obtain explicit analytical results. It is, however, unlikely that such an analysis would essentially change the theoretical predictions of the linear description. Since the nonlinearity of the noise is the dominating effect, the two transition points should almost coincide and we expect their separation to be proportional to τ_{cor}/τ as in the two previous models.

8.8 Turbulence and External Noise

Kai et al. [8.25] studied the influence of external noise not only on the transition to the Williams domain regime but also on the transition to the "dynamic scattering mode", the turbulent regime. Their experimental results are represented in Fig. 8.4 by the curve V_1. The outcome of this experiment is downright amazing: contrary to all intuition, noise *delays* the onset of turbulence. External noise, i.e., environmental stochasticity, *stabilizes* the system against a transition to chaotic behavior, i.e., intrinsic stochasticity. This result is even more astonishing if we take into account that we deal with nonlinear noise. Remember that $E\{V_t^2\} = V^2 + \sigma^2/2$, so that the effect trivially connected with quadratic noise would lead to a lowering of the threshold value by $\sigma^2/2$, analogous to (8.190, 207). Thus the experimental results clearly demonstrate the presence of a nontrivial, genuine noise effect.

In order to appreciate the full significance of these experimental findings, let us briefly summarize the presently known facts about the role of external noise in the different routes to chaos [1.29]. Of the three common scenarios, only the Ruelle-Takens scenario is not affected by (small) noise. This surprising result is demonstrated by *Kifer* [8.38]. Kifer proves, roughly speaking, that for systems with an Axiom A attractor the stationary probability of the system perturbed by colored external noise converges (weakly) to the invariant measure of the attractor as the intensity of the noise goes to zero. The question of external noise with finite, nonvanishing intensity remains open as well as the case of a non-Axiom A attractor. The influence of noise on the Lorenz strange attractor has been studied numerically [8.39]. The results indicate that here noise has the intuitively expected effect, namely it lowers the threshold of the transition to turbulence. The road to chaos via period doubling, the Feigenbaum scenario, and the road via intermittency, the Pomeau-Manneville scenario, are both influenced even by small noise. Again this is the intuitively expected behavior. In the Feigenbaum scenario high periods are wiped out leading to a "bifurcation gap" [8.40]. For a detailed study of the influence of noise on a system with a period-doubling sequence, namely the Verhulst or logistic model, see [8.41]. These authors also find that the behavior of the system in a deterministic periodic environment differs markedly from that in a noisy environment. For systems following the Feigenbaum scenario, scaling laws can be obtained for the influence of external noise at the onset of chaos [8.42, 43]. In systems which display intermittent behavior, the effect of noise was investigated by *Mayer-Kress* and *Haken* [8.44], *Eckmann* et al. [8.45], and *Hirsch* et al. [8.46]. The laminar time, i.e., the time between successive turbulent bursts, obeys a scaling law and decreases with increasing noise intensity. We can summarize this by saying that in the three commonly used scenarios for roads to chaos noise has either no influence or the expected effect of making the transition to turbulence easier. These scenarios offer therefore not even a qualitative understanding of the experimental findings of Kai and co-workers.

In the absence of a simple theoretical model for the transition to the "dynamic scattering mode" in nematic liquid crystals, an explanation of the counterintuitive behavior found by Kai et al. seems at the moment hard to come by. However, some light might be shed on the situation by comparison with another hydrodynamic system, where a simple but satisfactory phenomenological theory for the transition to turbulence exists, namely in the thermal counterflow of helium II. The growth of vortex line turbulence is very well described by the modified *Vinen* equation [8.47, 48]:

$$\dot{L} = Av\left(1 - \frac{\alpha}{d}L^{-1/2}\right)L^{3/2} - C\varkappa L^2 . \tag{8.218}$$

Here A, C and α are dimensionless constants ($\alpha \approx 1$), d is the tube diameter, $\varkappa$ the circulation quantum, v the counterflow velocity and L the vortex line density. For a detailed discussion of the phenomenology of the Vinen equation and its astonsihingly satisfactory description of superfluid turbulence see [8.48]. The counterflow velocity v is quite naturally a fluctuating quantity. This observation has recently led *Moss* and *Welland* [8.49] to use (8.218) for a theoretical analysis of the effect of noise in the counterflow velocity at the onset of superfluid turbulence. They found the remarkable result that noise will delay the onset of turbulence. In the rest of this section, we shall present their theoretical analysis in more detail, which interestingly points in the same direction as the experiment by Kai et al. Setting $\nu = v/v_c$ with $v_c = 4C\varkappa\alpha/Ad$, $x = L/\bar{L}(v_c)$, where $\bar{L}(v)$ is the steady state of (8.218), $\tau = 4C\varkappa(\alpha/d)^2 t$, we obtain (8.218) in dimensionless form

$$\frac{dX}{d\tau} = 2\nu X\left(X^{1/2} - \frac{1}{2}\right) - X^2 . \tag{8.219}$$

The steady-state solutions are $\bar{x}_1 = 0$, which is (linearly) stable for all $\nu > 0$, and $\sqrt{\bar{x}_{+/-}} = \nu[1 \pm \sqrt{1-(1/\nu)}]$ which exist for $\nu \geqslant 1$ and where $\bar{x}_+$ is stable and $\bar{x}_-$ unstable. The onset of turbulence occurs thus at $\nu = 1$, i.e., $v = v_c$. Moss and Welland consider the case of rapid fluctuations in the counterflow velocity and adopt the white-noise idealization $\nu_t = \sigma + \sigma\xi_t$. This yields the S SDE

$$dX_\tau = [2\nu X_\tau(X_\tau^{1/2} - \tfrac{1}{2}) - X_\tau^2]\,d\tau + 2\sigma X_\tau(X_\tau^{1/2} - \tfrac{1}{2}) \circ dW_\tau \,. \tag{8.220}$$

Equation (8.220) has three intrinsic boundaries, namely $b_1 = 0$ and $b_2 = 1/4$, where $g(x) = 0$, and $b_3 = \infty$, where $|f(x)| = \infty$. However, the middle boundary b_2 is nongeneric. As Moss and Welland remark, the Vinen phenomenology does not include any mechanism for the appearance of a nonzero line density in an initially circulation-free superfluid. This is quite natural: (8.218) is a deterministic equation, whereas the nucleation of vortex lines in a circulation-free superfluid is an essentially random event, caused for instance by vibrations in the cryostat. We shall model this mechanism by a second independent noise source and modify (8.220) by a (small) additive noise term:

$$dX_\tau = [2\nu X_\tau(X_\tau^{1/2} - \tfrac{1}{2}) - X_\tau^2]\,d\tau + 2\sigma X_\tau(X_\tau^{1/2} - \tfrac{1}{2}) \circ dW_\tau + \mu\, d\tilde{W}_\tau \,, \tag{8.221}$$

where W_τ and $\tilde{W}_\tau$ are independent Wiener processes. (For a different approach to remedy the shortcomings of (8.220) see [8.50].) Note that this destroys the intrinsic boundary $b_2 = 1/4$ (as well as $b_1 = 0$) and avoids the singularities connected with b_2. However, since the new noise term is additive, it in no way modifies the analysis done by Moss and Welland as far as the transition phenomena are concerned. To write down the FPE, let us change over to an Ito equation equivalent to (8.221):

$$\begin{aligned} dX_\tau = {} & [2\nu X_\tau(X_\tau^{1/2} - \tfrac{1}{2}) - X_\tau^2 + \tfrac{1}{2}\sigma^2 2X_\tau(X_\tau^{1/2} - \tfrac{1}{2})(3X_\tau^{1/2} - 1)]\,d\tau \\ & + 2\sigma X_\tau(X_\tau^{1/2} - \tfrac{1}{2})\,dW_\tau + \mu\, d\tilde{W}_\tau \,. \end{aligned} \tag{8.222}$$

Thus the FPE reads

$$\begin{aligned} \partial_\tau p(x,\tau) = {} & -\partial_x[2\nu x(x^{1/2} - \tfrac{1}{2}) - x^2 + \sigma^2 x(x^{1/2} - \tfrac{1}{2})(3x^{1/2} - 1)]\,p(x,\tau) \\ & + \tfrac{1}{2}\partial_{xx}[4\sigma^2 x^2(x^{1/2} - \tfrac{1}{2})^2 + \mu^2]\,p(x,\tau) \,. \end{aligned} \tag{8.223}$$

A word of caution is in order here. The noise-induced drift is

$$\tfrac{1}{2}\sigma^2 g' g = \sigma^2(3x^2 - \tfrac{5}{2}x^{3/2} + \tfrac{1}{2}x) \,. \tag{8.224}$$

Thus we have

$$f + \tfrac{1}{2}\sigma^2 g' g = (-1 + 3\sigma^2)x^2 + O(x^{3/2}) \,,$$

i.e., the noise-induced drift modifies the highest power in the phenomenological drift f; its coefficient $3\sigma^2 - 1$ will become positive for large σ^2. Clearly this would not happen if (8.221) were interpreted as an Ito equation.

This model is not robust against a change in interpretation of the SDE as far as the behavior near infinity is concerned. In the Stratonovich interpretation adopted here, $b_3 = \infty$ changes its character at $\sigma^2 = 1$ from natural to regular; thus the existence of a stationary probability density is not compromised. In a certain sense a "noise-induced runaway" occurs. This obviously does not happen if (8.221) is interpreted as an Ito equation. On the other hand, from a physical point of view (8.219, 221) should hardly be taken seriously for arbitrarily large values of x. We shall thus restrict the state space of x to $[0, x_{max}]$, where x_{max} is the highest line density that is physically realizable in the system. In the models considered previously, such a procedure is unnecessary, since the stationary probability density decreases at least exponentially whenever $b_2 = \infty$ is the upper boundary of the state space. This implies that extremely large values of the state variable occur only with an extremely small probability. In fact, this probability is exponentially small and can be considered to be zero for all practical purposes; the excursions to large values of x are harmless from a physical point of view in those models. If a large but finite upper boundary with instantaneous reflection were imposed, thereby reducing the probability of extremely large excursions to exactly zero, only the normalization constant of $p_s(x)$ would be modified by an exponentially small amount but no other changes would be caused in $p_s(x)$. However, the present model does not possess this feature, i.e., $p_s(x)$ does not decrease exponentially for $x \to \infty$. Therefore we shall consider (8.222) on $[0, x_{max}]$ and use instantaneous reflection at both boundaries as the appropriate boundary condition. The stationary probability density and its extrema can be found from the usual formulae:

$$p_s(x) = \mathcal{N}[4\sigma^2 x^2(\sqrt{x} - \tfrac{1}{2})^2 + \mu^2]^{-1}$$
$$\times \exp\left(2\int^x \frac{2\nu z(\sqrt{z} - \frac{1}{2}) - z^2 + \sigma^2 z(\sqrt{z} - \frac{1}{2})(3\sqrt{z} - 1)}{4\sigma^2 z^2(\sqrt{z} - \frac{1}{2})^2 + \mu^2} dz\right), \quad (8.225)$$

where $\mathcal{N}$ is determined by $\int_0^{x_{max}} p_s(x)\,dx = 1$, and

$$2\nu x_m(\sqrt{x_m} - \tfrac{1}{2}) - x_m^2 - \tfrac{1}{2}\sigma^2 2x_m(\sqrt{x_m} - \tfrac{1}{2})(3\sqrt{x_m} - 1) = 0\,. \quad (8.226)$$

The roots of (8.226) are:

$$x_{mo} = 0 \quad \text{for all } \sigma \text{ and } \nu^2 \quad (8.227)$$

and

$$\sqrt{x_{m+/-}} = \frac{1}{3\sigma^2 + 1}\left(\nu + \frac{5}{4}\sigma^2 \pm \sqrt{\nu^2 - \frac{1}{2}\nu\sigma^2 + \frac{1}{16}\sigma^4 - \nu - \frac{\sigma^2}{2}}\right). \quad (8.228)$$

It is easily verified that $x_{mo} = 0$ is always a maximum. Therefore x_{m-} is a minimum and x_{m+} a maximum if they exist and are in $[0, x_{max}]$. Further, $x_{m+/-}$ are real if

$$\nu^2 - \frac{1}{2}\nu\sigma^2 + \frac{1}{16}\sigma^4 - \nu - \frac{\sigma^2}{2} \geqslant 0 .$$

This implies that for σ^2 given there is a threshold ν, namely

$$\nu_{\text{th}} = \tfrac{1}{2}(1 + \tfrac{1}{2}\sigma^2 \pm \sqrt{1 + 3\sigma^2}) , \tag{8.229}$$

and vice versa, that for ν given there is a threshold σ^2, namely

$$\sigma^2_{\text{th}} = 4(\nu + 1 \pm \sqrt{3\nu + 1}) . \tag{8.230}$$

Taking the limit $\sigma^2 \to 0$ in (8.229), we see that the prolongation of the deterministic threshold for the onset of turbulence is given by the plus sign, thus

$$\nu_{\text{turb}}(\sigma^2) = \tfrac{1}{2}(1 + \tfrac{1}{2}\sigma^2 + \sqrt{1 + 3\sigma^2}) . \tag{8.231}$$

This corresponds to the minus sign in (8.230):

$$\sigma^2_{\text{turb}}(\nu) = 4(\nu + 1 - \sqrt{3\nu + 1}) , \qquad \nu \geqslant 1 . \tag{8.232}$$

For $\nu > \nu_{\text{turb}}$, σ^2 fixed, or $\sigma^2 < \sigma^2_{\text{turb}}$, ν fixed, superfluid turbulence occurs. Then (8.231) is the result obtained in [8.49]: the onset of turbulence is delayed by the noise. For small noise intensities, $\sigma^2 \ll 1$, we have

$$\nu_{\text{turb}}(\sigma^2) \approx 1 + 2\sigma^2 . \tag{8.233}$$

Furthermore the critical vortex line density at the onset of turbulence decreases. For small σ^2

$$x^{\text{c}}_m \approx 1 - \tfrac{1}{4}\sigma^2 . \tag{8.234}$$

The second root of (8.229, 230) corresponds to a pure noise-induced phenomenon. If $\sigma^2 \geqslant \sigma^2_+$ where

$$\sigma^2_+(\nu) = 4(\nu + 1 + \sqrt{3\nu + 1}) , \tag{8.235}$$

the system becomes turbulent. As is clear from (8.235), this can occur for any value of ν, even $0 < \nu < 1$.[3] In the extreme case of zero mean counterflow velocity for instance, turbulence sets in if the noise intensity crosses the threshold value $\sigma^2_+(0) = 8$. This implies that in addition to the shift of the deterministic transition point to turbulence as expressed by (8.231), there is a genuine noise-induced transition to turbulent behavior which occurs for any ν. This noise-induced phenomenon was also found already by Moss and Welland. The fluc-

[3] Here $\sqrt{x_m(\nu^2_+)} \approx 0.4$ for $0 < \nu < 1$, i.e., the average distance between vortex lines $L^{-1/2}$ is slightly bigger than the tube diameter. Effects from the tube walls might interfere with this pure noise-induced transition and its observability is at present unclear.

tuating Vinen equation (8.222) thus predicts some very interesting behavior for the thermal counterflow of helium II: for $\nu < 1$, turbulence occurs for $\sigma^2 \geqslant \sigma_+^2(\nu)$. For $\nu \geqslant 1$, turbulence occurs for $\sigma^2 < \sigma_{\text{turb}}^2(\nu)$ and $\sigma^2 > \sigma_+^2(\nu)$. The relative importance of the turbulence free region, i.e., $|\sigma_+^2 - \sigma_{\text{turb}}^2|/\sigma_{\text{turb}}^2$, decreases like $1/\sqrt{\nu}$ for large ν.

9. Markovian Dichotomous Noise: An Exactly Soluble Colored-Noise Case

We have emphasized that in general it is impossible to obtain an exact solution of, for instance, the stationary density of the system if an arbitrary form of noise is considered. This holds true even in such a convenient case as that of external Markovian Gaussian noise. Therefore, the general case of external colored noise can be tackled only by approximative methods. The methods developed in Chap. 8 allow us to treat the two limiting cases of very slow and very rapid external noise. In particular, for the latter case of short correlation times, we have a systematic perturbation expansion at our disposal. This method was used to show that the transitions induced by external noise with a short correlation time can be identified with the transitions found for the white-noise idealization. Due to the gap between the two approximation methods for rapid and slow noise it is, however, not clear how the transitions predicted for "fast" noise are related with the transitions occurring under a "slow" external noise. It is thus desirable to supplement the insight gained from the general approximation procedures by information obtained from the study of special classes of external colored noise. In other words, it is useful to look for types of colored noise such that for an arbitrary one-variable system at least the stationary probability density can be evaluated exactly for *any* value of the correlation time. As mentioned above, Gaussian noise does not belong to this class. We have to direct our search to random processes with a simpler structure and quite naturally Markov processes with a discrete state space come to mind. The simplest process of this kind is the dichotomous Markov noise, also known as the random telegraph signal. We shall show in this chapter that it indeed enables us to obtain exact results and a complete picture of the influence of correlations.

9.1 Markovian Dichotomous Noise: Formalism

The state space of the Markovian dichotomous noise I_t consists only of two levels $\{\Delta_-, \Delta_+\}$. This noise is a time-homogeneous Markov process and is therefore completely characterized by the following transition probability

$$P_{ij}(t) = \mathrm{Prob}(I_t = i | I_0 = j)\,, \quad i,j \in \{\Delta_+, \Delta_-\}\,. \tag{9.1}$$

The temporal evolution of the noise is given by the Kolmogorov forward equation, in the physical literature known as master equation:

$$\frac{d}{dt}\begin{pmatrix} P_{\Delta_-j}(t) \\ P_{\Delta_+j}(t) \end{pmatrix} = \begin{pmatrix} -\beta & \alpha \\ \beta & -\alpha \end{pmatrix}\begin{pmatrix} P_{\Delta_-j}(t) \\ P_{\Delta_+j}(t) \end{pmatrix}, \tag{9.2}$$

where α and β are the mean frequencies of passage from Δ_+ to Δ_- and from Δ_- to Δ_+. It is easily seen that the eigenvalues of the matrix appearing in (9.2) are

$$\omega_1 = 0\,, \qquad \omega_2 = -\gamma\,, \qquad \text{where} \tag{9.3}$$

$$\gamma = \alpha + \beta\,. \tag{9.4}$$

Thus the time-dependent solution of the master equation (9.2) is

$$P_{ij}(t) = \frac{1}{\gamma}\begin{pmatrix} \alpha + \beta e^{-\gamma t} & \alpha(1 - e^{-\gamma t}) \\ \beta(1 - e^{-\gamma t}) & \beta + \alpha e^{-\gamma t} \end{pmatrix}. \tag{9.5}$$

The stationary probability is given by

$$P_s(\Delta_+) = \frac{\beta}{\gamma}\,, \qquad P_s(\Delta_-) = \frac{\alpha}{\gamma}\,. \tag{9.6}$$

As usual we shall suppose that the external noise is a stationary random process and hence the dichotomous noise has to be started with (9.6) as initial condition. It then has the mean value

$$E\{I_t\} = \frac{1}{\gamma}(\alpha\Delta_- + \beta\Delta_+)\,, \tag{9.7}$$

and the correlation function $C(t)$ is

$$\begin{aligned} C(t) &= E\{I_t I_0\} - (E\{I_t\})^2 \\ &= \sum_{ij} ij P_{ij}(t) P_s(j) - (E\{I_t\})^2 \\ &= \frac{\alpha\beta}{\gamma^2}(\Delta_+ - \Delta_-)^2 \exp(-\gamma t)\,. \end{aligned} \tag{9.8}$$

A system subjected to this kind of colored noise is described by the following stochastic differential equation:

$$\dot{X}_t = h(X_t) + I_t g(X_t) \equiv f(X_t, I_t)\,, \tag{9.9}$$

which in the deterministic case reduces to

$$\dot{X}(t) = h(X(t)) + I g(X(t))\,, \qquad \text{with} \quad I = E\{I_t\}\,. \tag{9.10}$$

It is obvious that the pair process (X_t, I_t) is Markovian and obeys therefore a forward Kolmogorov equation for its transition probability density. From the particular structure of the pair process [more precisely (X_t, I_t) is a degenerate pair process whose component I_t fluctuates independently of X_t] it is intuitively clear that this equation should consist of the following two terms. One term describing the evolution of X_t for a fixed value of I_t, i.e., $I_t = \Delta_+$ or $I_t = \Delta_-$, and the other one containing the matrix of (9.2), governing the evolution of I_t. Hence [9.1, 2]:

$$\begin{aligned}\partial_t p(x, \Delta_-, t\,|x_0, i_0, 0) = &-\partial_x\{[h(x) + \Delta_- g(x)]\,p(x, \Delta_-, t\,|x_0, i_0, 0)\}\\ &- \beta p(x, \Delta_-, t\,|x_0, i_0, 0) + \alpha p(x, \Delta_+, t\,|x_0, i_0, 0)\,,\end{aligned} \tag{9.11}$$

$$\begin{aligned}\partial_t p(x, \Delta_+, t\,|x_0, i_0, 0) = &-\partial_x\{[h(x) + \Delta_+ g(x)]\,p(x, \Delta_+, t\,|x_0, i_0, 0)\}\\ &- \alpha p(x, \Delta_+, t\,|x_0, i_0, 0) + \beta p(x, \Delta_-, t\,|x_0, i_0, 0)\,.\end{aligned} \tag{9.12}$$

The starting point for a derivation of these two equations is, as usual, the Chapman-Kolmogorov equation, which reads for the pair process (X_t, I_t):

$$p(x, i, t+h\,|x_0, i_0, 0) = \sum_j \int dy\, p(x, i, t+h\,|y, j, t)\, p(y, j, t\,|x_0, i_0, 0)\,. \tag{9.13}$$

To abbreviate the notation we define

$$p(x, i, t\,|0) \equiv p(x, i, t\,|x_0, i_0, 0)\,. \tag{9.14}$$

In order not to burden the presentation, we shall present the derivation for the case $i = \Delta_-$, i.e., (9.11). The proof of (9.12), i.e., $i = \Delta_+$, is completely analogous. Let $g(z)$ be an infinitely often differentiable real-valued function with compact support, i.e., it vanishes identically outside a certain bounded interval. Then it follows from (9.13) with $i = \Delta_-$ that

$$\int dx\, g(x)\, p(x, \Delta_-, t+h\,|0) = \sum_j \int dy\, p(y, j, t\,|0)\,[\int dx\, g(x)\, p(x, \Delta_-, t+h\,|y, j, t)]$$

where $j \in \{\Delta_-, \Delta_+\}$.

Substracting on both sides $\int p(x, \Delta_-, t\,|0)\, g(x)\, dx$, we obtain:

$$\begin{aligned}&\int dx\, g(x)\,[p(x, \Delta_-, t+h\,|0) - p(x, \Delta_-, t\,|0)]\\ &\quad = \sum_j \int dy\, p(y, j, t\,|0)\,[\int dx\, g(x)\, p(x, \Delta_-, t+h\,|y, j, t)]\\ &\qquad - \int dx\, g(x)\, p(x, \Delta_-, t\,|0)\,.\end{aligned} \tag{9.15}$$

Dividing both sides by h and taking the limit $h \to 0$, we obtain for the lhs of (9.15):

$$\lim_{h\to 0}\frac{1}{h}\int dx\, g(x)\,[p(x,\Delta_-,t+h\,|\,0)-p(x,\Delta_-,t\,|\,0)]=\int dx\, g(x)\,\partial_t p(x,\Delta_-,t\,|\,0)^1\,. \tag{9.16}$$

To evaluate the rhs we have to determine the transition probability $p(x,\Delta_-,t+h\,|\,y,j,t)$ for infinitesimally small h. The random telegraph signal is independent of X_t. By its definition as a Markov process we have therefore the following transition probabilities at small time intervals for the three basic types of transitions which can occur:

1) $j=\Delta_-$, no jump $\Delta_-\to\Delta_-$, $p_{\Delta_-\Delta_-}(h)=1-\beta h+o(h)$, (9.17)

2) $j=\Delta_+$, exactly one jump $\Delta_+\to\Delta_-$, $p_{\Delta_-\Delta_+}(h)=\alpha h+o(h)$, (9.18)

3) two or more jumps: $o(h)$.

We need consider only the first two cases, since $\lim_{h\to 0} o(h)/h=0$. If the random telegraph signal does not jump and $X_t=y$ at time t, then the process X_t ends up at

$$x=y+f(y,\Delta_-)h+o(h)$$

at time $t+h$. This implies that

$$p(x,\Delta_-,t+h\,|\,y,\Delta_-,t)=\delta(x-[y+f(y,\Delta_-)h])(1-\beta h)+o(h)\,. \tag{9.19}$$

Consider now the second case that only one jump will occur in the interval $[t,t+h)$. The fraction of time v at which the process will jump is uniformly distributed on $[0,1]$ to dominant order in h, i.e., $p(v)=1+o(h)$. Then, starting at time t at y, the process X_t will end up at time $t+h$ at the value:

$$x=y+vf(y,\Delta_+)h+(1-v)f(y,\Delta_-)h+o(h)\,.$$

Hence we obtain:

$$p(x,\Delta_-,t+h\,|\,y,\Delta_+,t)=\delta(x-[y+vf(y,\Delta_+)h+(1-v)f(y,\Delta_-)h]) \times \alpha h+o(h)\,. \tag{9.20}$$

Again we neglect the $o(h)$ terms and the rhs of (9.15) reads with (9.19, 20):

---

[1] We suppose as usual that $\partial_t p(x,\Delta_-,t\,|\,0)$ exists and is continuous. The interchange of the integration and the limiting procedure is then permitted.

$$\lim_{h\to 0}\frac{1}{h}\{\int dy\, p(y,\Delta_-,t|0)\int dx\, g(x)\,\delta(x-[y+f(y,\Delta_-)h])(1-\beta h)$$
$$+\int dy\, p(y,\Delta_+,t|0)\int dx\, g(x)\,\delta(x-[y+vf(y,\Delta_+)h+(1-v)f(y,\Delta_-)h])\,\alpha h$$
$$-\int dx\, g(x)p(x,\Delta_-,t|0)\}$$
$$=\lim_{h\to 0}\frac{1}{h}[[\int dy\, p(y,\Delta_-,t|0)\, g(y+f(y,\Delta_-)h)(1-\beta h)$$
$$+\int dy\, p(y,\Delta_+,t|0)\, g(y+vf(y,\Delta_+)h+(1-v)f(y,\Delta_-)h)\,\alpha h$$
$$-\int dx\, g(x)p(x,\Delta_-,t|0)]\ .$$

Expanding $g(x)$ in a Taylor series, one has

$$\lim_{h\to 0}\frac{1}{h}\{\int dy\, p(y,\Delta_-,t|0)\,[g(y)+f(y,\Delta_-)hg'(y)+o(h)](1-\beta h)$$
$$+\int dy\, p(y,\Delta_+,t|0)\,[g(y)+(vf(y,\Delta_+)+(1-v)f(y,\Delta_-))hg'(y)+o(h)]\,\alpha h$$
$$-\int dx\, g(x)p(x,\Delta_-,t|0)\}\ .$$

Since terms in $o(h)$ have no bearing on the limit, we neglect them and obtain:

$$\lim_{h\to 0}\frac{1}{h}\{\int dy\, p(y,\Delta_-,t|0)[-\beta hg(y)+f(y,\Delta_-)hg'(y)]$$
$$+\int dy\, p(y,\Delta_+,t|0)\,g(y)\,\alpha h\}$$
$$=\int dy\, g(y)[-\beta p(y,\Delta_-,t|0)+\alpha p(y,\Delta_+,t|0)]$$
$$+\int dy\, p(y,\Delta_-,t|0)f(y,\Delta_-)g'(y)\ .$$

We integrate the last term by parts and take into account that $g(x)$ has compact support, i.e., the boundary terms vanish. Furthermore, we rename the integration variable y and combine the above expression for the rhs with (9.16) for the lhs and obtain

$$\int dx\, g(x)\,\partial_t p(x,\Delta_-,t|0)=\int dx\, g(x)[-\partial_x f(x,\Delta_-)p(x,\Delta_-,t|0)$$
$$-\beta p(x,\Delta_-,t|0)+\alpha p(x,\Delta_+,t|0)]\ . \qquad (9.21)$$

Since $g(x)$ is an arbitrary function, (9.11) follows from (9.21). If in this derivation the roles of Δ_- and Δ_+ and of α and β are interchanged, (9.12) is obtained.

This completes the proof that the Kolmogorov forward equation of the pair process (X_t, I_t) is indeed given by (9.11, 12). Note that no step of the demonstration requires the function $f(x,i)$ to be linear in the second argument. We shall come back in detail to the case of phenomenological equations nonlinear in the external parameter in Sect. 9.3.

So far we have derived an evolution equation for the joint transition probability $p(x,i,t\,|\,x_0,i_0,0)$ of the pair process. Of course the Kolmogorov forward equations (9.11, 12) also hold for the unconditioned joint probability $p(x, i, t)$ as is easily seen by integrating out the initial condition. However, the quantity we are really interested in is not the joint probability for the state variable and the fluctuating parameter but the probability density for the state variable X_t alone, i.e., $p(x, t)$. Obviously,

$$p(x,t) = p(x,\Delta_-,t) + p(x,\Delta_+,t) , \tag{9.22}$$

and we will now derive an evolution equation involving only this quantity, starting from (9.11, 12). To this aim, it is convenient to define the following function

$$q(x,t) \equiv \alpha p(x,\Delta_+,t) - \beta p(x,\Delta_-,t) . \tag{9.23}$$

Then the joint probability density can be written as

$$p(x,\Delta_-,t) = \gamma^{-1}(\alpha p(x,t) - q(x,t)) , \tag{9.24}$$

and

$$p(x,\Delta_+,t) = \gamma^{-1}(\beta p(x,t) + q(x,t)) . \tag{9.25}$$

The Kolmogorov forward equations (9.11, 12) can be written as

$$\begin{aligned} \gamma^{-1}\partial_t(\alpha p(x,t) - q(x,t)) = & -\{\partial_x[h(x) + \Delta_- g(x)]\gamma^{-1}(\alpha p(x,t) - q(x,t))\} \\ & + q(x,t) , \end{aligned} \tag{9.26}$$

and

$$\begin{aligned} \gamma^{-1}\partial_t(\beta p(x,t) + q(x,t)) = & -\{\partial_x[h(x) + \Delta_+ g(x)]\gamma^{-1}(\beta p(x,t) + q(x,t))\} \\ & - q(x,t) . \end{aligned} \tag{9.27}$$

Adding and substracting (9.26, 27) and rearranging terms, we obtain the following equations for $p(x,t)$ and $q(x,t)$ $[\tilde{I} = \gamma^{-1}(\beta\Delta_- + \alpha\Delta_+)]$:

$$\partial_t p(x,t) = -\partial_x\{[h(x) + I g(x)]p(x,t)\} + \gamma^{-1}(\Delta_- - \Delta_+)\partial_x g(x) q(x,t) , \tag{9.28}$$

$$\partial_t q(x,t) = -\{\partial_x[h(x) + \tilde{I} g(x)] + \gamma\}q(x,t) + \alpha\beta\gamma^{-1}(\Delta_- - \Delta_+)\partial_x g(x) p(x,t) , \tag{9.29}$$

where $I = E\{I_t\}$. A natural assumption for the initial condition is that in the infinite past the processes X_t and I_t are statistically independent, i.e.,

$$\begin{aligned} p(x,i,-\infty) &= p(x,-\infty)p(i,-\infty) \\ &= p(x,-\infty)p_s(i) . \end{aligned} \tag{9.30}$$

From (9.23) we have

$$\begin{aligned} q(x,-\infty) &= p(x,-\infty)[\alpha p_s(\Delta_+) - \beta p_s(\Delta_-)] \\ &= p(x,-\infty)[\alpha\beta/\gamma - \beta\alpha/\gamma] \\ &= 0 . \end{aligned} \tag{9.31}$$

The formal solution (9.29), obeying this initial condition, is given by

$$\begin{aligned} q(x,t) = \int_{-\infty}^{t} dt' \exp\{-[\gamma + \partial_x(h(x) + \tilde{I}g(x))](t-t')\}\alpha\beta\gamma^{-1}(\Delta_- - \Delta_+) \\ \times \partial_x g(x) p(x,t') , \end{aligned} \tag{9.32}$$

where the operator ∂_x, also the one in the exponential, acts on all x-dependent functions written to the right of it. Injecting this expression in (9.28) we attain our aim to find a closed evolution equation for $p(x,t)$:

$$\begin{aligned} \partial_t p(x,t) = &- \partial_x[h(x) + Ig(x)]p(x,t) \\ &+ \alpha\beta\gamma^{-2}(\Delta_- - \Delta_+)^2 \partial_x g(x) \int_{-\infty}^{t'} dt' \exp\{-[\gamma + \partial_x(h(x) \\ &+ \tilde{I}g(x))](t-t')\}\partial_x g(x) p(x,t') . \end{aligned} \tag{9.33}$$

Note that this equation contains derivatives of $p(x,t')$ of infinitely high order, namely $\exp(-\partial_x)$. Since the process X_t is non-Markovian as is clearly displayed by the memory kernel in (9.33), this is compatible with the property that X_t has continuous realizations. It is clear from the results of the preceding chapter that the random telegraph signal $I_t^\varepsilon = (I_{t/\varepsilon^2} - I)/\varepsilon$, where

$$I = \gamma^{-1}(\alpha\Delta_- + \beta\Delta_+) , \tag{9.34}$$

converges for $\varepsilon \to 0$ to Gaussian white noise with intensity

$$\sigma^2 = 2\alpha\beta\gamma^{-3}(\Delta_+ - \Delta_-)^2 . \tag{9.35}$$

If $I = 0$, this limit amounts to the rescaling

$$\alpha \to \alpha/\varepsilon^2 , \tag{9.36}$$

$$\beta \to \beta/\varepsilon^2 , \tag{9.37}$$

$$\Delta_\pm \to \Delta_\pm/\varepsilon . \tag{9.38}$$

As was also etablished in the last chapter, in the white-noise limit the probability density $p(x,t)$ obeys a FPE corresponding to the Stratonovich interpretation of the SDE (8.72). This result can be verified directly for the case of dichotomous

Markov noise. Employing the scaling (9.36 – 38) it is easily established that the memory kernel reduces to a Dirac delta function in the limit $\varepsilon \to 0$:

$$\lim_{\varepsilon \to 0} \frac{\alpha\beta}{\gamma^2} \frac{(\Delta_- - \Delta_+)^2}{\varepsilon^2} \exp\{-[\gamma/\varepsilon^2 + \partial_x f(x)](t-t')\} = \frac{\sigma^2}{2}\delta(t-t'), \tag{9.39}$$

and (9.33) turns into the following FPE:

$$\begin{aligned}\partial_t p(x,t) &= -\partial_x f(x) p(x,t) + \alpha\beta\gamma^{-3}(\Delta_+ - \Delta_-)^2 \partial_x g(x)\partial_x g(x) p(x,t)\\ &= -\partial_x \left[f(x) + \frac{\sigma^2}{2} g(x) g'(x)\right] p(x,t) + \frac{\sigma^2}{2}\partial_{xx} g^2(x) p(x,t).\end{aligned} \tag{9.40}$$

The most convenient way to calculate the stationary probability density $p_s(x)$, i.e., the time-independent solution of (9.33), is to go back to (9.28, 27):

$$0 = -\partial_x\{[h(x) + Ig(x)]p_s(x)\} + \gamma^{-1}(\Delta_- - \Delta_+)\partial_x g(x) q_s(x), \tag{9.41}$$

$$0 = -\partial_x\{[h(x) + \Delta_+ g(x)][\beta p_s(x) + q_s(x)] - \gamma q_s(x). \tag{9.42}$$

From (9.41) it follows that

$$[h(x) + Ig(x)]p_s(x) + \gamma^{-1}(\Delta_+ - \Delta_-) g(x) q_s(x) = C. \tag{9.43}$$

As explained in Sect. 6.2, we consider only systems that are deterministically stable, i.e., the solution $X(t)$ of the deterministic equation $\dot{X} = h(X) + \lambda g(X)$ is bounded away from infinity for all times t:

$$|X(t)| \nrightarrow \infty \quad \text{for} \quad t \to \infty$$

(no explosion). If the state space consists of a finite interval on the real line, then the system is trivially deterministically stable. If the state space is unbounded to the right or to the left, then recall that a necessary and sufficient condition for a system to be deterministically stable is that a positive K exists, possibly dependent on λ, such that

$$h(x) + \lambda g(x) < 0 \quad \text{for all} \quad x > K(\lambda) \tag{9.44}$$

or

$$h(x) + \lambda g(x) > 0 \quad \text{for all} \quad x < -K(\lambda). \tag{9.45}$$

This implies that for finite initial values $x(0)$ the state of the system lies with certainty in the interval $[-K(\lambda), K(\lambda)]$ for time tending to infinity. It follows that for deterministically stable systems the joint stationary probability $p_s(x,i)$ has a compact support. This is trivially true if the state space is finite. If it is, say, unbounded to the right, the most frequent case, then indeed:

$$p_s(x,\Delta_-) \equiv 0 \quad \text{for all} \quad x > \max(K(\Delta_-), K(\Delta_+)) ,$$
$$p_s(x,\Delta_+) \equiv 0 \quad \text{for all} \quad x > \max(K(\Delta_-), K(\Delta_+)) . \tag{9.46}$$

For deterministically stable systems we can conclude that the constant C in (9.43) equals zero, since we can always choose x such that the lhs vanishes. Therefore (9.43) implies

$$q_s(x) = -a[h(x)+Ig(x)]p_s(x)/g(x) , \tag{9.47}$$

where we have set

$$a^{-1} = \gamma^{-1}(\Delta_+ - \Delta_-) .$$

We obtain for the derivative of $q_s(x)$[2]:

$$q_s'(x) = \left[-a\frac{h'(x)+Ig'(x)}{g(x)} + a\frac{h(x)+Ig(x)}{g^2(x)}g'(x) \right] p_s(x) - a\frac{h(x)+Ig(x)}{g(x)}p_s'(x) . \tag{9.48}$$

Injecting these expressions for $q_s(x)$ and $q_s'(x)$ in (9.42) and rearranging terms, we obtain the following differential equation for $p_s(x)$:

$$p_s'(x) = \left[-\gamma\frac{h(x)+Ig(x)}{(h(x)+\Delta_+ g(x))(h(x)+\Delta_- g(x))} - \frac{h'(x)+\Delta_+ g'(x)}{h(x)+\Delta_+ g(x)} + \frac{1}{g(x)}\frac{h(x)g'(x)-g(x)h'(x)}{h(x)+\Delta_- g(x)} \right] p_s(x) .$$

This equation can be written in the equivalent form

$$p_s'(x) = \left\{ -\gamma\frac{h(x)+Ig(x)}{[h(x)+\Delta_+ g(x)][h(x)+\Delta_- g(x)]} - \ln'[h(x)+\Delta_+ g(x)] - \ln'[h(x)+\Delta_- g(x)] + \ln' g(x) \right\} p_s(x) , \tag{9.49}$$

which is easily integrated and yields the following expression for the stationary probability density of the system:

[2] Derivation with respect to x is indicated by a prime in the expressions to follow.

$$p_s(x) = N \frac{g(x)}{[h(x)+\Delta_+ g(x)][h(x)+\Delta_- g(x)]} \times \exp\left(-\gamma \int^x \frac{h(z)+Ig(z)}{[h(z)+\Delta_+ g(z)][h(z)+\Delta_- g(z)]} dz\right). \tag{9.50}$$

It vanishes identically outside the support $U(\Delta_+, \Delta_-)$, which is the closed interval, or the union of closed intervals, whose boundaries are given by the stable stationary states of the deterministic equation

$$\dot{X} = h(X) + \lambda g(X)$$

for $\lambda = \Delta_-$ and $\lambda = \Delta_+$. The significance of the support is explained in more detail in Fig. 9.1. The extrema of the stationary probability density $p_s(x)$ are the roots of the following equation:

$$h(x_m) + Ig(x_m) + \frac{\Delta_+ \Delta_-}{\gamma} g'(x_m) g(x_m) + \frac{2}{\gamma} h'(x_m) h(x_m) - \frac{1}{\gamma} \frac{h^2(x_m)}{g(x_m)} g'(x_m) + \frac{1}{\gamma} h'(x_m) g(x_m)(\Delta_- + \Delta_+) = 0, \tag{9.51}$$

if they lie inside the support U.

For the sake of clarity, we shall henceforth restrict our discussion to the case of symmetric dichotomous noise (D noise for short):

$$\Delta_+ = -\Delta_- = \Delta \tag{9.52}$$

and

$$\alpha = \beta = \gamma/2. \tag{9.53}$$

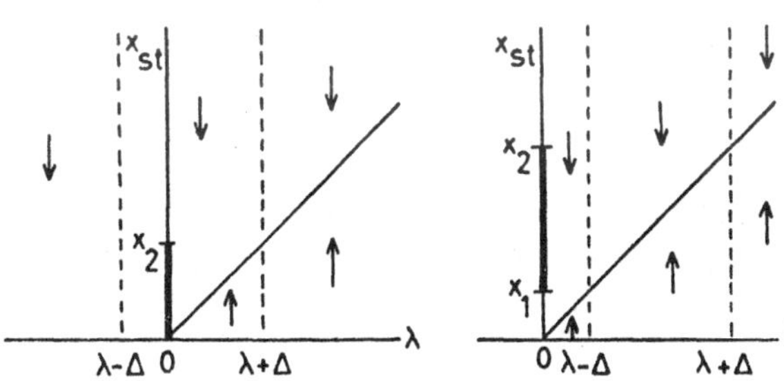

Fig. 9.1. Plot of the curve of the deterministic steady states of the Verhulst model as a function of λ. The state space of the random process (X_t, I_t) (for the case $\lambda - \Delta < 0$, respectively $\lambda - \Delta > 0$) is given by the two broken lines; the arrows indicate the direction of evolution of X_t. The support of $p_s(x)$ is indicated on the x_{st} axis by the bold-face line

Equations (9.2, 7, 8) read in this case

$$\frac{d}{dt}\begin{pmatrix} P_{-\Delta,j}(t) \\ P_{\Delta,j}(t) \end{pmatrix} = \frac{\gamma}{2}\begin{pmatrix} -1 & 1 \\ 1 & -1 \end{pmatrix}\begin{pmatrix} P_{-\Delta,j}(t) \\ P_{\Delta,j}(t) \end{pmatrix} \tag{9.54}$$

$$E\{I_t\} = 0 \tag{9.55}$$

and

$$C(t) = \Delta^2 \exp(-\gamma t) . \tag{9.56}$$

Consider now the following stochastic differential equation

$$\dot{X}_t = h(X_t) + \lambda_t g(X_t) \tag{9.57}$$

where the external parameter fluctuates according to a symmetric random telegraph signal around its mean value λ, i.e.,

$$\lambda_t = \lambda + I_t . \tag{9.58}$$

We define again

$$f_\lambda(x) \equiv h(x) + \lambda g(x) \tag{9.59}$$

and suppress the dependence on λ in the notation. As is easily derived from (9.50) the stationary probability density $p_s(x)$ for a system subjected to symmetric dichotomous noise has the form

$$p_s(x) = N\frac{g(x)}{\Delta^2 g^2(x) - f^2(x)} \exp\left[-\frac{\gamma}{2}\int^x dz\left(\frac{1}{f(z)-\Delta g(z)} + \frac{1}{f(z)+\Delta g(z)}\right)\right] . \tag{9.60}$$

One recovers the expression obtained by Pawula using the "extended Fokker-Planck equation" for the stationary probability density [4.5, 9.3]. The zeros of $\Delta^2 g^2 - f^2 = -(f-\Delta g)(f+\Delta g)$ constitute of course, roughly speaking, the boundaries of the support U. Let b_i denote the lower (respectively upper) boundary of U, i.e.,

$$\lambda \pm \Delta = -h(b_i)/g(b_i) . \tag{9.61}$$

Then $p_s(x)$ in the vicinity of, say, the lower boundary is given by

$$p_s(b_1+\delta x) = N\frac{g(b_1+\delta x)}{F_1(b_1+\delta x)F_2(b_1+\delta x)} \times \exp\left[-\frac{\gamma}{2}\int d\delta x\left(\frac{1}{F_1(b_1+\delta x)} + \frac{1}{F_2(b_1+\delta x)}\right)\right], \tag{9.62}$$

where

$$F_{1,2}(x) = f(x) \mp \Delta g(x) . \tag{9.63}$$

Using a Taylor expansion of the various functions and keeping only first-order terms in δx, we obtain

$$p_s(b_1 + \delta x) \simeq \bar{N} \frac{1}{F_1'(b_1)\delta x} \exp\left(-\frac{\gamma}{2}\int d\delta x \frac{1}{F_1'(b_1)\delta x}\right),$$

if $F_2(b_1) \neq 0$ and $g(b_1) \neq 0$. Carrying out the integration we have finally

$$p_s(b_1 + \delta x) \simeq \bar{\bar{N}} |\delta x|^{-([\gamma/2F_1'(b_1)]+1)} , \tag{9.64}$$

and similarly for the other boundary of the support U

$$p_s(b_2 + \delta x) \simeq \bar{N} |\delta x|^{-([\gamma/2F_2'(b_2)]+1)} . \tag{9.65}$$

For a deterministically stable system

$$F_1'(b_1) < 0 , \quad F_2'(b_2) < 0 , \tag{9.66}$$

since the boundaries of U are the stable steady states of the deterministic equation for $\lambda - \Delta$ respectively $\lambda + \Delta$. It thus follows from (9.64, 65) that

$$\int_U dx p_s(x) < \infty , \tag{9.67}$$

i.e., $p_s(x)$ is normalizable for a deterministically stable system. Furthermore, we conclude from (9.64, 65) that near the boundary b_i the stationary probability density $p_s(x)$ behaves as follows:

i) it diverges to infinity (in an integrable manner) if

$$-\frac{\gamma}{2F_i'(b_i)} - 1 < 0 ; \tag{9.68}$$

ii) it vanishes with divergent tangent, if

$$0 < -\frac{\gamma}{2F_i'(b_i)} - 1 < 1 ; \tag{9.69}$$

iii) it vanishes with a horizontal tangent, if

$$1 < -\frac{\gamma}{2F_i'(b_i)} - 1 . \tag{9.70}$$

Note that all the above formulae hold only if $F_j(b_i) \neq 0$ for $i \neq j$. In the following section, we shall deal with an example where this condition does not hold. Physically it is interesting to see that for $p_s(x)$ to diverge or to vanish with vertical or horizontal tangent at the boundaries of the support the conditions (9.68 – 70) can be related to the linear deterministic stability analysis around the steady states b_i. Indeed, according to (3.4) the inverse of the relaxation time $\tau(b_i)$ determined that way is given by $\tau^{-1}(b_i) = -F_i'(b_i)$. So conditions (9.68, 69) can be written as

$$\tau(b_i) = -1/F_i'(b_i) < 2/\gamma\,, \tag{9.71}$$

and

$$2/\gamma < \tau(b_i) < 4/\gamma\,. \tag{9.72}$$

As remarked at the beginning of this section, α and β are the mean frequencies of passage from one state into the other. It is easily seen that for symmetric dichotomous noise, $2/\gamma$ is the mean waiting time in either one of the two states. Thus condition (9.68) means that only if the relaxation time of X_t is smaller than the mean waiting time between transitions from one noise level to the other can the process X_t relax to the deterministic steady state b_i, constituting the boundary of U, and spend a finite amount of time in its neighborhood, leading to a divergence of $p_s(x)$ at $x = b_i$. If the relaxation time is larger than the mean waiting time between transitions in the noise, but does not exceed it by a factor two, the process X_t accumulates still sufficient probability mass in the neighborhood of b_i to produce a divergent tangent.

The extrema of $p_s(x)$ for the symmetric dichotomous noise are given by the equation:

$$f(x_m) - \frac{\Delta^2}{\gamma} g(x_m) g'(x_m) + \frac{2}{\gamma} f(x_m) f'(x_m) - \frac{1}{\gamma} f^2(x_m) \frac{g'(x_m)}{g(x_m)} = 0\,, \tag{9.73}$$

if $g(x_m) \neq 0$ and $x_m \in U$. This equation has a very interesting structure. The first term equal to zero yields the deterministic steady state. As is clear from the scaling (9.36 – 38), the white-noise limit for the symmetric dichotomous noise corresponds to $\Delta \to \infty$, $\gamma \to \infty$ such that $\Delta^2/\gamma = \sigma^2/2$. In this limit only the first two terms survive. It indeed yields the equation for the extrema used in the white-noise limit, (6.36) with $\nu = 1$, in the Stratonovich interpretation. This is a good illustration of the theorem of Wong and Zakai and the general result presented in Chap. 8. The last terms are corrections due to a nonvanishing correlation time, i.e., γ finite. Note, however, that these corrections are not equal to the correction terms obtained in Chap. 8 for the O-U noise. This is not astonishing since the nature of the two noises is very different despite the fact that they have the same correlation function. Since the O-U process is Gaussian, the mean value and most probable value coincide. On the contrary, for the D noise the mean value does not even lie in the state space.

9.2 Phase Diagrams for D Noise-Induced Transitions

9.2.1 The Verhulst Model

Assuming that the fluctuations of the parameter λ of the Verhulst equation are given by a symmetrical D noise

$$\lambda_t = \lambda + I_t , \tag{9.74}$$

we consider the following stochastic differential equation

$$\dot{X}_t = \lambda X_t - X_t^2 + I_t X_t . \tag{9.75}$$

We shall restrict ourselves to the case that in a nonfluctuating environment genuine growth is possible, i.e., $\lambda > 0$. From expression (9.60) we obtain for the stationary probability density [9.1]:

$$p_s(x) = N x^{(\gamma\lambda/[\Delta^2-\lambda^2]-1)} |x-(\lambda-\Delta)|^{-(\gamma/[2(\Delta-\lambda)]+1)} |x-(\lambda+\Delta)|^{(\gamma/[2(\lambda+\Delta)]-1)} . \tag{9.76}$$

The support of $p_s(x)$ is given by the interval

$$U = [\max(0, \lambda-\Delta), \lambda+\Delta] . \tag{9.77}$$

Inspecting (9.76) for the stationary probability density $p_s(x)$, we see that $\int_U dx p_s(x) < \infty$ for all positive λ.

The behavior of the stationary probability density in the neighborhood of the boundaries of U is given by (9.68–70). For the upper boundary $x = \lambda + \Delta$ we have

$$\begin{array}{lll} p_s(\lambda+\Delta) = \infty & & \Delta+\lambda > \gamma/2 , \\ p_s(\lambda+\Delta) = 0 , & p_s'(\lambda+\Delta) = \infty & \gamma/4 < \lambda+\Delta < \gamma/2 , \\ p_s(\lambda+\Delta) = 0 , & p_s'(\lambda+\Delta) = 0 & \lambda+\Delta < \gamma/4 . \end{array} \tag{9.78}$$

For $\lambda - \Delta > 0$, the lower boundary is given by $x = \lambda - \Delta$ and we have the following situation:

$$\begin{array}{lll} p_s(\lambda-\Delta) = \infty & & \lambda-\Delta > \gamma/2 , \\ p_s(\lambda-\Delta) = 0 & p_s'(\lambda-\Delta) = \infty & \gamma/4 < \lambda-\Delta < \gamma/2 , \\ p_s(\lambda-\Delta) = 0 & p_s'(\lambda-\Delta) = 0 & \lambda-\Delta < \gamma/4 . \end{array} \tag{9.79}$$

For $\lambda - \Delta < 0$, the lower boundary is given by $x = 0$. Since $\bar{x} = 0$ is a deterministic steady state for all $\lambda \in \mathbb{R}$, $F_1(0) = F_2(0) = 0$ for $\lambda - \Delta < 0$ and (9.68–70) cannot be used to discuss the behavior of $p_s(x)$ near the lower boundary of the support in this case. Working directly with the explicit expression (9.76) for $p_s(x)$, we obtain

$$\begin{aligned} &p_s(0)=\infty && \text{if} \quad \gamma\lambda+\lambda^2<\Delta^2\,, \\ &p_s(0)=0\,, \quad p_s'(0)=\infty && \text{if} \quad \frac{\gamma}{2}\lambda+\lambda^2<\Delta^2<\gamma\lambda+\lambda^2\,, \\ &p_s(0)=0\,, \quad p_s'(0)=0 && \text{if} \quad \Delta^2<\frac{\gamma}{2}\lambda+\lambda^2\,. \end{aligned} \tag{9.80}$$

To characterize the stationary probability density in more detail and to obtain further insight into the modifications of the steady-state behavior of the Verhulst system induced by D noise, we determine the extrema of the stationary probability density, using (9.73):

$$\lambda x_m - x_m^2 - \frac{\Delta^2}{\gamma}x_m + \frac{2}{\gamma}x_m(\lambda - x_m)(\lambda - 2x_m) - \frac{1}{\gamma}x_m(\lambda - x_m)^2 = 0\,. \tag{9.81}$$

In the white-noise case, the degree of the polynomial is not changed compared to that for the deterministic steady states; a transition is induced due to the fact that the position of the maximum is shifted to $\lambda - \Delta^2/\gamma$ (6.47). The D noise leads to additional modifications, since the last two terms in (9.81), which stem from the nonvanishing correlation time, increase the degree of the polynomial by one. The roots of (9.81) are given by:

$$x_0 = 0\,, \quad x_{1,2} = \tfrac{1}{6}[4\lambda + \gamma \pm ((2\lambda - \gamma)^2 + 12\Delta^2)^{1/2}]\,. \tag{9.82}$$

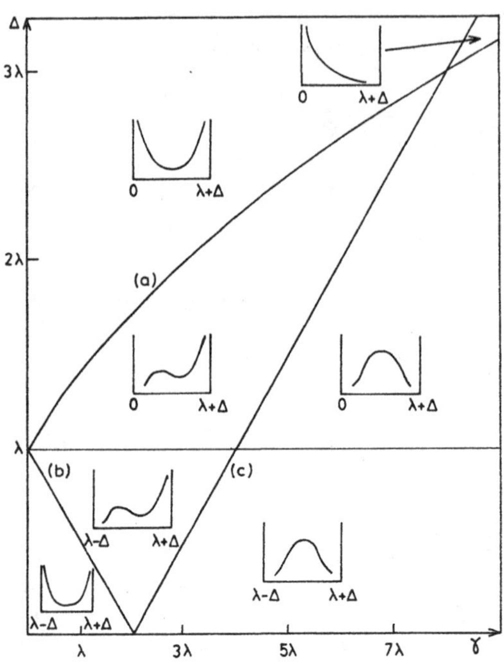

Fig. 9.2. The (Δ, γ) phase diagram for the steady-state behavior of X_t. *Curves* $(a-c)$ correspond to conditions (9.80, 79, 78) respectively. The shape of $p_s(x)$ for the different regions confined by these curves is sketched

Analysis of this expression together with those concerning the behavior of $p_s(x)$ near the boundaries of the support U (9.78 – 80) permits us to construct a phase diagram of the steady-state behavior of the Verhulst system in the (Δ, γ)-plane, displayed in Fig. 9.2. This phase diagram shows that either by increasing the amplitude Δ of the external noise or by changing its correlation time $\tau_{cor} = 1/\gamma$, the system undergoes a series of transitions which are absent from the usual phenomenological description. The neighborhood of white noise corresponds to the upper right corner of the phase diagram. Here Curve (a) is given by $\lambda = \sigma^2/2 - O(1/\gamma)$, i.e., the results of the white-noise analysis are indeed explicitly recovered if the correlation time of the external noise is sufficiently small. From Fig. 9.2 it can be seen that approximately $\gamma \gtrsim 10\lambda$, i.e., the correlation time less than one-tenth of the typical macroscopic time, is necessary for the white-noise idealization to give a satisfactory qualitative picture of the steady-state behavior of the system in a fluctuating environment.

9.2.2 The Genetic Model

Let us now analyze how the genetic system behaves if it is subjected to external D noise [9.2]. Suppose that the selection coefficient λ is modulated by the symmetrical dichotomous Markov process: $\lambda_t = \lambda + I_t$. The system is then described by the SDE

$$\dot{X}_t = 1/2 - X_t + \lambda X_t(1 - X_t) + I_t X_t(1 - X_t) . \tag{9.83}$$

The stationary probability density $p_s(x)$ can of course be explicitly calculated. All information needed to construct the phase diagram of this model in the (Δ, γ) plane can be obtained from (9.68 – 70) describing the behavior of $p_s(x)$ near the boundaries of U, and from the extrema of $p_s(x)$ given by (9.73). Note that (9.68 – 70) can be used for this system in every case, since the curve of the deterministic steady states is single valued for $-\infty < \lambda < \infty$, implying that indeed $F_i(b_j) \neq 0$ for $i \neq j$. Let us first discuss the case $\lambda = 0$, i.e., no allele is favored on the average. The support U of $p_s(x)$ is given in this case by the solution of

$$1/2 - x \pm \Delta x(1 - x) = 0 , \tag{9.84}$$

i.e.,

$$U = \left[\frac{1}{2\Delta} [\Delta + 1 - (\Delta^2 + 1)^{1/2}] , \quad \frac{1}{2\Delta} [\Delta - 1 + (\Delta^2 + 1)^{1/2}] \right] . \tag{9.85}$$

Since for $\lambda = 0$, $p_s(x)$ is obviously symmetric around $x = 1/2$, the change of behavior near the boundaries occurs simultaneously for both of them. It is thus sufficient to consider the upper boundary b_2. According to condition (9.68) $p_s(x)$ changes from $p_s(b_2) = \infty$ to $p_s(b_2) = 0$ on the line

$$-\gamma [2F_2'(b_2)]^{-1} = 1 ,$$

which yields

$$\gamma = 2(\Delta^2+1)^{1/2} \tag{9.86}$$

and from $p_s'(b_2) = \infty$ to $p_s'(b_2) = 0$ on the line

$$-\gamma[2F_2'(b_2)]^{-1} = 2, \quad \text{i.e.,} \quad \gamma = 4(\Delta^2+1)^{1/2}. \tag{9.87}$$

As already explained, further information on the steady-state behavior of the system can be obtained from the extrema of $p_s(x)$. Equation (9.73) yields in the case $\lambda = 0$:

$$(1/2-x_m)x_m(1-x_m) - \frac{\Delta^2}{\gamma}x_m^2(1-x_m)^2(1-2x_m) - \frac{2}{\gamma}x_m(1-x_m)(1/2-x_m) - \frac{1}{\gamma}(1/2-x_m)^2(1-2x_m) = 0. \tag{9.88}$$

Obviously,

$$x_0 = 1/2 \tag{9.89}$$

is a root of (9.88). Dividing by $(1-2x_m)$ we obtain the fourth-order polynomial

$$-2\frac{\Delta^2}{\gamma}x_m^4 + 4\frac{\Delta^2}{\gamma}x^3 - \left(2\frac{\Delta^2}{\gamma}+1\right)x_m^2 + x_m - 1/2\gamma = 0. \tag{9.90}$$

The fact that for $\lambda = 0$ $p_s(x)$ is symmetric around $x_0 = 1/2$ implies that the roots of (9.90), i.e., the extrema of $p_s(x)$, are of the form

$$x_{2/3} = 1/2 \pm a, \quad x_{4/5} = 1/2 \pm b. \tag{9.91}$$

The standard form of (9.90) is obtained by division with $-2\Delta^2/\gamma$. Then (9.90) can be written as

$$(x_m - 1/2 - a)(x_m - 1/2 + a)(x_m - 1/2 - b)(x_m - 1/2 + b) = 0, \tag{9.92}$$

and a comparison of coefficients yields:

$$1 + \gamma/2\Delta^2 = 3/2 - a^2 - b^2, \tag{9.93}$$

$$-\gamma/2\Delta^2 = -1/2 + a^2 + b^2, \tag{9.94}$$

$$1/4\Delta^2 = 1/16 - b^2/4 - a^2/4 + b^2a^2, \tag{9.95}$$

where (9.93, 94) are linearly dependent. Solving for b^2, we obtain

$$b_{+/-}^2 = [\Delta^2 - \gamma \pm (\gamma^2 - 4\Delta^2)^{1/2}]/4\Delta^2. \tag{9.96}$$

Obviously a^2 should be equal to b^2, which is indeed easily verified to hold. Further, b^2 is real if

$$\gamma \geqslant 2\Delta . \tag{9.97}$$

This implies that for $\gamma < 2\Delta$ the stationary probability density has only one extremum, namely at $x_0 = 1/2$. It is a minimum, since according to (9.86) $p_s(b_i) = \infty$ for $\gamma < 2(\Delta^2 + 1)^{1/2}$. To continue gathering information about $p_s(x)$ that will eventually enable us to construct the phase diagram in the (Δ, γ) plane for $\lambda = 0$, we now turn our attention to the critical points. Remember that in the white-noise analysis, $x_0 = 1/2$ is a critical point for $\lambda = 0$ if $\sigma^2 = 4$. So let us look for the conditions under which x_0 is a triple root of (9.88). This implies that $b^2 = 0$ and is the case for

$$\Delta^2 = 2\gamma - 4 \quad \text{and} \quad \gamma > 2 . \tag{9.98}$$

Since $\Delta^2/\gamma = \sigma^2/2$ in the white-noise limit, the above condition means:

$$\sigma_c^2 = 4 - 8/\gamma = 4 - 8\tau_{cor} , \qquad \tau_{cor} < 1/2 . \tag{9.99}$$

It is worthwhile mentioning here that in Chap. 8 the bandwidth perturbation expansion for Ornstein-Uhlenbeck noise with $0 \neq \tau_{cor} \ll 1$ gave the following result, see below [8.7]:

$$\sigma_c^2 = 4(1 + \tau_{cor})^{-1} = 4 - 4\tau_{cor} + O(\tau_{cor}^2) .$$

Though no explicit upper limit for τ_{cor} can be determined by this method, (9.99) indicates that our result presumably does not depend qualitatively on the exact nature of the colored noise. The expression (9.99) for the critical variance, which we want to emphasize is an exact result, is very interesting for two reasons:

i) The intensity σ^2 necessary to induce critical behavior in the model, which does not show any instability at all in a deterministic environment, decreases as the correlation time of the colored noise increases.

ii) There is an upper limit for the correlation time of the noise, beyond which the critical behavior disappears. In this case, the upper limit $\tau_{cor} = 0.5$ is of the order of the macroscopic relaxation time, which for $\lambda = 0$ is $\tau_{macro} = 1$.

To continue the construction of the phase diagram, we remark that on the critical line (9.98) the second solution for b^2 reads

$$b^2 = \frac{2(\gamma - 4)}{4(2\gamma - 4)} . \tag{9.100}$$

This is positive for $\gamma > 4$ and vanishes at the point $\gamma = 4$. So at this point $x_0 = 1/2$ is a fivefold root of (9.88) for the extrema of $p_s(x)$ in the case $\lambda = 0$. This constitutes a strong indication that at this point $\gamma = 4$, $\Delta = 2$ on the critical line something happens. Indeed a numerical solution of the general equation for the extrema of $p_s(x)$, i.e., not specifying $\lambda=0$, shows that at the point $\gamma=4$, $\Delta=2$ two symmetric branches of critical points bifurcate which have $|\lambda|\neq 0$ and which are symmetric with respect to $\lambda=0$.

Result (9.100) implies furthermore that for $\lambda=0$, on the critical line $\Delta^2=2\gamma-4$ the stationary probability density $p_s(x)$ has in addition to the triple extremum at $x=1/2$ an extremum to the right and left of the critical point, as long as $x=1/2\pm b$ is in the support U. Since this is an important question for all roots of (9.90), calculated via (9.96), we shall now derive the corresponding conditions. Combining (9.85, 96) we have that $x_i \in U$ ($i=2$-5) if $b^2 \geqslant 0$ and

$$\frac{1}{2\Delta}[\Delta^2-\gamma\pm(\gamma^2-4\Delta^2)^{1/2}]^{1/2} \leqslant -\frac{1}{2\Delta}[1-(1+\Delta^2)^{1/2}]\,.$$

Thus

$$x_{2/3}=1/2\pm b_+ \in U \quad \text{if} \quad (\gamma^2-4\Delta^2)^{1/2} \leqslant 2+\gamma-2(1+\Delta^2)^{1/2} \tag{9.101}$$

and

$$x_{4/5}=1/2\pm b_- \in U \quad \text{if} \quad -(\gamma^2-4\Delta^2)^{1/2} \leqslant 2+\gamma-2(1+\Delta^2)^{1/2}\,. \tag{9.102}$$

Since γ has to fulfill the inequality (9.97) for b^2 to be real and since furthermore $2(1+\Delta^2)^{1/2}-2 \leqslant 2\Delta$, the right-hand side of (9.101, 102) is nonnegative. This implies that $x_{4/5} \in U$ if they exist. On the other hand, from the inequality (9.101) we conclude that $x_{2/3} \in U$, existence provided, if

$$\gamma \leqslant 2(1+\Delta^2)^{1/2}\,. \tag{9.103}$$

This is identical to (9.86), i.e., the roots $x_{2/3}=1/2\pm b_+$ leave the support of $p_s(x)$ on the same line, where the behavior of the stationary probability density changes from $p_s(b_i)=0$ to $p_s(b_i)=\infty$. Putting together all the information obtained on the extrema of the stationary probability density as well as on its behavior near the boundaries of the support U, we can construct the phase diagram of the steady-state behavior of the genetic model under the influence of external D noise for $\lambda=0$. The result is displayed in Fig. 9.3. Most remarkable is that not only are there transitions from monomodal to bimodal behavior, but also a region, albeit small, where $p_s(x)$ displays two maxima and three minima appears.

The various domains of this diagram and the manner in which the system makes the transition from one to the other suggest that two distinct mechanisms, simultaneously at work, determine the form of the stationary probability density. First we have a *peak-damping mechanism* which the multiplicative noise considered here has in common with the additive noise appearing in the usual Langevin treatment. This merely reflects the fact that the randomness of the environment tends to disorganize the system. Second, we have a *peak-splitting mechanism* which is specific of multiplicative noise and corresponds to a noise-

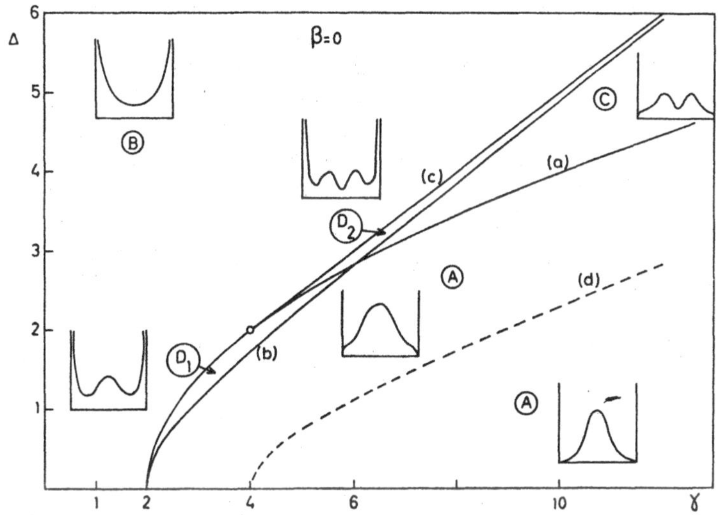

Fig. 9.3. The (Δ, γ) phase diagram for the steady-state behavior of the genetic model. *Curves* $(a-d)$ correspond to the conditions: (a) $\Delta^2 = 2\gamma - 4$; (b) $\gamma = 2(1+\Delta^2)^{1/2}$; (c) $\gamma = 2\Delta$; (d) $\gamma = 4(\Delta^2+1)^{1/2}$. The shape of $p_s(x)$ for the different regions A, B, C, D_1 and D_2 confined by these curves is sketched

induced phase transition. Regions A and B are characterized by the dominance of the peak-damping mechanism. In Region A however, since the correlation time is short, the system still adjusts to the mean value. In Region B on the contrary, the noise has a large correlation time and the system has time to relax to the "instantaneous" values of the external fluctuations, i.e., it is most probably near the deterministic steady-state corresponding to $\lambda + \Delta$ and $\lambda - \Delta$. This is of course a particularity of the D noise, namely a consequence of the fact that the state space of the noise consists of a finite number of levels. In this sense the environment is not as random as in the case of, say, an Ornstein-Uhlenbeck process where the state space is a continuum of states. As we have seen in Chap. 8, in this case no divergence of the probability density occurs at the boundary of the support, which coincides for the O-U noise with the physical state space. In Region C the behavior of the probability density is dominated by the peak-splitting mechanism. Clearly this bimodal density does not correspond to a compromise between the behavior of Regions A and B. The two maxima never move outwards sufficiently far in order to reach the boundaries of the support and to make the value of $p_s(x)$ diverge there. Indeed, $p_s(x)$ changes already from non-divergent to divergent behavior when the two maxima are still at a finite distance from the boundaries of the support. This transition is observed in domain D_2, which may thus be viewed as a compromise between the behavior of C and B. It is most interesting to note that at the "bifurcation point" $\gamma = 4$, $\Delta = 2$, where $x_{m0} = 1/2$ is a fivefold root and two symmetric branches of critical point with $|\lambda| \neq 0$ appear, something similar to an exchange of stability in deterministic bifurcations occurs: namely for γ to the left of this point, the critical behavior is caused by the peak-damping mechanism. For γ to the right of this point, the peak-splitting mechanism is responsible for inducing a critical point.

9.2.3 Hongler's Model

To conclude this section let us briefly consider the behavior of Hongler's model which, as explained in Chap. 6, is related to the genetic model. Starting with the SDE

$$\dot{X}_t = -\frac{1}{2\sqrt{2}}\tanh(2\sqrt{2}X_t) + \frac{1}{4}\,\frac{I_t}{\cosh(2\sqrt{2}X_t)} \tag{9.104}$$

one finds for the stationary probability density

$$p_s(x) = N\cosh(2\sqrt{2}x)\left[\frac{\Delta^2}{2} - \sinh^2(2\sqrt{2}x)\right]^{(\gamma/2)-1}. \tag{9.105}$$

Its support consists of the interval

$$\left[-\frac{1}{2\sqrt{2}}\operatorname{arc\,sinh}\left(\frac{\sqrt{2}}{2}\Delta\right),\quad \frac{1}{2\sqrt{2}}\operatorname{arc\,sinh}\left(\frac{\sqrt{2}}{2}\Delta\right)\right]. \tag{9.106}$$

The extrema of (9.105) are the zeros of

$$\sinh(2\sqrt{2}x_{\rm m}) = 0\,, \quad \text{i.e.,} \quad x_{\rm m1} = 0\,, \tag{9.107}$$

and

$$\frac{\Delta^2}{2} - \sinh^2(2\sqrt{2}x_{\rm m}) - 2(\gamma/2-1)\cosh^2(2\sqrt{2}x_{\rm m}) = 0\,. \tag{9.108}$$

Using the relation $\cosh^2 z - \sinh^2 z = 1$, (9.108) implies that

$$\cosh^2(2\sqrt{2}x_{\rm m}) = -\frac{\Delta^2+2}{2(1-\gamma)}\,. \tag{9.109}$$

Since the lhs is always bigger than or equal to one, it follows that $x_{\rm m1} = 0$ is a triple root if

$$\Delta^2 = 2\gamma - 4 \quad \text{with} \quad \gamma \geqslant 2\,. \tag{9.110}$$

We see that the line of noise-induced critical points in Hongler's model coincides with that of the genetic model (9.98, 99). It can be verified that the extrema

$$x_{\rm m2,3} = \frac{1}{2\sqrt{2}}\operatorname{arcosh}\left(\frac{\Delta^2+2}{2(\gamma-1)}\right)^{1/2} \tag{9.111}$$

lie in the support, if and only if

$$\gamma > 2\,. \tag{9.112}$$

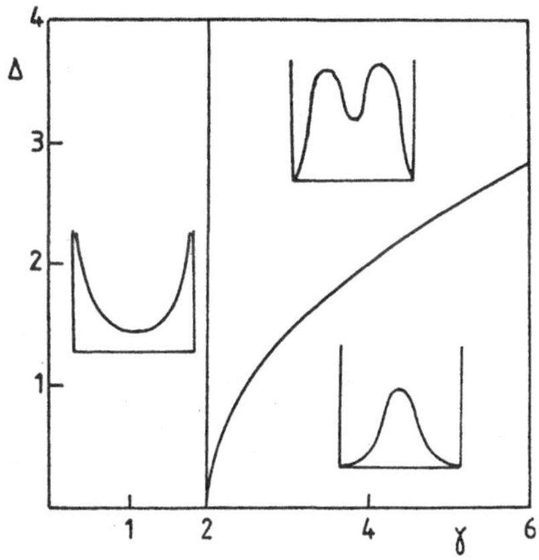

Fig. 9.4. The (Δ, γ) phase diagram of Hongler's model

The value of γ for which the probability density diverges at the boundaries of the support is given by (9.68) which reads in this case

$$-F'(b_i) = \frac{1}{\cosh^2(2\sqrt{2}\,b_i)} + 2\sqrt{2}\,\Delta\frac{\sinh(2\sqrt{2}\,b_i)}{4\cosh^2(2\sqrt{2}\,b_i)} = \gamma/2\,. \tag{9.113}$$

Substituting the expression for b_i from (9.106) one obtains

$$\gamma/2 = 1\,. \tag{9.114}$$

The fact that (9.114) does not depend on Δ reflects the property that in Hongler's model the normal mode $\omega(\bar{x})$ is independent of $\bar{x}$. This is a particular feature common to all models belonging to the class defined in Sect. 8.3.

The phase diagram of Hongler's model displayed in Fig. 9.4 consists of three domains and compared with the genetic model is much simpler. This is a good illustration of the fact that deterministic models which are closely related present markedly different macroscopic responses when subjected to external noise. In particular, there is no point on the line of critical points at which the peak-splitting mechanism which operates at large values of γ is replaced by a peak-damping mechanism dominating the behavior of $p_s(x)$ for small γ.

On the other hand, it is now possible to compare on related models, namely the genetic model and Hongler's model, the influence of the correlation time of the noise on the location of the pure noise-induced transition point. This is done in Table 9.1. It reveals that an increase in the correlation time of the noise can either increase or decrease the value of σ_c, depending on the nature of the noise and certain features of the system. The different behavior of the genetic model and Hongler's model for O-U noise is in our opinion due to the fact that the latter does not possess built-in finite asymptotes to which the switching curve tends for values of $\lambda \to \pm\infty$. This feature combined with the coincidence of the average state of the O-U noise with its most probable state, i.e., the state in the neighborhood of which it spends most of its time, clearly makes the transition to a bimodal distribution more difficult when $\tau_{\rm cor}$ increases. This interpretation is

Table 9.1. Influence of τ_{cor} on the critical variance σ_c^2

Model	O-U noise	Dichotomous noise
Genetic model	$\sigma_c^2 = 4 - 4\tau_{cor}$	$\sigma_c^2 = 4 - 8\tau_{cor}$
Hongler's model	$\sigma_c^2 = 4 + 4\tau_{cor}$	$\sigma_c^2 = 4 - 8\tau_{cor}$

supported by the finding that with dichotomous noise, Hongler's model behaves qualitatively like the genetic model. This is a consequence of the structure of the dichotomous noise. It always imposes a finite support to $p_s(x)$ and furthermore as τ_{cor} increases it favors the sojourn of the realizations of X_t in the neighborhood of $\bar{x}(\lambda \pm \Delta)$, since for a D noise the fluctuating parameter is always in one of its extremes, i.e., $\lambda \pm \Delta$.

9.2.4 Dichotomous Periodic Forcing

We shall close this section by pointing out the essential differences between the behavior of the genetic model in the presence of a dichotomous noise and its response to regular variations of the external parameter amounting to periodic jumps between two well-defined values. We thus consider that $\lambda_t = \lambda_{t+nT}$ is a time periodic function of period T such that it takes each of the values $+\Delta$ and $-\Delta$ for one half of the period (Fig. 9.5). Obviously in the course of time, the state variable $X(t)$ will evolve, whatever its initial condition, towards a stationary periodic regime of period T: $X(t) = X(t+nT)$. We want to compare this asymptotic situation which corresponds to the simplest form of *deterministic* external forcing with the stationary regime realized in the presence of dichotomous Markovian noise. For this purpose, we define the "stationary probability density" $\tilde{p}_s(x)$ of finding the system in the interval $(x, x+dx)$ as

$$\begin{aligned}\tilde{p}_s(x)\,dx &= \frac{1}{T}\left|\frac{dt}{dx}\right| dx \\ &= \frac{1}{T}\left(\left|\frac{dt}{dx}\right|_{\lambda=\Delta} + \left|\frac{dt}{dx}\right|_{\lambda=-\Delta}\right) dx \\ &= \frac{1}{T}\left(1/\left|1/2 - x + \Delta x(1-x)\right| + 1/\left|1/2 - x - \Delta x(1-x)\right|\right) dx\end{aligned}$$

and investigate its behavior as a function of Δ and T. As in the case of the dichotomous Markovian noise Δ is the amplitude of the jumps around the average value $\lambda = 0$ and T is the analog of the correlation time γ^{-1}.

The extrema x_m of $\tilde{p}_s(x)$ are the zeros of the equation:

$$[2x_m(1-x_m) + \tfrac{1}{2}(1-2x_m)^2 + 2\Delta^2 x_m^2(1-x_m)^2](1-2x_m) = 0$$

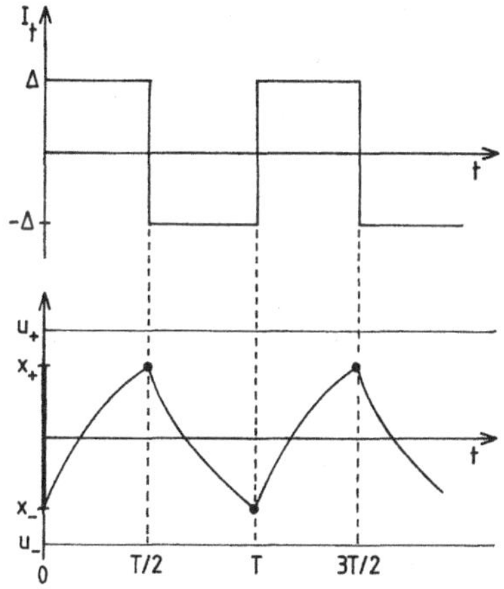

Fig. 9.5. Sketch of the response of the genetic model to a periodic forcing

which immediately shows that $x_m = 1/2$ is the only acceptable solution for all values of Δ and T. Furthermore, contrary also to what happens in the presence of noise, the nature of this extremum, i.e., maximum or minimum, is totally insensitive both to the amplitude of the forcing Δ and to its period T. Indeed one finds that

$$\left.\frac{d^2\tilde{p}_s(x)}{dx^2}\right|_{x=1/2} = \frac{64}{\Delta}\left(1+\frac{4}{\Delta^2}\right)$$

is always positive definite: *the extremum* $x_m = 1/2$ *is always a minimum.* The mechanism of peak splitting described above has completely disappeared. For all values of Δ and T the deterministic stationary state $x = 1/2$ is "destabilized", i.e., $x = 1/2$ has the lowest stationary probability density. This clearly results from the fact that $X(t)$ and the sign of its time derivative are strictly slaved to the jumps of λ; these jumps being completely discrete, the sign of dX/dt must necessarily change at each jump so that the harmonics of the noise play no role. For the same reason the support of $\tilde{p}_s(x)$ is necessarily smaller than (9.85). It is easily seen that it is restricted to the range $U' = [x_- = 1/2 - s, x_+ = 1/2 + s]$ (Figs. 9.5, 6) where

$$s = \frac{1}{2}\left\{-\left(\frac{1+k}{k-1}\right)\left(\frac{\Delta^2+1}{\Delta^2}\right)^{1/2} - \left[\left(\frac{1+k}{k-1}\right)^2\left(\frac{\Delta^2+1}{\Delta^2}\right) - 1\right]^{1/2}\right\}$$

with

$$k = \exp[-(1+\Delta^2)^{1/2}T/2]\,.$$

For a fixed value of intensity Δ, the more rapid the forcing, the smaller the support of $\tilde{p}_s(x)$ becomes. Furthermore transitions corresponding to a change in behavior of $\tilde{p}_s(x)$ near the boundaries of U' are excluded since always

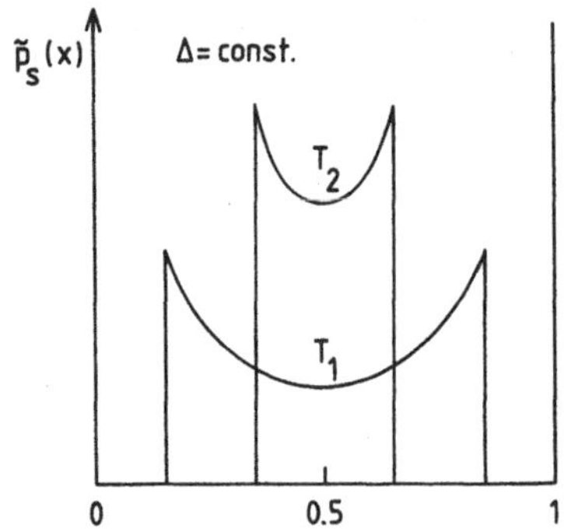

Fig. 9.6. Sketch of $\tilde{p}_s(x)$ for $T_2 < T_1$

$$\left.\frac{d\tilde{p}_s(x)}{dx}\right|_{x=x_+} > 0 \quad \text{and} \quad \left.\frac{d\tilde{p}_s(x)}{dx}\right|_{x=x_-} < 0 \, .$$

In the limit $\Delta \to \infty$, $T^{-1} \to \infty$ such that $\Delta^2/T^{-1} = \text{const}$, the support U' tends to zero but nevertheless keeps its "bimodal" shape. The behavior of the "probability density" $\tilde{p}_s(x)$ reflects the fact that contrary to what happens for the dichotomous noise, this is a noiseless limit. From the expressions of k and s it is clear that the appropriate scaling which keeps the support constant in the limit of an infinitely fast forcing ($T^{-1} \to \infty$) is $\Delta \to \infty$ such that $\Delta/T^{-1} = \text{const}$.

In conclusion: (i) the shape of $p_s(x)$ is bimodal and is *qualitatively independent* of the intensity Δ and period T; (ii) only the support $U' = (x_-, x_+)$ is affected by changes of the parameters describing the forcing.

9.3 Electrically Excitable Membranes

Electrically excitable nerve membranes play a primordial role in the study of complex nonequilibrium phenomena displayed by biological systems. For most biological systems, it is their very complexity which has made a quantitative theoretical description an elusive goal so far. Nerve membranes are one of the few exceptions in biology; they have been analyzed quantitatively in great detail experimentally as well as theoretically. One of the most extensively studied systems is the giant squid axon. It has the advantage of having a diameter of approximately 500 μm. This makes it far easier than in other systems to insert microelectrodes into the axon and stimulate or record its electrical activity. The understanding of electrically excitable membranes received a tremendous boost by the work of *Hodgkin* and *Huxley* [9.4] on the giant squid axon using the voltage clamp method (Figs. 9.7, 8). They developed a satisfactory phenomenological description of the dynamical properties of nerve membrane which still occupies a central place in electrophysiology.

Let us briefly summarize the relevant facts about nerve membranes. A nerve cell consists of a spherical cell body, the soma, and a number of branched extensions which ensure connections with other nerve cells. One extension is relatively large and long and called the axon. In consists of a membrane filled with an

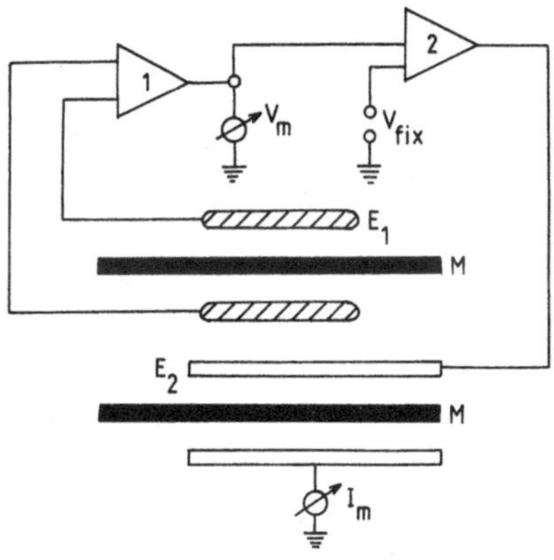

Fig. 9.7. Voltage clamp experimental setup

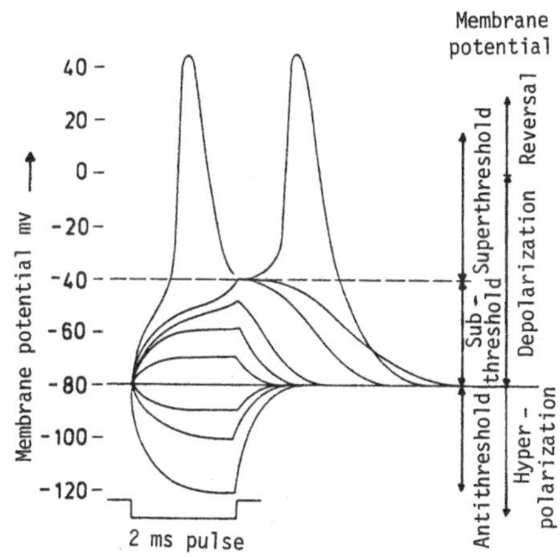

Fig. 9.8. Response of the axon membrane to a stimulus of 2 ms of increasing amplitude

electrolyte, the axoplasm, which is embedded in the extracellular fluid. We shall consider here only nerve membranes which are not surrounded by an insulating sheath, i.e., so-called unmyelinated axons. Measurements reveal that in the equilibrium or rest state an electrical potential exists across the membrane. It is negative inside the cell and of the order of -70 mV. The origin of this potential is the different composition of the axoplasm and the extracellular fluid as shown in Fig. 9.9.

The rest potential of the membrane does not correspond to the potential for which thermodynamic equilibrium exists between the extracellular and axoplasmic concentrations of the ionic species. The deviation is largest for the sodium ions. The concentration gradients are maintained against the passive diffusion of the ions across the membrane by an energy-dependent metabolic process, the so-called sodium pump. This process, which regulates the composition of the axoplasm and the extracellular fluid, is largely independent of the phenomenon of electrical excitation. Electrical excitation occurs if the membrane potential is perturbed beyond a certain threshold value. To be precise, if a small

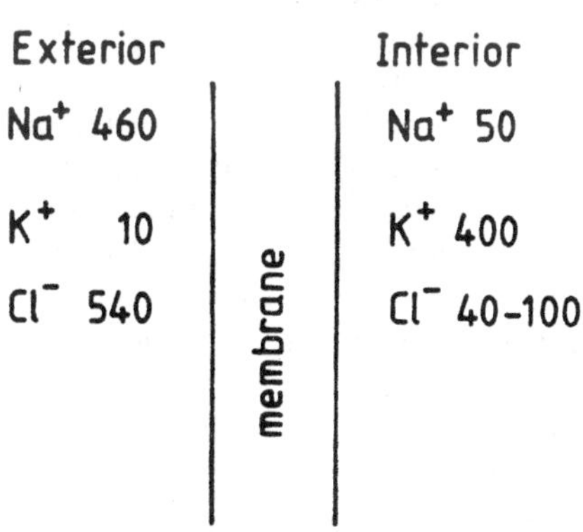

Fig. 9.9. Composition of the interior and exterior bulk phase of the axon membrane in mmol/l

stimulus is applied to the nerve membrane which displaces the membrane potential to a higher value, i.e., reduces the difference between the two sides of the cell membrane, so-called depolarization, then the potential rapidly returns to its initial value at the end of stimulus. Small stimulus means there is a linear relation between the strength of the stimulus and depolarization. This linearity breaks down if the stimulus is strong enough to cause a depolarization larger than some well-defined threshold value in general of the order of 30 – 40 mV. The depolarization continues to grow rapidly and nonlinearly and the membrane potential reaches a maximum of about + 30 mV before decaying back to the rest value. This excitation propagates along the axon. It is said that the nerve has undergone an action potential (Fig. 9.8).

The mechanism that gives rise to an action potential has been studied by Hodgkin and Huxley using the voltage clamp method (for a simple survey, see [9.5]). The voltage clamp technique corresponds to the setup shown in Fig. 9.7. The membrane potential V_m is measured with the shaded electrodes and electronically compared to V_{fix} which is imposed across the membrane. Amplifier 2 is used to drive a current across the membrane to compensate any deviations of V_m from V_{fix}. The voltage clamp allows the study of the response of the membrane to a fixed externally imposed displacement of the membrane potential from its rest value. Hodgkin and Huxley demonstrated that the mechanism, which generates an action potential, functions via the voltage dependence of the ionic permeabilities of the membrane. When the membrane potential is close to its rest potential, the conductivity g of the nerve membrane for sodium ions is very small. However, once the depolarization exceeds the threshold value, i.e., V_m reaches about – 30 mV, the permeability of the membrane for sodium ions increases explosively. As a result, sodium ions flow into the axoplasm and this influx enhances the depolarization further. This "autocatalytic" process is very fast, on a time scale of ms, and operates up to the point where the potential difference reaches the approximate equilibrium potential of the sodium ions. This flare, due to the change in the Na^+ permeability, is followed by two slower processes, on a time scale of 10 ms; one process decreases the sodium permeability and the other corresponds to an increase of the potassium permeability. The combined action of these two processes brings the membrane potential back to its rest value; the action potential is "regenerative". It is now widely believed that the sodium and potassium ion pass through distinct and specific so-called sodium and potassium channels.

Hodgkin and Huxley have provided a satisfactory quantitative description of the electric excitability of nerve membranes. We shall use their equations in the next section as a starting point for a theoretical analysis of nerve behavior under the influence of Markovian dichotomous voltage noise. This type of external noise is an appropriate choice here for two principal reasons. First, it allows us to obtain exact results in the case of a nonlinearly fluctuating parameter for any value of the variance and correlation time. Second, from an experimental point of view, this kind of noise requires only minor modifications of the traditional voltage clamp setup.

Up to now, studies of noise in membranes have been exclusively devoted to the spontaneous intrinsic noise of these systems, as for instance generated by the passage of the ions through the channels. However, as discussed in Chap. 1, these internal fluctuations can be safely neglected in the study of the macroscopic properties of axon patches, which contain a large number of ionic channels. We shall show here that a study of external voltage noise is worthwhile in view of the new phenomena which are predicted to exist [9.6]. Furthermore, it would constitute an experimental test of the Hodgkin-Huxley equations under conditions quite different from those for which they were initially derived.

9.3.1 The Hodgkin-Huxley Axon and the Dichotomous Voltage Noise

The data obtained from voltage clamp experiments led Hodgkin and Huxley to develop the following phenomenological description of nerve-membrane dynamics. To describe the potassium activation process, they introduced an auxiliary variable n which obeys the kinetic equation

$$\dot{n} = \alpha_4(1-n) - \beta_4 n\,, \qquad n \in [0,1] \tag{9.115}$$

where α_4 and β_4 are known functions of the membrane potential V (V in mV), namely

$$\begin{aligned} \alpha_4 &= \frac{V+10}{100}\left[\exp\left(\frac{V}{10}+1\right)-1\right]^{-1}, \\ \beta_4 &= \exp(V/80)/8\,. \end{aligned} \tag{9.116}$$

The potassium conductance g is related to the auxiliary variable by

$$g = \tilde{g}_k n^4\,. \tag{9.117}$$

We set the maximal conductance equal to one, $\tilde{g}_k = 1$, and obtain for the temporal evolution of the potassium conductance

$$\dot{g} = 4\alpha_4(V_t)(g^{3/4}-g) - 4\beta_4(V_t)g\,. \tag{9.118}$$

The change in sodium permeability involves an activation and an inactivation process, which necessitates the use of two auxiliary variables, m to describe the activation of the sodium channels, and h to take inactivation into account. They obey the kinetic equations:

$$\begin{aligned} \dot{m} &= \alpha_3(1-m) - \beta_3 m\,, \qquad m \in [0,1] \\ \dot{h} &= \alpha_1(1-h) - \beta_1 h\,, \qquad h \in [0,1] \end{aligned} \tag{9.119}$$

with

$$\begin{aligned} \alpha_3 &= \frac{(V+25)}{10}\left[\exp\left(\frac{V}{10}+\frac{5}{2}\right)-1\right]^{-1}, \\ \beta_3 &= 4\exp(V/18)\,, \end{aligned} \tag{9.120}$$

$$\alpha_1 = \frac{7}{100} \exp(V/20) ,$$

$$\beta_1 = \left[\exp\left(\frac{V}{10} + 3 \right) + 1 \right]^{-1} .$$

The sodium conductance is given by

$$g = \tilde{g}_{\text{Na}} m^3 h .$$

In the following, we shall analyze separately the behavior of the sodium and potassium activation processes. In other words, we consider only situations in which the nerve membranes have been treated with chemical or pharmacological agents blocking selectively either the sodium or the potassium channels. Experimentally this can be realized as follows: (i) the sodium channels are blocked by tetrodotoxin so that only the behavior of the potassium channels is followed [9.7]; (ii) the potassium channels are blocked with tetraethylammonium [9.8]. In addition, we should like to avoid any interference between the sodium activation and inactivation processes. We therefore suppose that the membrane has been treated with pronase which inhibits inactivation [9.9]. The merit of these artificial situations is that it enables us to avoid getting bogged down by the full complexity of the behavior of electrically excitable membranes. It permits us to study the basic building blocks of electrical excitation separately which is in any case essential for a subsequent interpretation of the nerve response as a whole to externally fluctuating constraints. Furthermore, these conditions exclude any deterministic instability. No threshold behavior is possible, regardless of the value of the membrane potential. This implies that any transition induced by external voltage noise is a pure noise effect. Since the sodium inactivation process can be disregarded in the situations considered here, the sodium conductance is a function of m only, $\tilde{g}_{\text{Na}} m^3$, and its temporal evolution is given by (maximal sodium conductance has been normalized to one)

$$\dot{g} = 3\,\alpha_3(V_t)(g^{2/3} - g) - 3\,\beta_3(V_t)\,g . \tag{9.121}$$

In summary, the sodium and potassium conductances obey here an equation of the general form

$$\dot{g} = \nu\alpha_\nu(V_t)(g^{(\nu-1)/\nu} - g) - \nu\beta_\nu(V_t)\,g \equiv F(g, V_t) , \tag{9.122}$$

with

$$\nu = 4 \text{ for } \mathrm{K}^+ \quad \text{and} \quad \nu = 3 \text{ for } \mathrm{Na}^+ .$$

In the following, we shall analyze the behavior of these conductances for the case where V_t is a random process fluctuating between two well-defined values (Fig. 9.10):

$$V_t = V + I_t . \tag{9.123}$$

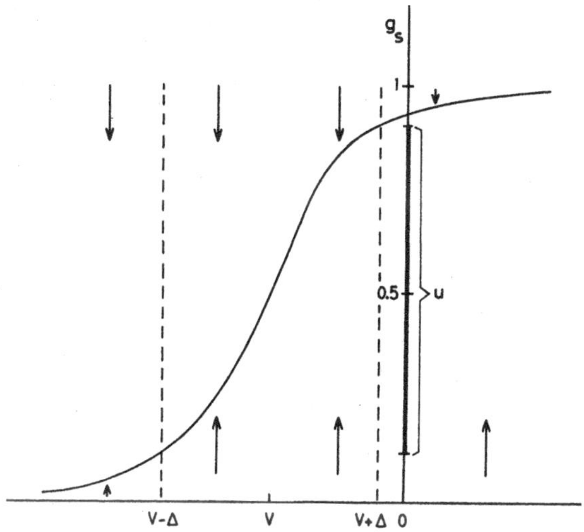

Fig. 9.10. Schematic plot of the curve of the deterministic steady states of axon conductance as a function of V. The state space of the random process (g_t, V_t) is given by the two broken lines. The arrows indicate the direction of the evolution of g_t. It is obvious that the whole probability mass will be inside the support U, indicated by the bold-face line on the g_s axis, for $t \to \infty$

The dichotomous Markov process is I_t with state space $\{\Delta, -\Delta\}$. It is easily verified that in the case of nonlinear D noise the stationary probability density is given by the expression:

$$p_s(g) = N\left(\frac{1}{F(g, V+\Delta)} - \frac{1}{F(g, V-\Delta)}\right)$$
$$\times \exp\left[-\frac{\gamma}{2}\int^g dg' \left(\frac{1}{F(g', V+\Delta)} + \frac{1}{F(g', V-\Delta)}\right)\right]. \tag{9.124}$$

The support of $p_s(g)$ is given by

$$U = [\bar{g}(V-\Delta), \bar{g}(V+\Delta)] , \tag{9.125}$$

where $\bar{g}(V \pm \Delta)$ is the steady-state solution of (9.122) corresponding to the value $V \pm \Delta$ of the potential.

The local extrema of the stationary probability density $p_s(g)$ are the zeros of

$$-\left[F'(g, V+\Delta) + \frac{\gamma}{2}\right] \cdot F^2(g, V-\Delta) + \left[F'(g, V-\Delta) + \frac{\gamma}{2}\right] \cdot F^2(g, V+\Delta) = 0$$
$$F' \equiv \partial_g F(g, V) . \tag{9.126}$$

The behavior of the probability density $p_s(g)$ near the boundaries of the support can change from divergent to nondivergent. It is easily verified that the probability density diverges at the boundaries when the inverse of the correlation time γ satisfies the following condition

$$-2F'(\bar{g}(V\pm\Delta), V\pm\Delta) > \gamma_{\pm} \,. \tag{9.127}$$

Here γ_+ refers to the conditions for the upper boundary, γ_- for the lower boundary. When (9.127) is not fulfilled, $p_s(g)$ vanishes at the boundary. On the basis of these formulae, a complete picture of the steady-state properties of the Hodgkin-Huxley axon under fluctuating voltage clamp conditions can now be obtained.

9.3.2 Phase Diagrams for Sodium and Potassium Conductance of the Hodgkin and Huxley Axon

Using the expression for $F(g, V\pm\Delta)$ given in (9.127), and the relations (9.116, 117) for α and β, the potassium and sodium stationary probability density can straightforwardly be calculated from (9.124). In the potassium case it reads:

$$\begin{aligned} p_s(g) = {} & N g^{-3/4}\{\alpha(V-\Delta)-\alpha(V+\Delta)+[\alpha(V-\Delta)+\beta(V-\Delta) \\ & -\alpha(V+\Delta)-\beta(V+\Delta)]\, g^{1/4}\} \\ & \times |\alpha(V+\Delta)-[\alpha(V+\Delta)+\beta(V+\Delta)]\, g^{1/4}|^{(\gamma/2)[\alpha(V+\Delta)+\beta(V+\Delta)]-1} \\ & \times |\alpha(V-\Delta)-[\alpha(V-\Delta)+\beta(V-\Delta)]\, g^{1/4}|^{(\gamma/2[\alpha(V-\Delta)+\beta(V-\Delta)]-1} \,. \end{aligned}$$

Its extrema are given by (9.126):

$$\begin{aligned} & -\left[\alpha(V+\Delta)(3/4-n)-\beta(V+\Delta)n+\frac{\gamma}{2}n\right][\alpha(V-\Delta)(1-n)-\beta(V-\Delta)n]^2 \\ & +\left[\alpha(V-\Delta)(3/4-n)-\beta(V-\Delta)n+\frac{\gamma}{2}n\right][\alpha(V+\Delta)(1-n)-\beta(V+\Delta)n]^2 = 0 \,, \\ & n = g^{1/4} \,. \end{aligned} \tag{9.128}$$

The stationary probability density diverges at the lower boundary of the support U when

$$\gamma_+ = 2[\alpha(V+\Delta)+\beta(V+\Delta)] \,,$$

and at the upper boundary when

$$\gamma_- = 2[\alpha(V-\Delta)+\beta(V-\Delta)] \,.$$

Similarly, the corresponding expressions for the sodium system can be obtained.

Using these expressions, we can construct phase diagrams which display concisely the response of the conductances to external voltage fluctuations of amplitude Δ and correlation time γ^{-1}. For both potassium and sodium the mean value of the applied potential corresponds to g equal to one half under deterministic conditions (Fig. 9.10); that is $E\{V_t\} = -46.1$ mV for potassium and $E\{V_t\} = -38.4$ mV for sodium.

Figure 9.11 reports the results for the potassium system. For a given value of Δ, the probability density diverges at the upper (lower) boundary of the support U, if γ is chosen to the left of γ_+ (γ_-). This accounts for the existence of Regions A and D, in which the probability density exhibits only one extremum. In Region A, the extremum is a maximum, in D it is a minimum. If the value of Δ is chosen above 42.5 mV, and γ increases from zero and acrosses the γ_- line, $p_s[\bar{g}(V-\Delta)]$ vanishes and thus the probability density acquires an additional extremum which of course is a maximum. When it crosses the line γ_+, the disappearance of the divergence at the upper boundary of U leads to the disappearance of the minimum. Below 42.5 mV, the probability density has a saddle point for $\gamma = \gamma_s$ at which the minimum and maximum of Region B coalesce. As a result, inside γ_s one has a Domain C in which the probability density is monotone increasing from the lower to the upper boundary of the support. Figure 9.12 represents for each of these regions the typical behavior of the probability density $p_s(g)$. The influence of the correlation time on the position of the extrema is given in Fig. 9.15a for two typical values of the intensity of the external noise. One is chosen below the threshold value $\Delta \approx 42.5$ mV and the other is chosen above this value. Below, provided γ is not taken in Region C (Fig. 9.11), there is always an extremum in the neighborhood of $g = 0.5$. This extremum is a minimum to the left of Region C and a maximum to the right of C. The position of the second extremum, which exists in Region B, increases rapidly with γ from values which lie near the lower boundary of the support to values near the upper boundary. Above $\Delta = 42.5$ mV, the minimum which starts out near $g = 0.5$ for small values of γ, moves rapidly to the upper boundary which it attains at $\gamma = \gamma_+$. The maximum which appears at $\gamma = \gamma_-$ approaches $g = 0.5$ when $\gamma \to \infty$.

In Fig. 9.13 we report the results for the sodium system. Since the characteristic time of the sodium channels is much shorter than for the potassium system as can be seen from the coefficients α_ν and β_ν, the region of interest is shifted to higher values of γ, i.e., to shorter correlation times. Remarkably, despite the fact that sodium and potassium activations obey equations which are very similar, differing only in the power ν of the auxiliary variable, there is an essential difference between the phase diagrams of Figs. 9.11 and 13. This is basically due to the fact that for the sodium system the γ_+ and γ_- curves cross each other for a depolarization value which remains physiologically acceptable. This has the following three consequences: (i) contrary to what happens in the K^+ system, Domain C in Fig. 9.13, where $p_s(g)$ is a monotone function, is restricted to the value of Δ larger than 45 mV; (ii) the divergence occurs at the lower boundary of the support, whereas for the potassium case it is at the upper boundary; (iii) Region B, where two extrema occur, now occupies a finite domain in the (Δ, γ) plane. The behavior of the probability density in each of these regions is displayed in Fig. 9.14. Though the behavior of the probability density in Region C (Fig. 9.13) is qualitatively identical in both systems, it is important to underline that the mechanism of transition to this region is basically different in the sodium and potassium systems. In the case of the potassium system one deals with a soft transition corresponding to a local event inside the support, namely

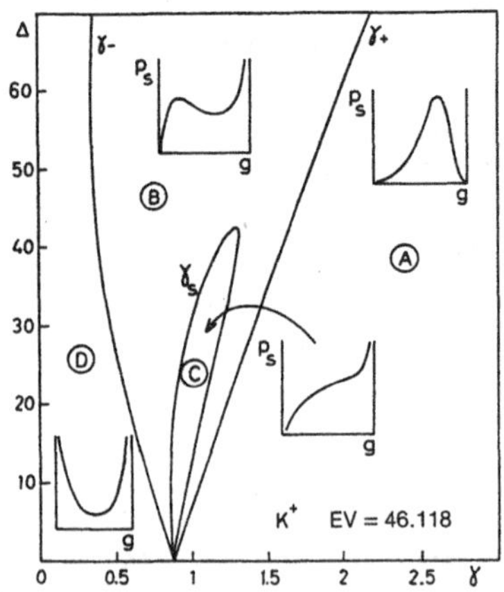

Fig. 9.11

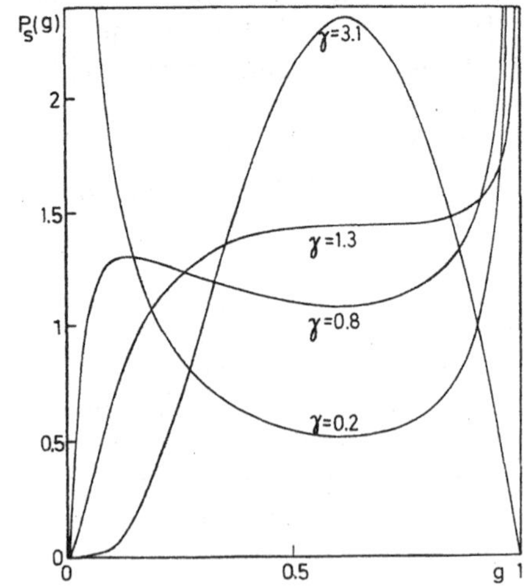

Fig. 9.12

Fig. 9.11. The (Δ, γ) phase diagram of the potassium system. The value chosen for the average potential is -46.1 mV for K^+. It corresponds to $g_K = 1/2$ under deterministic conditions

Fig. 9.12. Probability density of the potassium system for the values of γ indicated and chosen in regions A, B, C, D (Fig. 9.11) and $\Delta = 40$ mV. The support $U = (2.94 \times 10^{-2}, 0.904)$ has been uniformly stretched to $(0, 1)$

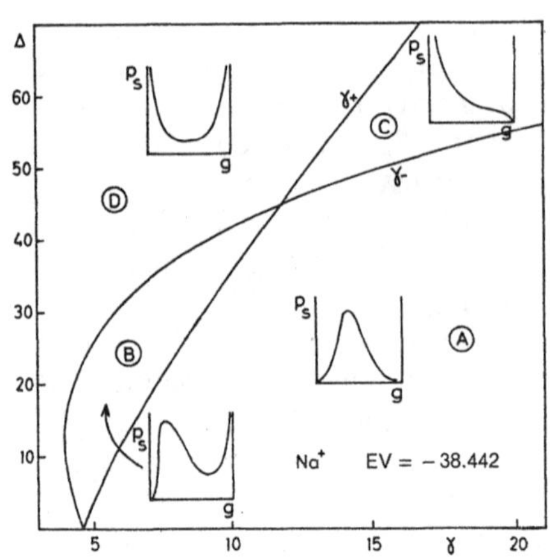

Fig. 9.13

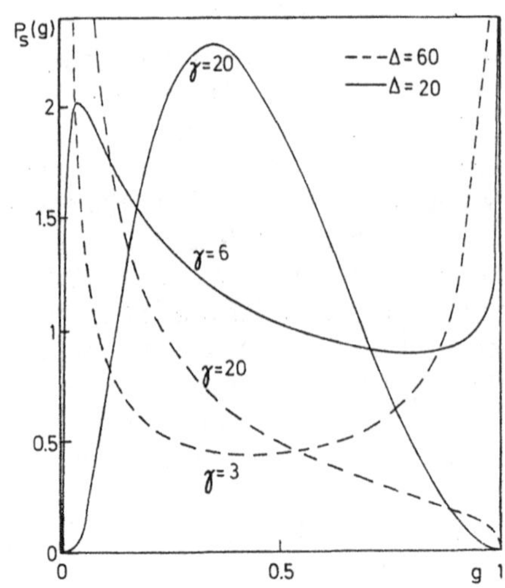

Fig. 9.14

Fig. 9.13. The (Δ, γ) phase diagram of the sodium system. The value chosen for the average potential is -38.442 mV. It corresponds to $g_{Na} = 1/2$ under deterministic conditions

Fig. 9.14. Probability density of the sodium system for values of the parameters as indicated in the figure and chosen in the Regions A, B, C and D (Fig. 9.13). The support, $U = (3.599 \times 10^{-2}, 0.876)$ for $\Delta = 20$ and $U = (3.789 \times 10^{-8}, 0.993)$ for $\Delta = 60$, has been uniformly stretched to $(0, 1)$

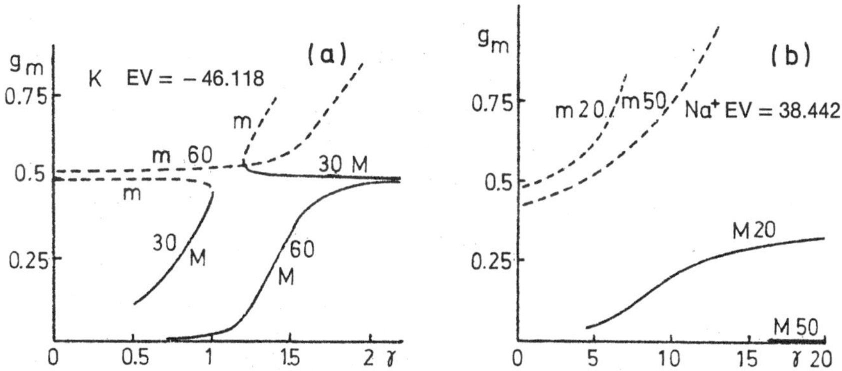

Fig. 9.15. **(a)** The location g_m of the extrema of $p_s(g)$ for the potassium system is displayed as a function of γ for two typical values of the amplitude Δ. Broken line: minima; full line: maxima. **(b)** The location g_m of the extrema of $p_s(g)$ for the sodium system is displayed as a function of γ for two tyical values of the amplitude Δ. For $\Delta = 50$ mV, the unique extremum in Region D coincides, for the values of γ reported, almost with the lower boundary of U. However, for γ tending to infinity it will approach the determinstic steady state $g = 0.5$. Broken line: minima; full line: maxima

the coalescence of two extrema into a saddle point. This phenomenon does not affect the behavior of the probability density near the boundaries of the support. This aspect connects it with the noise-induced phase transitions of the *peak-splitting type* described in the genetic model, Hongler's model or the Nitzan-Ross system. In contradistinction, in the sodium system Region C appears via a hard transition caused by the abrupt change of behavior at one of the boundaries of the support.

The marked difference between the potassium and the sodium channels is further illustrated in Fig. 9.15 which gives the location of the extrema of $p_s(g)$ as a function of γ for Δ fixed. Whatever the value of Δ, for small γ there is always a minimum near $g = 0.5$. It moves towards the upper boundary of the support which it attains for $\gamma = \gamma_+$. For $\gamma \geqslant \gamma_-$ the location of the additional extremum of $p_s(g)$ (maximum), which appears at the lower boundary, increases slowly towards $g = 0.5$. The larger the intensity of the noise, the slower the rate of approach to $g = 0.5$. As is to be expected from the phase diagram, the two curves overlap only if $\Delta < 45$ mV. The results of this section are a good illustration of the fact that systems, the deterministic phenomenological description of which closely resemble each other, can respond to externally fluctuating constraints in quite different ways.

These results show some features which are quite interesting from a physiological point of view. (i) A relation between the existence of Domain C and the threshold depolarization capable of triggering an action potential seems to exist: indeed the lower boundary of this domain in the sodium case, respectively upper boundary in the potassium case, typically corresponds to values of Δ of the order of 40 mV. (ii) In the regions where the probability density diverges only at one boundary of the support U, the potassium and sodium systems behave in the same way below this "threshold" Δ, i.e., both channels are preferentially in the open state. In contrast, above this "threshold" Δ the behavior of the sodium system changes in the sense that now *it favors the closed state*. This is a strong indication that in the complete Hodgkin-Huxley system, i.e., where the nerve membrane has not been treated with the chemical and pharmacological agents mentioned above, external voltage noise could modify the threshold value and the generation of the action potential. (iii) Compared with the genetic model which also has no instability under deterministic environmental conditions, the phase diagram of the nerve system is less complex. This suggests that the nerve system is less sensitive to the influence of external noise.

Let us now briefly discuss some experimental aspects. We believe that an experimental observation of the predicted phenomena is feasible for the following reasons: (i) dichotomous Markov noise with a well-defined amplitude Δ and correlation time γ^{-1} can be generated by simple electronics; (ii) the typical switching time for voltage clamp experiments is microseconds whereas the typical time scale of the conductances is $1-10$ ms, so that the correlation time of the noise can range over a sufficiently large interval (down to one-tenth or less of the characteristic time of the conductances); (iii) since the interesting range of correlation times is shorter or of the order of the typical duration of an action potential, the mean frequency γ is high enough for the experiments to be carried out over a sufficiently long time to obtain reliable conductance statistics.

To compare our theoretical predictions directly with experimental results, it is necessary to disregard in the current measurements the capacitive currents that follow each jump of the potential. It is also known that nerve membranes have a leakage current which was not explicitly taken into account in the above analysis. This is justified since the leakage conductance is not voltage dependent. This has the consequence that the leakage current takes only two well-defined values corresponding to the values of the potential. The currents measured under fluctuating potential conditions can thus be corrected by subtracting the leakage component corresponding to the actual value of the potential.

To close this section, let us remark that the use of different kinetic schemes for the nerve membrane, e.g., two-state channel models [9.10–14] or models assuming cooperative interactions in the membrane [9.15–17], which agree reasonably well with each other and the Hodgkin-Huxley model under deterministic environmental conditions, lead to distinctly different predictions for the behavior of excitable membranes under fluctuating external constraints [9.6]. Therefore, any essential deviation in the kinetics of real nerve membranes from the Hodgkin-Huxley axon would be clearly revealed in a fluctuating environment. From an experimental point of view, probing the nerve membrane behavior with external dichotomous noise should thus be a convenient method to explore its molecular properties and to discriminate between different kinetic descriptions.

10. The Symbiosis of Noise and Order – Concluding Remarks

The major theme of Chap. 1 was the omnipresence of noise in natural systems. In view of the noise-induced phenomena described in the preceding chapters this fact gains considerably in importance. The question arises as to how far the macroscopic behavior observed in natural systems is determined by environmental variability. The more complex the system is, the larger the number of noise sources and the more pressing this question becomes. The example par excellence is certainly the brain. Not only does it function with strongly varying sensory input, but its activity involves a great deal of intrinsic stochasticity, as, for instance, the random firing of neurons. On a different level, this astonishing symbiosis of order and randomness occurs in the behavior of populations of social insects. To cope with a fluctuating food supply, it seems to be necessary for the survival of the insect societies to incorporate an appreciable amount of behavioral randomness in the food exploration mechanism. On the other hand, highly complex systems may also be influenced by fairly simple types of random variability. Amazingly, even essential properties of complex systems can be critically dependent on fluctuating environmental conditions. For example, in many species of turtles the sex of the embryos is not genetically fixed but primarily decided by the temperature at which the eggs are incubated [10.1]. A temperature change of only a few degrees will change a clutch from male to female or vice versa. Situations similiar to these three examples can be found in quite a few natural systems. The results obtained in the preceding chapters lead us to speculate that the effect of noise will be fundamental for an understanding of the behavior of complex natural systems. Contrary to intuitive belief, external noise might well play an active role in the organization of such systems. A more precise theoretical investigation of this question will remain very much an open problem until satisfactory phenomenological descriptions of such complex systems can be achieved.

However, already in simpler laboratory systems many intriguing questions pertaining to the role of noise in self-organizing systems arise. A prime example is the role of external noise in structured hydrodynamic flows. To be specific, it is observed that for Bénard or Couette systems with large aspect ratios, turbulent behavior occurs at values of the Rayleigh or Reynolds number above, but astonishingly close to the critical value [10.2–4]. These observations led *Gorman* et al. [10.5] to pose the question whether this low R chaos is of intrinsic origin or caused by external noise. The most interesting aspect of this question stems from the fact that if the answer turns out to be the latter alternative, then it would demonstrate that even an external noise of small amplitude can significantly change the behavior of nonlinear systems. The question of turbulence and external noise has already been investigated theoretically in a different hydrodynamic system, namely superfluid helium [8.49]. Here it is predicted that finite amplitude noise delays the onset of turbulence (Sect. 8.8).

In this monograph we treated only one-variable systems. This was motivated by the fact that for two- or more variable systems explicit analytical results are the exception and not the rule. This restriction to one-variable systems excludes, of course, certain types of behavior at noise-induced transition points. A new behavior which is frequent in two- or more variable systems is the occurrence of self-sustained oscillations. The influence of external noise on this behavior has been investigated by *Arnold* et al. [10.11] for the case of conservative oscillatory systems of the Lotka-Voltera type. It has been found that the existence of a stationary solution critically depends on whether or not the noise power spectrum contains the eigenfrequencies of the Lotka-Voltera system. It is also interesting that while studying a model for the oxidation of CO on a platinum surface, *de la Rubia* et al. [10.6] recently predicted that colored external noise can induce limit cycle behavior. In addition, to the restriction to one-variable systems, we also assumed in general, except in Sect. 8.7, that the system was spatially homogeneous, i.e., of an effective spatial dimension zero. An attempt to go beyond systems of zero spatial dimensionality and to study the influence of external noise on spatial structures in reaction-diffusion systems has been made by *Mikhailov* [10.7 – 10].

In its present stage the theoretical part of the field of noise-induced transitions is clearly more developed than its experimental side. Though the few experiments described in Chaps. 7 and 8 give qualitative support to the theory, much remains to be done from an experimental point of view. We hope that this book will encourage further experiments on noise-induced transitions. We have proposed a few new experiments in the three preceding chapters. The systems considered there have the merit that a quantitative comparison between theory and experiment is possible. Those systems are certainly not the only ones suitable for an experimental study. Hopefully this monograph will help experts in other fields to identify situations in which an investigation of the influence of external noise would be profitable. More experimental results on the role of external noise are desirable for two main reasons: (i) To gain a better understanding of the implications of noise-induced phenomena in complex natural systems. The proposed experiment on electrically excitable membranes falls in this category. (ii) To advance and refine theoretical modeling. Experiments on laser and other quantum optical systems, electrical circuits, photochemical reactions and hydrodynamic systems belong to this class. We expect from such experiments information about the white-noise idealization, time-dependent behavior, the interplay of chaos and noise, and the influence of noise on spatial structures. For the last two fields, experiments on nematic liquid crystals seem to be an ideal choice, as already stressed earlier.

While the theory of the influence of external noise on one-variable systems seems to us reasonably complete, the field of noise-induced transitions is still rather new and quite a few open questions remain, not only on the experimental side. Hopefully this monograph will provide the encouragement to study these questions.

Appendix

A. Generalized Stochastic Processes

Consider the following experimental situation: a certain time-dependent quantity $f(t)$ has to be measured at some instant s with say an electric device. Due to its inertia, the measuring device will however not immediately respond to the "input" $f(t)$ at the instant of time we are interested in. Instead, some time averaging will be carried out by the instrument and if it is a linear device, then its output can be represented by

$$\int_{\mathbb{R}} \varphi(t) f(t)\, dt = \phi_f(\varphi) , \tag{A.1}$$

where $\varphi(t)$ is the transfer function, a characteristic of the instrument. From the above remarks, it is obvious that is should have a peak at $t=s$ and then drop rapidly to zero. The narrower $\varphi(t)$, the better the instrument. Obviously the instrument effects a smoothing of the input $f(t)$. Hence even if $f(t)$ is a very irregular object, the integral and the quantity $\phi_f(\varphi)$ still exist. This observation indicates how to handle functions which do not exist in any ordinary sense, e.g., the derivative of the sample paths of a Wiener process. These generalized functions are defined via their action on a suitably large class of ordinary functions. To be specific: let $\varphi(t)$ be an arbitrary real-valued function which is infinitely often differentiable and which has compact support, i.e., vanishes identically outside some bounded interval on the time axis. Denote the space of all such functions by $C_K^\infty(\mathbb{R})$. By a functional on this space we understand the following: if φ is a function belonging to $C_K^\infty(\mathbb{R})$, then $\phi(\varphi)$ is a real (or generally a complex) number. For instance

$$\phi(\varphi) = \int_{\mathbb{R}} \varphi(t)\, g(t)\, dt \tag{A.2}$$

with a fixed continuous $g(t)$ and

$$\phi(\varphi) = \varphi(s) \tag{A.3}$$

are functionals on $C_K^\infty(\mathbb{R})$. They are also examples for a continuous linear functional. Here ϕ is called linear, if

$$\phi(a\varphi_1 + b\varphi_2) = a\phi(\varphi_1) + b\phi(\varphi_2) , \tag{A.4}$$

where a, b are real (complex) numbers. The functional ϕ is said to be continuous, if

$$\lim \varphi_n = \varphi \quad \text{implies} \quad \lim \phi(\varphi_n) = \phi(\varphi)\,. \tag{A.5}$$

Here $\varphi_n \to \varphi$ if $\varphi_n \equiv 0$ for $x \notin [t_1, t_2]$ independent of n and if φ_n and its derivatives converge in the usual sense uniformly to φ. Every continuous linear functional ϕ on $C_K^\infty(\mathbb{R})$ is called a generalized function or distribution. In order to avoid confusion with probability distributions we shall not use the latter name. Hence a generalized function is characterized by the set of real numbers $\phi(\varphi)$, $\varphi \in C_K^\infty(\mathbb{R})$. The second example of a generalized function given by (A.3), namely

$$\phi(\varphi) = \varphi(s) \tag{A.6}$$

corresponds to the Dirac delta function $\delta(t-s)$. This linear functional is formally written as an integral

$$\phi(\varphi) = \int_{\mathbb{R}} \varphi(t)\,\delta(t-s)\,dt = \varphi(s)\,. \tag{A.7}$$

A convenient property of generalized functions is that they possess derivatives of any order, which are again generalized functions. Indeed, inspired by the rule of integration by parts, the derivative of a generalized function $\dot{\phi}$ is defined as

$$\dot{\phi}(\varphi) = -\phi(\dot{\varphi})\,. \tag{A.8}$$

A generalized random process is defined in a similar way. We associate with every function φ a random variable $\phi(\varphi)$, i.e., $\phi(\varphi)$ is a stochastic process with the parameter set $C_K^\infty(\mathbb{R})$ and the real numbers as state space. Furthermore, $\phi(\varphi)$ has to be linear with probability one, i.e., fulfill (A.4) almost surely. In addition, $\phi(\varphi)$ has to be continuous in the following sense: if φ_{in}, $i = 1, \ldots, m$ converges to φ_i for $n \to \infty$ then the probability distribution of $(\phi(\varphi_{1n}), \ldots \phi(\varphi_{in}), \ldots \phi(m_{mn}))$ converges to the probability distribution of $(\phi(\varphi_1), \ldots, \phi(\varphi_i), \ldots, \phi(\varphi_m))$. A generalized stochastic process is said to be Gaussian if for arbitrary linearly independent functions $\varphi_1, \ldots, \varphi_n$ the random variables $(\phi(\varphi_1), \ldots, \phi(\varphi_n))$ have a joint Gauss distribution. Such a process is, like an ordinary Gaussian process, completely characterized by the mean-value functional

$$E\{\phi(\varphi)\} = m(\varphi) \tag{A.9}$$

and the covariance functional

$$\operatorname{Cov}(\phi) = E\{[\phi(\varphi) - m(\varphi)]\,[\phi(\psi) - m(\psi)]\} = C(\varphi, \psi)\,. \tag{A.10}$$

Obviously the derivative of a Gaussian generalized process is again a Gaussian generalized process with

$$E\{\dot{\phi}(\varphi)\} = -m(\dot{\varphi}) \tag{A.11}$$

and

$$\mathrm{Cov}(\dot{\phi}) = C(\dot{\varphi}, \dot{\psi}) \,. \tag{A.12}$$

This allows us to establish that the derivative of the Wiener process is the Gaussian white noise. Indeed, with any ordinary process X_t may be associated a generalized process ϕ_X by

$$\phi_X(\varphi) = \int_{\mathbb{R}} \varphi(t) X_t dt \,. \tag{A.13}$$

In particular, for the Wiener process W_t we have

$$\phi_w(\varphi) = \int_0^\infty \varphi(t)\, W_t dt \tag{A.14}$$

which clearly is a Gaussian generalized process with

$$m_w(\varphi) = 0 \tag{A.15}$$

and

$$\begin{aligned} C_w(\varphi, \psi) &= E\left\{ \int_0^\infty \varphi(t)\, W_t dt \cdot \int_0^\infty \psi(s)\, W_s ds \right\} \\ &= \int_0^\infty \int_0^\infty \varphi(t)\, \psi(s) E\{W_t W_s\} dt\, ds \\ &= \int_0^\infty \int_0^\infty \varphi(t)\, \psi(s) \min(t,s)\, dt\, ds \,. \end{aligned} \tag{A.16}$$

Let us consider the derivative of the generalized process $\phi_w(\varphi)$ associated with the Wiener process W_t.

$$E\{\dot{\phi}_w(\varphi)\} = -m_w(\dot{\varphi}) = 0$$

$$E\{\dot{\phi}_w(\varphi)\, \dot{\phi}_w(\psi)\} = E\{\phi_w(\dot{\varphi})\, \phi_w(\dot{\psi})\}$$

$$= \int_0^\infty \int_0^\infty \dot{\varphi}(t)\, \dot{\psi}(s) \min(t,s)\, dt\, ds$$

$$= \int_0^\infty dt\, \dot{\varphi}(t) \int_0^t ds\, s\, \dot{\psi}(s) + \int_0^\infty dt\, t\, \dot{\varphi}(t) \int_t^\infty ds\, \dot{\psi}(s)$$

$$= \int_0^\infty dt\, \dot{\varphi}(t) \left[t\, \psi(t) - \int_0^t \psi(s)\, ds \right] + \int_0^\infty dt\, t\, \dot{\varphi}(t) [\psi(\infty) - \psi(t)]$$

(since $\psi \in C_K^\infty(\mathbb{R})$ we have $\psi(\infty) = 0$)

$$= -\int_0^\infty dt\,\dot{\varphi}(t)\int_0^t \psi(s)\,ds$$

$$= \varphi(\infty)\int_0^\infty \psi(s)\,ds - \varphi(0)\cdot 0 + \int_0^\infty dt\,\varphi(t)\,\psi(t)$$

$$= \int_0^\infty \int_0^\infty dt\,ds\,\delta(t-s)\,\varphi(t)\,\psi(s)\,. \tag{A.17}$$

Hence the derivative of the Wiener process W_t is a generalized Gaussian process with mean-value zero

$$E\{\dot{W}_t\} = 0\,, \tag{A.18}$$

and a Dirac delta function as correlation function

$$E\{\dot{W}_t \dot{W}_s\} = \delta(t-s)\,. \tag{A.19}$$

B. Markov Property of Solutions of Ito SDE's

We shall represent here the essential steps in the proof of theorem (5.52). The measurability of X_s with respect to $\mathscr{F}_s$ is important, i.e., $X_s^{-1}(B)\in\mathscr{F}_s$ for all $B\in\mathscr{B}$. Furthermore, the family of σ fields $\mathscr{F}_s$ increases: $\mathscr{F}_u \subset \mathscr{F}_s$ for $u<s$. This means that all random variables X_u with $t_0\leqslant u\leqslant s$ are measurable with respect to $\mathscr{F}_s$. On the other hand, by definition, we have that $X_u^{-1}(B)\in\mathscr{A}[t_0,s]$ for all $t_0\leqslant u\leqslant s$ and all $B\in\mathscr{B}$ and $\mathscr{A}[t_0,s]$ is the smallest sub-σ-field for which this is true. It is the smallest sub-σ-field with respect to which all random variables X_u, $t_0\leqslant u\leqslant s$, are measurable. From these properties we conclude that all elements in $\mathscr{A}[t_0,s]$ belong also to $\mathscr{F}_s$. In other words, $\mathscr{F}_s$ contains $\mathscr{A}[t_0,s]$: $\mathscr{F}_t \supset \mathscr{A}[t_0,s]$.

It suffices thus to prove instead of (4.2) the stronger condition:

$$P(X_t\in B\,|\,\mathscr{F}_s) = P(X_t\in B\,|X_s) \quad \text{a.s.}\,. \tag{B.1}$$

To establish this relation, consider bounded measurable functions $h(x,\omega)$, for which $h(x,\cdot)$ is a random variable that is independent of $\mathscr{F}_s$ for every fixed x. For this class of functions the following equalities hold:

$$E\{h(X_s,\omega)\,|\,\mathscr{F}_s\} = E\{h(X_s,\omega)\,|X_s\} = H(X_s)\,, \tag{B.2}$$

where $H(x) = E\{h(x,\omega)\}$. The proof of this fact is rather technical and gives no further insight. Consequently it will be skipped here. To exploit (B.2), choose

$$h(x,\omega) = I_B(X_t(s,x)(\omega))$$

where $X_t(s,x)$ is the (unique) solution of the Ito SDE with the initial condition $X_s = x$. Then $X_t(s,x)$ is a function of the increments of the Wiener process, $W_{s+h} - W_s$. Since the initial condition is a nonrandom constant, $X_t(s,x)$ is independent of $\mathscr{F}_s$. (Recall that $\mathscr{F}_s$ and $\mathscr{M}_s^+$ are independent.) Therefore $h(x,\cdot) = I_B(X_t(s,x))$ is also independent of $\mathscr{F}_s$ and exploiting (5.42), i.e., $X_t = X_t(s, X_s(t_0, x_0)) = X_t(s, X_s)$, we have $h(X_s, \omega) = I_B(X_t)$. Theorem (5.52) follows now via the chain of equalities:

$$\begin{aligned}
P(X_t \in B \mid \mathscr{F}_s) &= E\{I_B(X_t) \mid \mathscr{F}_s\} \\
&= E\{h(X_s, \omega) \mid \mathscr{F}_s\} \\
&= E\{h(X_s, \omega) \mid X_s\} \\
&= E\{I_B(X_t) \mid X_s\} \\
&= P(X_t \in B \mid X_s)\,.
\end{aligned}$$

Furthermore we have

$$\begin{aligned}
E\{I_B(X_t) \mid X_s\} &= E\{h(X_s, \omega) \mid X_s\} \\
&= H(X_s) \\
&= E\{h(x, \omega)\}\big|_{x=X_s} \\
&= E\{I_B(X_t(s,x))(\omega)\}\big|_{x=X_s} \\
&= P(X_t(s,x) \in B)\big|_{x=X_s}\,,
\end{aligned}$$

which proves the asserted relation for the transition function.

C. The Stratonovich Calculus Obeys Classical Rules

Let X_t be a solution of the S SDE

$$dX_t = f(X_t)\,dt + \sigma g(X_t) \circ dW_t \tag{C.1}$$

and $U(x)$ a bijective real-valued, twice continuously differentiable function. The process $Y_t = U(X_t)$ obeys then the S SDE

$$\begin{aligned}
dY_t &= U'(X_t) f(X_t)\,dt + U'(X_t)\sigma g(X_t) \circ dW_t \\
&= U'(U^{-1}(Y_t))\,[f(U^{-1}(Y_t))\,dt + \sigma g(U^{-1}(Y_t)) \circ dW_t]\,.
\end{aligned} \tag{C.2}$$

The easiest way to prove this is using the transformation formula (5.75) and the Ito rule. According to (5.75) the following Ito SDE is equivalent to (C.1)

$$dX_t = \left[f(X_t) + \frac{\sigma^2}{2} g'(X_t) g(X_t) \right] dt + \sigma g(X_t)\,dW_t\,.$$

Using the Ito rule (5.45) we have

$$dY_t = \left[U'(X_t)f(X_t) + U'(X_t)\frac{\sigma^2}{2}g'(X_t)g(X_t) + \frac{1}{2}U''(X_t)\sigma^2 g^2(X_t) \right] dt$$
$$+ U'(X_t)\sigma g(X_t)dW_t, \qquad \text{(C.3)}$$

or writing it as a SDE

$$dY_t = U'(U^{-1}(Y_t))[f(U^{-1}(Y_t))dt + \sigma g(U^{-1}(Y_t))dW_t]$$
$$+ \frac{\sigma^2}{2}U''(U^{-1}(Y_t))g^2(U^{-1}(Y_t))dt$$
$$+ \frac{\sigma^2}{2}U'(U^{-1}(Y_t))g'(U^{-1}(Y_t))g(U^{-1}(Y_t))dt. \qquad \text{(C.4)}$$

To obtain the equivalent S SDE we have to apply (5.75) in the inverse sense, i.e., subtract

$$\frac{\sigma^2}{2}U'(U^{-1}(Y_t))g(U^{-1}(Y_t)) \cdot \frac{\partial}{\partial y}[U'(U^{-1}(y))g(U^{-1}(y))]\Big|_{y=Y_t}$$
$$= \frac{\sigma^2}{2}U'(U^{-1}(Y_t))g(U^{-1}(Y_t)) \cdot [U''(U^{-1}(Y_t))U^{-1'}(Y_t)g(U^{-1}(Y_t))$$
$$+ U'(U^{-1}(Y_t))g'(U^{-1}(Y_t))U^{-1'}(Y_t)]$$
$$= \frac{\sigma^2}{2}g(U^{-1}(Y_t))[U''(U^{-1}(Y_t))g(U^{-1}(Y_t))$$
$$+ U'(U^{-1}(Y_t))g'(U^{-1}(Y_t))] \qquad \text{(C.5)}$$

since $U'(U^{-1}(y)) \cdot U^{-1'}(y) = 1$. Hence (C.5) compensates the second term in (C.4) and Y_t obeys the S SDE (C.2) as claimed.

D. Critical Exponents of the Mean Field Theory

Though the mean-field theory does not agree with the experimental results for equilibrium critical points precisely because it neglects spatial fluctuations and so has been replaced by renormalization group theory [6.6] it is found that non-equilibrium critical points, under deterministic conditions at least, are often well described by it. This is a consequence of the fact that for nonequilibrium phase transitions, as in the laser or in well-stirred chemical systems, spatial fluctuations indeed play *no* role. Though the classical mean-field theory predicts the same critical exponents for all equilibrium critical points, as the liquid-gas critical point, the ferromagnetic critical point, etc., its concepts are best illustrated for the transition from paramagnetic to ferromagnetic behavior. The order parameter for this transition is the magnetization m of the sample. It is well known that the source of magnetization is the spin of electrons in incomplete atomic shells. The spins attain their lowest level of energy if they are parallel as a result of a quantum phenomenon known as the "exchange effect." If the temperature of the sample is zero, all spins are parallel and there is a finite magnetization, the material is ferromagnetic. The direction of m is undefined; in the isotropic case all directions are possible. As the temperature T is increased, thermal motion will perturb the perfect alignment of spins. However, for not too high temperatures there is still a net fraction of spins pointing in the same direction. Hence the magnetization will be reduced but still positive. When the temperature T reaches the so-called critical temperature T_c, there will be no net fraction of parallel spins any more and the magnetization disappears; the material becomes paramagnetic. Thus an equilibrium transition, a critical point, occurs as the temperature T is varied. This phenomenon is modeled by the mean-field theory which despite its shortcomings describes the important features of critical points in a qualitatively correct way. For the ferromagnetic critical point, the theory goes as follows [6.6]. Each electron spin is subjected to a local magnetic field h', which is made up of two parts, the external field h assumed to be very small and the field generated by the surrounding spins. Supposing that the average value m of a spin follows a Curie law, we have

$$m = ch'/T, \tag{D.1}$$

where c is a constant. The cornerstone of the mean-field theory is the assumption that the field due to the surrounding spins is a function of the average of all spins, i.e.,

$$h' = h + \mu(m). \tag{D.2}$$

Supposing that $\mu(m)$ can be written as a power series, we obtain

$$h' = h + am - bm^3, \tag{D.3}$$

where a and b are constants and even powers are absent for symmetry reasons. Combining (D.1, 3) we have

$$m\left(1 - \frac{ac}{T} + \frac{cb}{T}m^2\right) = \frac{ch}{T}. \tag{D.4}$$

In the absence of an external magnetic field h, we have a critical point at $T = T_c = ca$. Indeed $m = 0$ is a triple root of (D.4).

If we impose at $T = T_c$ a small external field h, then

$$m = b^{1/3} \cdot h^{1/3}. \tag{D.5}$$

Since for a ferromagnet magnetization is parallel to the external field, we conclude that the constant b has to be positive. Let us now define the critical exponents and determine their values as predicted by the mean-field theory:

i) The exponent β: order parameter m as a function of T for vanishing external field.

For $h = 0$ and T near but below the critical point T_c, the magnetization displays the power law behavior

$$m \sim (T_c - T)^{\beta} .$$

The exponent β which describes the behavior of the order parameter near the critical point is known as a critical exponent. Mean-field theory predicts

$$m = \left(\frac{T_c - T}{cb}\right)^{1/2} \quad \text{for} \quad T < T_c , \tag{D.6}$$

i.e., $\beta = 1/2$.

ii) The exponent δ: order parameter as a function of h at $T = T_c$. Again the magnetization obeys a power law

$$m \sim h^{1/\delta} ,$$

which the mean-field theory predicts to correspond to

$$\delta = 3 . \tag{D.7}$$

iii) The exponent γ: the magnetic susceptibility $\chi = (\partial m/\partial h)_T$ for $h = 0$ as a function of T. The susceptibility x follows the power law

$$\begin{aligned} \chi &\sim (T - T_c)^{-\gamma} \qquad T > T_c \\ \chi &\sim (T_c - T)^{-\gamma'} \qquad T < T_c . \end{aligned}$$

The mean-field theory predicts

$$\begin{aligned} \chi = (\partial m/\partial h)\,|_{h=0}(T) &\sim 1/(T_c - T) \qquad T < T_c , \\ &\sim 1/(T - T_c) \qquad T > T_c , \end{aligned}$$

i.e., $\gamma = \gamma' = 1$.

The above values for the critical exponents $\beta = 1/2$, $\delta = 3$, $\gamma = \gamma' = 1$ are known as the classical exponents.

References

Chapter 1

1.1 J. Kepler: *Mysterium Cosmographicum* (1609). Cited by C. Sagan: In *Cosmos* (Random House, New York 1980) p.45

1.2 C. Bernard: *Introduction à l'étude de la medecine experimentale* (1866); Reedited (Editions P. Beltond, Paris 1966) p.116

1.3 E. Durkheim: *Les formes élementaires de la vie religieuse* (Alcan, Paris 1912)

1.4 J. Moser: *Stable and Random Motions in Dynamical Systems* (University Press, Princeton 1973)

1.5 V.I. Arnold, A. Avez: *Ergodic Problems of Classical Mechanics* (Benjamin, New York 1968)

1.6 B.V. Chirikov: A universal instability of many-dimensional oscillator systems. Phys. Rep. *52*, 263 (1979)

1.7 R.H.G. Helleman: "Self-Generated Chaotic Behavior in Nonlinar Mechanics," in *Fundamental Problems in Statistical Mechanics*, Vol.5, ed. by E.G.D. Cohen (North-Holland, Amsterdam 1980) p.165

1.8 I. Prigogine: *From Being to Becoming* (Freeman, San Francisco 1980)

1.9 I. Prigogine: "Entropy, Time and Kinetic Description," in *Order and Fluctuation in Equilibrium and Nonequilibrium Statistical Mechanics*, ed. by G. Nicolis, G. Dewel, J.W. Turner (Wiley, New York 1981) p.35

1.10 I. Prigogine, I. Stengers: *La nouvelle alliance. Métamorphose de la science.* (Editions Gallimard, Paris 1979)

1.11 M. Eigen: Selforganization of matter and the evolution of biological macromolecules. Naturwissenschaften *64*, 541 (1971)

1.12 M. Eigen, P. Schuster: *The Hypercycle. A Principle of Natural Selection* (Springer, Berlin, Heidelberg, New York 1979)

1.13 M.W. Ho, P.T. Saunders: Beyond neo-Darwinism - an epigenetic approach to evolution. J. Theor. Biol. *78*, 573 (1979)

1.14 P. Glansdorff, I. Prigogine: *Thermodynamic Theory of Structure, Stability and Fluctuation* (Wiley, New York 1971)

1.15 G. Nicolis, I. Prigogine: *Self-Organization in Nonequilibrium Systems. From Dissipative Structure to Order Through Fluctuations* (Wiley, New York 1977)

1.16 P. Bergé: "Experiments on Hydrodynamic Instabilities and the Transition to Turbulence," in *Dynamical Critical Phenomena and Related Topics*, Lecture Notes in Physics, Vol.104, ed. by C.P. Enz (Springer, Berlin, Heidelberg, New York 1979) p.288

1.17 D. Ruelle, F. Takens: On the nature of turbulence. Commun. Math. Phys. *20*, 167 (1971)

1.18 E.N. Lorenz: Deterministic nonperiodic flow. J. Atmos. Sci. *20*, 130 (1963)

1.19 O.E. Rössler: An equation for continuous chaos. Phys. Lett. A*77*, 397 (1976)

1.20 O.E. Rössler: "Chaos and Strange Attractors in Chemical Kinetics," in *Synergetics Far from Equilibrium*, Springer Ser. Synergetics, Vol.3, ed. by A. Pacault, C. Vidal (Springer, Berlin, Heidelberg, New York 1979) p.107

1.21 M. Henon: A two-dimensional mapping with a strange attractor. Commun. Math. Phys. *50*, 69 (1976)

1.22 Y. Pomeau, P. Manneville: Intermittent transition to turbulence in dissipative dynamical systems. Commun. Math. Phys. *74*, 189 (1980)

1.23 P. Coullet, C. Tresser: Itérations d'endomorphismes et groupe de renormalisation. J. Phys. C*5*, 25 (1978)

1.24 M. Feigenbaum: Quantitative universality for a class of nonlinear transformations. J. Stat. Phys. *19*, 25 (1978)

1.25 M. Feigenbaum: The universal metric properties of nonlinear transformations. J. Stat. Phys. *21*, 669 (1979)
1.26 M. Feigenbaum: The transition to aperiodic behavior in turbulent systems. Commun. Math. Phys. *77*, 65 (1980)
1.27 P. Collet, J.P. Eckmann, O. Lanford: Universal properties of maps on an interval. Commun. Math. Phys. *76*, 211 (1980)
1.28 P. Collet, J.P. Eckmann, H. Koch: Period doubling bifurcations for families of maps on $\mathbb{R}^n$. J. Stat. Phys. *25*, 1 (1981)
1.29 J.P. Eckmann: Roads to turbulence in dissipative dynamical systems. Rev. Mod. Phys. *53*, 643 (1981)
1.30 E. Ott: Strange attractors and chaotic motions of dynamical systems. Rev. Mod. Phys. *53*, 655 (1981)
1.31 J.B. McLaughin, P.C. Martin: Transition to turbulence in a statistically stressed fluid system. Phys. Rev. A*12*, 186 (1975)
1.32 P. Manneville, Y. Pomeau: Different ways to turbulence in dissipative dynamical systems. Physica D*1*, 219 (1980)
1.33 M. Levi: Qualitative analysis of the periodically forced relaxation oscillations. Mem. Am. Math. Soc. *244*, 1 (1981)
1.34 C. Vidal, J.C. Roux, S. Bachelart, A. Rossi: Experimental study of the transition to turbulence in the Belousov-Zhabotinsky reaction. Ann. NY Acad. Sci. *357*, 377 (1980)
1.35 J.C. Roux, H.L. Swinney: "Topology of Chaos in a Chemical Reaction," in *Nonlinear Phenomena in Chemical Dynamics*, Springer Ser. Synergetics, Vol.12, ed. by C. Vidal, A. Pacault (Springer, Berlin, Heidelberg, New York 1981) p.38
1.36 J.S. Turner, J.C. Roux, W.D. McCormick, H. Swinney: Alternating periodic and chaotic regimes in a chemical reaction - experiment and theory. Phys. Lett. A*85*, 9 (1981)
1.37 G. Ahlers: Low-temperature studies of the Rayleigh-Bénard instability and turbulence. Phys. Rev. Lett. *33*, 1185 (1974)
1.38 J.P. Gollub, H.L. Swinney: Onset of turbulence in a rotating fluid. Phys. Rev. Lett. *35*, 927 (1975)
1.39 J.P. Gollub, S.V. Benson: Many routes to turbulent convection. J. Fluid Mech. *100*, 449 (1980)
1.40 H.L. Swinney, J.P. Gollub (eds.): *Hydrodynamic Instabilities and the Transition to Turbulence*. Topics Appl. Phys., Vol.45 (Springer, Berlin, Heidelberg, New York 1978)
1.41 A. Libchaber, J. Maurer: Une expérience de Rayleigh-Bénard de géométrie réduite; multiplication, accrochage et démultiplication de fréquence. J. Phys. Paris C*3*, 51 (1980)
1.42 P. Bergé, M. Dubois, P. Manneville, Y. Pomeau: Intermittency in Rayleigh-Bénard convection. J. Phys. Paris Lett. *41*, L341 (1980)
1.43 Y. Pomeau, J.C. Roux, A. Rossi, S. Bachelart, C. Vidal: Intermittent behaviour in the Belousov-Zhabotinski reaction. J. Phys. Paris Lett. *42*, L271 (1981)
1.44a C. Vidal, A. Pacault (eds.): *Nonlinear Phenomena in Chemical Dynamics*, Springer Ser. Synergetics, Vol.12 (Springer, Berlin, Heidelberg, New York 1981)
1.44b C. Vidal, S. Bachelart, A. Rossi: Bifurcations en cascade conduisant à la turbulence dans la réaction de Belousov-Zhabotinsky. J. Phys. Paris *43*, 7 (1982)
1.45 J.L. Hudson, J. Mankin, J. McCullough, P. Lamba: "Experiments on Chaos in a Continuous Stirred Reactor," in *Nonlinear Phenomena in Chemical Dynamics*, Springer Ser. Synergetics, Vol.12, ed. by C. Vidal, A. Pacault (Springer, Berlin, Heidelberg, New York 1981) p.44
1.46 H. Haken: *Laser Theory*, Handbuch der Physik/Encyclopedia of Physics Vol. XXV/2c (Springer, Berlin, Heidelberg, New York 1970)
1.47 H. Haken: *Laser Theory* (Springer, Berlin, Heidelberg, New York 1983)
1.48 H. Haken: "Introduction to Synergetics," in *Synergetics. Cooperative Phenomena in Multi-Component Systems*, ed. by H. Haken (Teubner, Stuttgart 1973) p.9
1.49 R. Graham: "Phase-Transition-Like Phenomena in Lasers and Nonlinear Optics," in *Synergetics-Cooperative Phenomena in Multi-Component Systems*, ed. by H. Haken (Teubner, Stuttgart 1973) p.71
1.50 R. Graham: "The Phase Transition Concept and Coherence in Atomic Emission," in *Progress in Optics*, Vol.12, ed. by E. Wolf (North-Holland, Amsterdam 1974)
1.51 F. Schlögl: Chemical reaction models for non-equilibrium phase transition. Z. Phys. *253*, 147 (1972)

1.52 A. Nitzan, P. Ortoleva, J. Deutch, J. Ross: Fluctuations and transitions at chemical instabilities: the analogy to phase transitions. J. Chem. Phys. *61*, 1056 (1974)

1.53 H. Haken: Cooperative phenomena in systems far from thermal equilibrium and in nonphysical systems. Rev. Mod. Phys. *47*, 67 (1975)

1.54 H. Haken: *Synergetics. An Introduction*, Springer Ser. Synergetics, Vol.1, 2nd ed. (Springer, Berlin, Heidelberg, New York 1977)

1.55 R. Graham: "Onset of Cooperative Behavior in Nonequilibrium Steady States," in *Order and Fluctuations in Equilibrium and Nonequilibrium Statistical Mechanics*, ed. by G. Nicolis, G. Dewel, J. Turner (Wiley, New York 1981) p.235

1.56 D. Walgraef, G. Dewel, P. Borckmans: Nonequilibrium phase transitions and chemical instabilities. Adv. Chem. Phys. *49*, 311 (1982)

1.57 S.J. Gould: The chance that shapes our ends. New Sci. *89*, 347 (1981)

1.58a S. Rosenblat, D.S. Cohen: Periodically perturbed bifurcation I. Simple bifurcation. Stud. Appl. Math. *63*, 1 (1980)

1.58b S. Rosenblat, D.S. Cohen: Periodically perturbed bifurcation II. Hopf bifurcation. Stud. Appl. Math. *64*, 143 (1981)

1.59 W.L. Kath: Resonance in periodically perturbed Hopf bifurcation. Stud. Appl. Math. *65*, 95 (1981)

1.60 F.H. Busse: Nonlinear properties of thermal convection. Rep. Prog. Phys. *41*, 1929 (1978); "Transition to Turbulence in Rayleigh-Bênard Convection," in *Hydrodynamic Instabilities and the Transition to Turbulence*, Topics Appl. Phys., Vol.45, ed. by H.L. Swinney, J.P. Gollub (Springer, Berlin, Heidelberg, New York 1981) p.97

1.61 G. Nicolis, R. Lefever (eds.): Membranes, dissipative structures, and evolution. Adv. Chem. Phys. *29* (1975)

1.62 T. Riste (ed.): *Fluctuations, Instabilities and Phase Transitions* (Plenum, New York 1975)

1.63 A. Pacault, C. Vidal (eds.): *Synergetics. Far from Equilibrium*, Springer Ser. Synergetics, Vol.3 (Springer, Berlin, Heidelberg, New York 1979)

1.64 O. Gurel, O.E. Rössler (eds.): Bifurcation theory and applications in scientific disciplines. Ann. NY Acad. Sci. *316* (1979)

1.65 P.C. Fife: *Mathematical Aspects of Reacting and Diffusing Systems*, Lecture Notes Biomath., Vol.28 (Springer, Berlin, Heidelberg, New York 1979)

1.66 J.P. Kernevez: *Enzyme Mathematics*, Studies Math. and Its Applications, Vol.10 (North-Holland, Amsterdam 1980)

1.67 H.G. Helleman (ed.): Nonlinear dynamics. Ann. NY Acad. Sci. *357* (1980)

1.68 G. Nicolis, G. Dewel, J.W. Turner (eds.): *Order and Fluctuations in Equilibrium and Nonequilibrium Statistical Mechanics*, Wiley Ser. Nonequilibrium Problems in the Physical Sciences and Biology, Vol.1 (Wiley, New York 1981)

1.69 I. Prigogine: Modération et transformations irréversibles des systèmes ouverts. Bull. Cl. Sci. Acad. R. Belg. *31*, 600 (1945)

1.70 I. Prigogine: "Structure, Dissipation and Life," in *Theoretical Physics and Biology*, ed. by M. Marois (North-Holland, Amsterdam 1969) p.23

1.71 I. Prigogine: *Etude thermodynamique des phénomènes irréversibles* (Desoer, Liège 1947)

1.72 I. Prigogine: *Introduction to Thermodynamics of Irreversible Processes* (Wiley, New York 1967)

1.73 N. Minorsky: *Nonlinear Oscillations* (Krieger, Huntington, NY 1962)

1.74 I. Prigogine, R. Balescu: Phénomènes critiques dans la thermodynamique des processus irreversibles. Bull. Cl. Sci. Acad. R. Belg. *42*, 256 (1956)

1.75 M. Malek Mansour, C. Van den Broeck, G. Nicolis, J.W. Turner: Asymptotic properties of Markovian master equations. Ann. Phys. (NY) *131*, 283 (1981)

1.76 G. Nicolis, J.W. Turner: Stochastic analysis of a nonequilibrium phase transition: some exact results. Physics A*89*, 326 (1977)

1.77 P.I. Kuznetsov, R.L. Stratonovich, V.I. Thikhonov: "The Effect of Electrical Fluctuations on a Valve Oscillator," in *Nonlinear Transformations of Stochastic Processes*, ed. by P.I. Kuznetsov, R.L. Stratonovich, V.I. Thikhonov (Pergamon, Oxford 1965) p.223

1.78 I.N. Amiantov, V.I. Thikhonov: "The Response of Typical Nonlinear Elements to Normally Fluctuating Inputs," in *Nonlinear Transformations of Stochastic Processes*, ed. by P.I. Kuznetsov, R.L. Stratonovich, V.I. Thikhonov (Pergamon, Oxford 1965) p.175

1.79 R.L. Stratonovich, P.S. Landa: "The Effect of Noise on an Oscillator with Fixed Excitation," in *Nonlinear Transformations of Stochastic Processes*, ed. by P.I. Kuznetsov, R.L. Stratonovich, V.I. Thikhonov (Pergamon, Oxford 1965) p.259
1.80 R.L. Stratonovich: *Topics in the Theory of Random Noise*, Vols.1,2 (Gordon and Breach, New York 1967)
1.81 R.M. May: *Stability and Complexity in Model Ecosystems* (University Press, Princeton, NJ 1973)
1.82 H.S. Hahn, A. Nitzan, P. Ortoleva, J. Ross: Threshold excitations, relaxation oscillations, and effect of noise in an enzyme reaction. Proc. Natl. Acad. Sci. USA *71*, 4067 (1974)
1.83 R. Kubo: "A Stochastic Theory of Line-Shape and Relaxation," in *Fluctuation, Relaxation and Resonance in Magnetic Systems*, ed. by D.Ter Haar (Oliver and Boyd, Edinbourgh 1962) p.23
1.84 R. Brout: *Phase Transitions* (Benjamin, New York 1965)
1.85 A.Z. Patashniskii, V.I. Pokrovskii: *Fluctuation Theory of Phase Transitions*, Int. Ser. in Natural Philosophy, Vol.98 (Pergamon, New York 1979)
1.86 K.L. Chung: *Elementary Probability Theory with Stochastic Processes* (Springer, Berlin, Heidelberg, New York 1979)
1.87 W. Feller: *An Introduction to Probability Theory and Its Applications*, Vol.1, 3rd ed. and Vol.2, 2nd ed. (Wiley, New York 1968 and 1971)
1.88 S. Karlin, H.M. Taylor: *A First Course in Stochastic Processes*, 2nd ed. (Academic, New York 1975)
1.89 R.E. Mortensen: Mathematical problems of modeling stochastic nonlinear dynamic systems. J. Stat. Phys. *1*, 271 (1969)

Chapter 2

2.1 D. Kannan: *An Introduction to Stochastic Processes* (North-Holland, New York 1979)
2.2 L. Arnold: *Stochastic Differential Equations: Theory and Applications* (Wiley, New York 1974)
2.3 Yu.V. Prohorov, Yu.A. Rozanov: *Probability Theory* (Springer, Berlin, Heidelberg, New York 1969)
2.4 I.I. Gihman, A.V. Skorohod: *Introduction to the Theory of Random Processes* (Saunders, Philadelphia 1969)
2.5 S. Chandrasekhar: Stochastic problems in physics and astronomy. Rev. Mod. Phys. *15*, 1 (1943); reprinted in *Selected Papers on Noise and Stochastic Processes*, ed. by N. Wax (Dover, New York 1954) p.3
2.6 T. Hida: *Brownian Motion* (Springer, Berlin, Heidelberg, New York 1980)
2.7 G.E. Uhlenbeck, L.S. Ornstein: On the theory of Brownian motion. Phys. Rev. *36*, 823 (1930)

Chapter 3

3.1 W. Horsthemke: "Nonequilibrium Transitions Induced by External White and Coloured Noise," in *Dynamics of Synergetic Systems*, Springer Ser. Synergetics, Vol.6, ed. by H. Haken (Springer, Berlin, Heidelberg, New York 1980) p.67
3.2a Y. Rozanov: *Processus aléatoires* (Editions Mir, Moscow 1975)
3.2b I.M. Gel'fand, N.J. Wilenkin: *Verallgemeinerte Funktionen (Distributionen)*, Vol.4 (VEB Deutscher Verlag der Wissenschaften, Berlin 1964)
3.3 I.I. Gihman, A.V. Skorohod: *The Theory of Stochastic Processes*, Vol.1 (Springer, Berlin, Heidelberg, New York 1974)

Chapter 4

4.1 I.I. Gihman, A.V. Skorohod: *The Theory of Stochastic Processes*, Vol.2 (Springer, Berlin, Heidelberg, New York 1975)
4.2 J.L. Doob: The Brownian movement and stochastic equations. Ann. Math. *43*, 351 (1942)
4.3 T. Maruyama: *Stochastic Problems in Population Genetics*, Lecture Notes in Biomath., Vol.17 (Springer, Berlin, Heidelberg, New York 1977)

4.4 P. Hänggi, H. Thomas, H. Grabert, P. Talkner: Note on time evolution of non-Markov processes. J. Stat. Phys. *18*, 155 (1978)
4.5 R.F. Pawula: Generalization and extensions of the Fokker-Planck-Kolmogorov equations. IEEE Trans. Inform. Theory IT-*13*, 33 (1967)
4.6 R.F. Pawula: Approximation of the linear Boltzmann equation by the Fokker-Planck equation. Phys. Rev. *162*, 186 (1967)
4.7 P. Hänggi: Correlation functions and master equations of generalized (non-Markovian) Langevin equations. Z. Phys. B*31*, 407 (1978)
4.8 H.C. Tuckwell: Synaptic transmission in a model for stochastic neural activity. J. Theor. Biol. *77*, 65 (1979)
4.9 H.C. Tuckwell: "Poisson Processes in Biology," in *Stochastic Nonlinear Systems in Physics, Chemistry, and Biology*, Springer Ser. Synergetics, Vol.8, ed. by L. Arnold, R. Lefever (Springer, Berlin, Heidelberg, New York 1981) p.162
4.10 A.V. Holden: *Models of the Stochastic Activity of Neurones*, Lecture Notes in Biomath., Vol.12 (Springer, Berlin, Heidelberg, New York 1976)
4.11 P. Hänggi, K.E. Shuler, I. Oppenheim: On the relations between Markovian master equations and stochastic differential equations. Physica A*107*, 143 (1981)

Chapter 5

5.1 J.L. Doob: *Stochastic Processes* (Wiley, New York 1953)
5.2 K. Ito: On a formula concerning stochastic differentials. Nagoya Math. J. *3*, 55 (1951)
5.3 R.L. Stratonovich: A new representation for stochastic integrals and equations. SIAM J. Control *4*, 362 (1966)
5.4 K. Ito: Stochastic differential equations on a differentiable manifold. Nagoya Math. J. *1*, 35 (1950)
5.5 A.H. Gray, T.K. Caughey: A controversy in problems involving random parametric excitation. J. Math. Phys. *44*, 288 (1965)
5.6 D.W. Strook, S.R.S. Varadhan: *Multidimensional Diffusion Processes* (Springer, Berlin, Heidelberg, New York 1979)
5.7 H.P. McKean: *Stochastic Integrals* (Academic, New York 1969)
5.8 E. Wong, M. Zakai: On the convergence of ordinary integrals to stochastic integrals. Ann. Math. Stat. *36*, 1560 (1965)
5.9 G. Blankenship, G.C. Papanicolaou: Stability and control of stochastic systems with wide-band noise disturbances. SIAM J. Appl. Math. *34*, 437 (1978)
5.10 N.G. van Kampen: Itô versus Stratonovich. J. Stat. Phys. *24*, 175 (1981)
5.11 A.H. Jazwinski: *Stochastic Processes and Filtering Theory* (Academic, New York 1970)
5.12 I.I. Gihman, A.V. Skorohod: *Stochastic Differential Equations* (Springer, Berlin, Heidelberg, New York 1972)
5.13 A.T. Bharucha-Reid: *Elements of the Theory of Markov Processes and Their Applications* (McGraw-Hill, New York 1960)
5.14 S. Karlin, H.M. Taylor: *A Second Course in Stochastic Processes* (Academic, New York 1981)

Chapter 6

6.1 M. Lavrentiev, B. Chabat: *Méthodes de la théorie des fonctions d'une variable complexe* (Mir, Moscow 1972)
6.2 G. Nicolis, R. Lefever: Comment on the kinetic potential and the Maxwell construction in nonequilibrium chemical phase transitions. Phys. Lett. A*62*, 469 (1977)
6.3 R. Landauer: Fluctuations in bistable tunnel diode circuits. J. Appl. Phys. *33*, 2209 (1962)
6.4 R. Thom: *Stabilité structurelle et morphogénèse* (Ediscience, Paris 1973); *Structural Stability and Morphogenesis* (Benjamin, Reading, MA 1975)
6.5 T. Poston, I. Stewart: *Catastrophe Theory and Its Applications* (Pitman, London 1978)
6.6 S.K. Ma: *Modern Theory of Critical Phenomena* (Benjamin, Reading, MA 1976)
6.7 W. Horsthemke, M. Malek Mansour: The influence of external noise on nonequilibrium phase transitions. Z. Phys. B*24*, 307 (1976)

6.8 L. Arnold, W. Horsthemke, R. Lefever: White and coloured external noise and transition phenomena in nonlinear systems. Z. Phys. B*29*, 867 (1978)
6.9 W. Ebeling: Nonequilibrium transitions and stationary probability distributions of stochastic processes. Phys. Lett. A*68*, 430 (1978)
6.10 N.S. Goel, N. Richter-Dyn: *Stochastic Models in Biology* (Academic, New York 1974)
6.11 S. Karlin, U. Lieberman: Random temporal variation in selection intensities: case of large population size. Theor. Popul. Biol. *6*, 355 (1974)
6.12 B. Levikson, S. Karlin: Random temporal variation in selection intensities acting on infinite diploid population: diffusion method analysis. Theor. Popul. Biol. *8*, 292 (1975)
6.13 J.H. Gillepsie: Polymorphism in patchy environments. Am. Nat. *108*, 145 (1974)
6.14 J.H. Gillepsie: The role of the environmental grain in the maintainance of genetic variation. Am. Nat. *108*, 831 (1974)
6.15 J.H. Gillepsie: The role of migration in the genetic structure of populations in temporally and spatially varying environments II. Island models. Theor. Popul. Biol. *10*, 227 (1976)
6.16 J.H. Gillepsie, C. Langley: Multi-locus behavior in random environments I. Random Levene models. Genetics *82*, 123 (1976)
6.17 W.J. Ewens: *Mathematical Population Genetics*. Biomathematics, Vol.9 (Springer, Berlin, Heidelberg, New York 1979)
6.18 M. Kimura, T. Ohta: *Theoretical Aspects of Population Genetics* (University Press, Princeton, NJ 1971)
6.19 M.O. Hongler: Exact time dependent probability density for a nonlinear non-Markovian stochastic process. Helv. Phys. Acta *52*, 280 (1979)
6.20 J. Elliott: Eigenfunction expansions associated with singular differential operators. Trans. Am. Math. Soc. *78*, 406 (1955)
6.21 R. Courant, D. Hilbert: *Methoden der mathematischen Physik I* (Springer, Berlin, Heidelberg, New York 1968)
6.22 K. Yosida: *Lectures on Differential and Integral Equations* (Interscience, New York 1960)
6.23 E.A. Coddington, N. Levinson: *Theory of Ordinary Differential Equations* (McGraw-Hill, New York 1955)
6.24 E. Hille: *Lectures on Ordinary Differential Equations* (Addison-Wesley, Reading, MA 1969)
6.25 E. Wong: The construction of a class of stationary Markoff processes. Proc. Symp. Appl. Math. *16*, 264 (1964)
6.26 A. Erdelyi: *Higher Transcendental Functions*, Vols.1 and 2 (McGraw-Hill, New York 1953)
6.27 Prajneshu: Time-dependent solution of the logistic model for population growth in random environment. J. Appl. Probab. *17*, 1083 (1980)
6.28 A. Schenzle, H. Brand: Multiplicative stochastic processes in statistical physics. Phys. Lett. A*69*, 313 (1979); Phys. Rev. A*20*, 1628 (1979)
6.29 M. Suzuki, K. Kaneko, F. Sasagawa: Phase transition and slowing down in non-equilibrium stochastic processes. Prog. Theor. Phys. *65*, 828 (1981)
6.30 W. Horsthemke: "Noise Induced Transitions," in *Stochastic Nonlinear Systems in Physics, Chemistry, and Biology*, Springer Ser. Synergetics, Vol.8, ed. by L. Arnold, R. Lefever (Springer, Berlin, Heidelberg, New York 1981) p.116
6.31 H. Fujisaka, S. Grossmann: External noise effects on the fluctuation line width. Z. Phys. B*43*, 69 (1981)
6.32 L. Brenig, V. Fairen: Analytical approach to initial-value problems in non-linear systems. J. Math. Phys. *22*, 649 (1981)
6.33 L. Brenig, N. Banai: Nonlinear dynamics of systems coupled with external noise: some exact results. Physica D*5*, 208 (1982)
6.34 R. Graham, A. Schenzle: Carleman imbedding of multiplicative stochastic processes. Phys. Rev. A*25*, 1731 (1982)
6.35 M.O. Hongler: Diffusion in a class of double-well potentials - Exact Results. Phys. Lett. *91*A, 396 (1982)
6.36 M.O. Hongler, W.M. Zheng: Exact solution for the diffusion in bistable potentials. J. Stat. Phys. *29*, 317 (1982)

Chapter 7

7.1 S. Kabashima, T. Kawakubo: Observation of a noise induced phase transition in a parametric oscillator. Phys. Lett. A*70*, 375 (1979)

7.2 S. Kabashima, S. Kogure, T. Kawakubo, T. Okada: Oscillatory to nonoscillatory transition due to external noise in parametric oscillator. J. Appl. Phys. *50*, 6296 (1979)

7.3 S. Kabashima: Observation of phase transition due to external fluctuations. (unpublished)

7.4 J.J. Tyson: *The Belousov-Zhabotinskii Reaction*, Lecture Notes in Biomath., Vol.10 (Springer, Berlin, Heidelberg, New York 1976)

7.5 J.J. Tyson, P.C. Fife: Target patterns in a realistic model of the Belousov-Zhabotinskii reaction. J. Chem. Phys. *73*, 2224 (1980)

7.6 M.L. Smoes: "Chemical Waves in the Oscillatory Zhabotinskii System. A Transition from Temporal to Spatio-Temporal Organization," in *Dynamics of Synergetic Systems*, Springer Ser. Synergetics, Vol.6, ed. by H. Haken (Springer, Berlin, Heidelberg, New York 1980) p.80

7.7 A. Winfree: *The Geometry of Biological Time*, Biomathematics, Vol.8 (Springer, Berlin, Heidelberg, New York 1980)

7.8 P. De Kepper, W. Horsthemke: Etude d'une réaction chimique périodique. Influence de la lumière et transitions induites par un bruit externe. C. R. Acad. Sci. Paris C*287*, 251 (1978)

7.9 P. De Kepper, W. Horsthemke: "Experimental Evidence of Noise-Induced Transition in an Open Chemical System," in *Synergetics. Far from Equilibrium*, Springer Ser. Synergetics, Vol.3, ed. by A. Pacault, C. Vidal (Springer, Berlin, Heidelberg, New York 1979) p.61

7.10 P. De Kepper: Contribution à l'étude expérimentale de systèmes dissipatifs chimique: réactions oscillantes de Briggs-Rauscher et de Belousov-Zhabotinskii. Thèse de Doctorat, Université de Bordeaux I (1978)

7.11 R. Graham, H. Haken: Laser light - first example of a second-order phase transition far away from thermal equilibrium. Z. Phys. *237*, 31 (1970)

7.12 V. De Giorgio, M.O. Scully: Analogy between the laser threshold region and a second-order phase transition. Phys. Rev. A*2*, 1170 (1970)

7.13 H. Haken: Generalized Ginzburg-Landau equations for phase transition like phenomena in lasers, nonlinear optics, hydrodynamics and chemical reactions. Z. Phys. B*21*, 105 (1975)

7.14 F.T. Arrechi: "Experimental Aspects of Transition Phenomena in Quantum Optics," in *Order and Fluctuations in Equilibrium and Nonequilibrium Statistical Mechanics*, ed. by G. Nicolis, G. Dewel, J.W. Turner (Wiley, New York 1981)

7.15 H.M. Gibbs, S.L. McCall, T.N.C. Venkatesan: Differential gain and bistability using a sodium-filled Fabry-Perot interferometer. Phys. Rev. Lett. *36*, 1135 (1976)

7.16 G.P. Agarwal, H.J. Carmichael: Optical bistability through nonlinear dispersion and absorption. Phys. Rev. A*19*, 2074 (1979)

7.17 R. Bonifacio, L.A. Lugiato: Optical bistability and cooperative effects in resonance fluorescence. Phys. Rev. A*18*, 1129 (1978)

7.18 A.R. Bulsara, W.C. Schieve, R.F. Gragg: Phase transitions induced by white noise in bistable optical systems. Phys. Lett. A*68*, 294 (1978)

7.19 R.F. Gragg: Stochastic Switching in Absorptive Optical Bistability. Ph. D. Thesis, University of Texas, Austin (1981)

7.20 F.T. Arrechi: "Transition Phenomena in Nonlinear Optics," in *Stochastic Nonlinear Systems in Physics, Chemistry, and Biology*, Springer Ser. Synergetics, Vol.8, ed. by L. Arnold, R. Lefever (Springer, Berlin, Heidelberg, New York 1981) p.222

7.21 J. de la Rubia, M.G. Velarde: Further evidence of a phase transition induced by external noise. Phys. Lett. A*69*, 304 (1978)

7.22 R. Levins: The effect of random variations of different types on population growth. Proc. Natl. Acad. Sci. USA *62*, 1055 (1969)

7.23 R.C. Lewontin, D. Cohen: On population growth in a randomly varying environment. Proc. Natl. Acad. Sci. USA *62*, 1056 (1969)

7.24 D. Ludwig: Optimal harvesting of a randomly fluctuating resource I: application of perturbation methods. SIAM J. Appl. Math. *37*, 166 (1979)

7.25 D. Ludwig, J.M. Varah: Optimal harvesting of a randomly fluctuating resource II: numerical methods and results. SIAM J. Appl. Math. *37*, 185 (1979)

7.26 B.S. White: The effect of a rapidly-fluctuating random environment on systems of interacting species. SIAM J. Appl. Math. *32*, 666 (1977)
7.27 R. Lefever, W. Horsthemke: Bistability in fluctuating environments. Implications in tumor immunology. Bull. Math. Biol. *41*, 469 (1979)
7.28 D. Ludwig, D.D. Jones, C.S. Holling: Qualitative analysis of insect outbreak systems: the spruce budworm and forest. J. Anim. Ecol. *47*, 315 (1978)
7.29 R.P. Garay, R. Lefever: A kinetic approach to the immunology of cancer: stationary state properties of effector-target cell reactions. J. Theor. Biol. *73*, 417 (1978)
7.30 R. Lefever, R. Garay: "Local Description of Immune Tumor Rejection", in *Biomathematics and Cell Kinetics*, ed. by A.J. Valleron, P.D.M. Macdonald (North-Holland, Amsterdam 1978) p.333
7.31 W. Horsthemke, R. Lefever: Phase transition induced by external noise. Phys. Lett. A*64*, 19 (1977)
7.32a R. Lefever: "Dynamics of Cell-Mediated Immune Response," in *Dynamics of Synergetic Systems*, Springer Ser. Synergetics, Vol.6, ed. by H. Haken (Springer, Berlin, Heidelberg, New York 1980) p.205
7.32b R. Lefever: "Noise-Induced Transitions in Biological Systems," in *Stochastic Nonlinear Systems in Physics, Chemistry, and Biology*, Springer Ser. Synergetics, Vol.8, ed. by L. Arnold, R. Lefever (Springer, Berlin, Heidelberg, New York 1981) p.127
7.33 J.C. Micheau, S. Boue, E. Van der Donckt: Theoretical kinetic analysis of biphotonic processes: evidence for the unusual but feasible occurrence of multistationary states and chemical oscillations. J. Chem. Soc. Faraday Trans. 2 *78*, 39 (1982)
7.34 J.C. Micheau, W. Horsthemke, R. Lefever: Sensitivity of biphotonic systems to light intensity fluctuations: experimental evidence in the thermoluminescence of fluorescein in boric acid glass. J. Chem Phys. (to appear)
7.35 F. Lesclaux, S. Ohayon, J. Joussot-Dubien: Contribution à l'étude de la luminescence différée de colorants en milieu rigide. Photoionisation de la fluoresceine dans l'acide borique par un processus à deux photons. Photochem. Photobiol. *11*, 401 (1970)
7.36 E. Walentynowicz: Thermoluminescence of fluorescein in boric acid glass. Acta Phys. Pol. *29*, 713 (1966)
7.37 M.H. Dung, J.J. Kozak: Analysis of mechanisms for the cyclic cleavage of water by visible light. J. Photochem. *16*, 121 (1981)
7.38 A. Nitzan, J. Ross: Oscillations, multiple steady states, and instabilities in illuminated systems. J. Chem. Phys. *59*, 241 (1973)
7.39 A. Nitzan, P. Ortoleva, J. Ross: Symmetry breaking instabilities in illuminated systems. J. Chem. Phys. *60*, 3194 (1974)
7.40 C.L. Creel, J. Ross: Multiple stationary states and hysteresis in a chemical reaction. J. Chem. Phys. *65*, 3779 (1976)
7.41 R. Lefever, W. Horsthemke: Multiple transitions induced by light intensity fluctuations in illuminated chemical systems. Proc. Natl. Acad. Sci. USA *76*, 2490 (1979)

Chapter 8

8.1 L. Arnold, W. Kliemann: "Qualitative Theory of Stochastic Systems," in *Probabilistic Analysis and Related Topics*, Vol.3, ed. by A. Bharucha-Reid (Academic, New York 1981)
8.2 R. Graham, H. Haken: Generalized thermodynamic potential for Markoff systems in detailed balance and far from thermal equilibrium. Z. Phys. *243*, 289 (1971)
8.3 R. Graham, H. Haken: Fluctuations and stability of stationary nonequilibrium systems in detailed balance. Z. Phys. *245*, 141 (1971)
8.4 R. Graham: "Stochastic Methods in Nonequilibrium Thermodynamics," in *Stochastic Nonlinear Systems in Physics, Chemistry, and Biology*, Springer Ser. Synergetics, Vol.8, ed. by L. Arnold, R. Lefever (Springer, Berlin, Heidelberg, New York 1981) p.202
8.5 H. Bunke: *Gewöhnliche Differentialgleichungen mit zufälligen Parametern* (Akademie, Berlin 1972)
8.6 W. Horsthemke, R. Lefever: A perturbation expansion for external wide band Markovian noise: application to transitions induced by Ornstein-Uhlenbeck noise. Z. Phys. B*40*, 241 (1980)

8.7 J.M. Sancho, M. San Miguel: External non-white noise and nonequilibrium phase transitions. Z. Phys. *B36*, 357 (1980)
8.8 M. San Miguel, J.M. Sancho: A colored-noise approach to Brownian motion in position space. Corrections to the Smoluchowski equation. J. Stat. Phys. *22*, 605 (1980)
8.9 M. San Miguel, J.M. Sancho: "Multiplicative Ornstein-Uhlenbeck Noise in Non-equilibrium Systems," in *Stochastic Nonlinear Systems in Physics, Chemistry, and Biology*, Springer Ser. Synergetics, Vol.8, ed. by L. Arnold, R. Lefever (Springer, Berlin, Heidelberg, New York 1981) p.137
8.10 M. San Miguel, J.M. Sancho: Theory of nonlinear Gaussian noise. Z. Phys. *B43*, 361 (1981)
8.11 R.Z. Has'minskii: *Stochastic Stability of Differential Equations* (Sijthoff and Noordhoff, Alphen aan den Rijn 1980)
8.12 N.G. van Kampen: Stochastic differential equations. Phys. Rep. *24*, 171 (1976)
8.13 R. Kubo: Stochastic Liouville equations. J. Math. Phys. *4*, 174 (1963)
8.14 R.F. Fox: Gaussian stochastic processes in physics. Phys. Rep. *48*, 179 (1978)
8.15 M. San Miguel: A class of exactly solvable Fokker Planck equations. Z. Phys. *B33*, 307 (1979)
8.16 E.A. Novikov: Functionals and the random force method in turbulence theory. Sov. Phys. JETP *20*, 1290 (1965)
8.17 P.C. Martin, E.D. Siggia, H.A. Rose: Statistical dynamics of classical systems. Phys. Rev. *A8*, 423 (1973)
8.18 L. Garrido, M. San Miguel: Response and correlation in Fokker-Planck dynamics I & II. Prog. Theor. Phys. *59*, 40 and 55 (1978)
8.19 H. Brand, A. Schenzle: Comment on "Effect of white noise on electrohydrodynamic instabilities in nematics". J. Phys. Soc. Jpn. *48*, 1382 (1980)
8.20 R. Lefever, W. Horsthemke: "Critical Exponents of a Pure Noise Induced Transition, Nonlinear Noise and Its Effect on an Electrohydrodynamic Transition in Nematics," in *Nonlinear Phenomena in Chemical Dynamics*, Springer Ser. Synergetics, Vol.12, ed. by C. Vidal, A. Pacault (Springer, Berlin, Heidelberg, New York 1981) p.120
8.21 S. Chandrasekhar: *Liquid Crystals* (University Press, Cambridge 1977)
8.22 H.J. Deuling: Elasticity of nematic liquid crystals. Solid State Phys. Suppl. *14*, 77 (1978)
8.23 E. Guyon: Second-order phase transitions: models and analogies. Am. J. Phys. *43*, 877 (1975)
8.24 P.G. De Gennes: *The Physics of Liquid Crystals* (Clarendon, Oxford 1974)
8.25 S. Kai, T. Kai, M. Takata, K. Hirakawa: Effect of the white noise on electrohydrodynamic transitions in nematics. J. Phys. Soc. Jpn. *47*, 1379 (1979)
8.26 T. Kawakubo, A. Yanagita, S. Kabashima: External noise effect on the onset of Williams domain in nematic liquid crystals. J. Phys. Soc. Jpn. *50*, 1451 (1981)
8.27 S. Chandrasekhar: *Hydrodynamic and Hydromagnetic Stability* (Clarendon, Oxford 1961)
8.28 G.Z. Gershuni, E.M. Zhukovitskii: *Convective Stability of Incompressible Fluids* (Keter, Jerusalem 1976)
8.29 E.L. Koschmieder: Bénard convection. Adv. Chem. Phys. *26*, 177 (1974)
8.30 E. Palm: Nonlinear thermal convection. Ann. Rev. Fluid Mech. *7*, 39 (1975)
8.31 C. Normand, Y. Pomeau, M.G. Velarde: Convective instability: a physicist's approach. Rev. Mod. Phys. *49*, 581 (1977)
8.32 F.H. Busse: Nonlinear properties of thermal convection. Rep. Prog. Phys. *41*, 1929 (1978)
8.33 F.H. Busse: "Transition to Turbulence in Rayleigh-Bénard Convection," in *Hydrodynamic Instabilities and the Transition to Turbulence*, Topics Appl. Phys., Vol.45, ed. by H.L. Swinney, J.P. Gollub (Springer, Berlin, Heidelberg, New York 1981) p.97
8.34 D.D. Joseph: *Stability of Fluid Motions*, Springer Tracts Nat. Philosophy, Vols.27,28 (Springer, Berlin, Heidelberg, New York 1976)
8.35 S. Kai, K. Hirakawa: Successive transitions in electrohydrodynamic instabilities of nematics. Suppl. Prog. Theor. Phys. *64*, 212 (1978)
8.36 R. Williams: Domains in liquid crystals. J. Chem. Phys. *39*, 384 (1963)
8.37 E. Dubois-Violette, P.G. de Gennes, O. Parodi: Hydrodynamic instabilities of nematic liquid crystals under a.c. electric fields. J. Phys. Paris *32*, 305 (1971)
8.38 J.I. Kifer: On small random perturbations of some smooth dynamical systems. Math. USSR Izv. *8*, 1083 (1974)

8.39 A. Zippelius, M. Lucke: The effect of external noise in the Lorenz model of the Bénard problem. J. Stat. Phys. *24*, 345 (1981)
8.40 J.P. Crutchfield, B.A. Huberman: Fluctuations and the onset of chaos. Phys. Lett. A*77*, 407 (1980)
8.41 G. Mayer-Kress, H. Haken: The influence of noise on the logistic model. J. Stat. Phys. *26*, 149 (1981)
8.42 J.P. Crutchfield, M. Nauenberg, J. Rudnicky: Scaling for external noise at the onset of chaos. Phys. Rev. Lett. *46*, 933 (1981)
8.43 B. Shraiman, C.E. Wayne, P.C. Martin: Scaling theory for noisy period-doubling transitions to chaos. Phys. Rev. Lett. *46*, 935 (1981)
8.44 G. Mayer-Kress, H. Haken: Intermittent behavior of the logistic system. Phys. Lett. A*82*, 151 (1981)
8.45 J.-P. Eckmann, L. Thomas, P. Wittwer: Intermittency in the presence of noise. J. Phys. A*14*, 3153 (1981)
8.46 J.E. Hirsch, B.A. Huberman, D.J. Scalapino: Theory of intermittency. Phys. Rev. A*25*, 519 (1982)
8.47 W.F. Vinen: "Vortex Lines in Liquid Helium II," in *Progress Low Temperature Physics*, Vol.3, ed. by C.J. Gorter (North-Holland, Amsterdam 1961) p.1
8.48 R.K. Childers, J.T. Tough: Helium II thermal counterflow: temperature- and pressure-difference data and analysis in terms of the Vinen theory. Phys. Rev. B*13*, 1040 (1976)
8.49 F. Moss, G.V. Welland: Multiplicative noise in the Vinen equation for turbulent superfluid ^{4}He. Phys. Rev. A*25*, 3389 (1982)
8.50 D. Brandt, F. Moss, G. Welland: A new model for superfluid helium turbulence: yet another example of bistability. (unpublished)

Chapter 9

9.1 K. Kitahara, W. Horsthemke, R. Lefever: Coloured-noise-induced transitions: exact results for external dichotomous Markovian noise. Phys. Lett. A*70*, 377 (1979)
9.2 K. Kitahara, W. Horsthemke, R. Lefever, Y. Inaba: Phase diagrams of noise induced transitions. Exact results for a class of external coloured noise. Prog. Theor. Phys. *64*, 1233 (1980)
9.3 R.F. Pawula: The probability density and level-crossings of first order nonlinear systems driven by the random telegraph signal. Int. J. Control *25*, 283 (1977)
9.4 A. Hodgkin, A.F. Huxley: Currents carried by sodium and potassium ions through the membrane of the giant axon of Loligo. J. Physiol. London *116*, 449 (1952)
9.5 B. Katz: *Nerve, Muscle and Synapse* (McGraw-Hill, New York 1966)
9.6 W. Horsthemke, R. Lefever: Voltage-noise induced transitions in electrically excitable membranes. Bioph. J. *35*, 415 (1981)
9.7 J. Narahashi, J.W. Moore, W.R. Scott: TTX blockage of sodium conductance increase in lobster giant axons. J. Gen. Physiol. *47*, 965 (1969)
9.8 J. Tasaki, S. Hagiwara: Demonstration of two stable potential states in squid giant axon under tetraethylammonium chloride. J. Gen. Physiol. *40*, 851 (1957)
9.9 C.M. Armstrong, F. Bezanilla, E. Rojas: Destruction of sodium inactivation in squid axons perfused with pronase. J. Gen. Physiol. *62*, 375 (1973)
9.10 F. Conti, E. Wanke: Channel noise in nerve membranes and lipid bilayers. Q. Rev. Biophys. *8*, 451 (1975)
9.11 C.F. Stevens: Study of membrane permeability changes by fluctuation analysis. Nature *270*, 391 (1977)
9.12 J. de Goede, A.A. Verveen: "Electrical Membrane Noise; Its Origin and Interpretation", in *Electrical Phenomena at the Biological Membrane Level* ed. E. Roux (Elsevier, Amsterdam 1977) p.337
9.13 I. Begenisich, C.F. Stevens: How many conductance states do potassium channels have? Biophys. J. *15*, 843 (1975)
9.14 F.J. Sigworth: Sodium channels in nerve apparently have two conductance states. Nature *270*, 265 (1977)
9.15 J.P. Changeux, J. Thiery, Y. Tung, C. Kittel: On the cooperativity of biological membranes. Proc. Natl. Acad. Sci. USA *57*, 335 (1967)
9.16 R. Blumenthal, J.P. Changeux, R. Lefever: Membrane excitability and dissipative instabilities. J. Membr. Biol. *2*, 351 (1970)
9.17 H. Kijima, S. Kijima: Cooperative response of chemically excitable membrane. J. Theor. Biol. *71*, 567 (1978)

Chapter 10

10.1 J.J. Bull, R.C. Vogt: Temperature-dependent sex determination in turtles. Science *206*, 1186 (1979)
10.2 R.J. Donnelly, K. Park, R. Shaw, R.W. Walden: Early non periodic transitions in Couette flow. Phys. Rev. Lett. *44*, 987 (1980)
10.3 G. Ahlers, R.P. Behringer: Evolution of turbulence from the Rayleigh-Bênard instability. Phys. Rev. Lett. *40*, 712 (1978)
10.4 G. Ahlers, R.W. Walden: Turbulence near onset of convection. Phys. Rev. Lett. *44*, 445 (1980)
10.5 M. Gorman, L.A. Reith, H.L. Swinney: Modulation patterns, multiple frequencies, and other phenomena in circular Couette flow. Ann. NY Acad. Sci. *357*, 10 (1980)
10.6 F.J. de la Rubia, J. Garcia Sanz, M.G. Velarde: "Role of Multiplicative Non-white Noise in an Nonlinear Surface Catalytic Reaction," in *Nonlinear Stochastic Problems*, ed. by R. Bucy, J.M.F. Moura (Reidel, Dordrecht 1982)
10.7 A.S. Mikhailov: Noise-induced phase transition in a biological system with diffusion. Phys. Lett. A*73*, 143 (1979)
10.8 A.S. Mikhailov: Effects of diffusion in fluctuating media: a noise-induced phase transition. Z. Phys. B*41*, 277 (1981)
10.9 A.S. Mikhailov, I.V. Uporov: Noise-induced phase transition and the percolation problem for fluctuating media with diffusion. Sov. Phys. JETP *52*, 989 (1980)
10.10 A.S. Mikhailov, I.V. Uporov: Diffusion on the fluctuating random surface. Physica A*112*, 588 (1982)
10.11 L. Arnold, W. Horsthemke, J.W. Stucki: The influence of external white noise on the Lotka-Volterra model. Biom. J. *21*, 451 (1979)

Subject Index

Springer Series in Synergetics

Synergetics An Introduction 3rd Edition
By H. Haken

Synergetics A Workshop
Editor: H. Haken

Synergetics Far from Equilibrium
Editors: A. Pacault, C. Vidal

Structural Stability in Physics
Editors: W. Güttinger, H. Eikemeier

Pattern Formation by Dynamic Systems and Pattern Recognition
Editor: H. Haken

Dynamics of Synergetic Systems
Editor: H. Haken

Problems of Biological Physics
By L. A. Blumenfeld

Stochastic Nonlinear Systems in Physics, Chemistry, and Biology
Editors: L. Arnold, R. Lefever

Numerical Methods in the Study of Critical Phenomena
Editors: J. Della Dora, J. Demongeot, B. Lacolle

The Kinetic Theory of Electromagnetic Processes By Yu. L. Klimontovich

Chaos and Order in Nature
Editor: H. Haken

Nonlinear Phenomena in Chemical Dynamics Editors: C. Vidal, A. Pacault

Handbook of Stochastic Methods for Physics, Chemistry, and the Natural Sciences 2nd Edition
By C. W. Gardiner

Concepts and Models of a Quantitative Sociology The Dynamics of Interacting Populations By W. Weidlich, G. Haag

Noise-Induced Transitions Theory and Applications in Physics, Chemistry, and Biology By W. Horsthemke, R. Lefever

Physics of Bioenergetic Processes
By L. A. Blumenfeld

Evolution of Order and Chaos in Physics, Chemistry, and Biology
Editor: H. Haken

The Fokker-Planck Equation
2nd Edition By H. Risken

Chemical Oscillations, Waves, and Turbulence By Y. Kuramoto

Advanced Synergetics
2nd Edition By H. Haken

Stochastic Phenomena and Chaotic Behaviour in Complex Systems
Editor: P. Schuster

Synergetics – From Microscopic to Macroscopic Order Editor: E. Frehland

Synergetics of the Brain
Editors: E. Başar, H. Flohr, H. Haken, A. J. Mandell

Chaos and Statistical Methods
Editor: Y. Kuramoto

Dynamics of Hierarchical Systems
An Evolutionary Approach
By J. S. Nicolis

Self-Organization and Management of Social Systems Editors: H. Ulrich, G. J. B. Probst

Non-Equilibrium Dynamics in Chemical Systems
Editors: C. Vidal, A. Pacault

Self-Organization Autowaves and Structures Far from Equilibrium
Editor: V. I. Krinsky

Temporal Order Editors: L. Rensing, N. I. Jaeger

Dynamical Problems in Soliton Systems
Editor: S. Takeno

Complex Systems – Operational Approaches in Neurobiology, Physics, and Computers Editor: H. Haken

Dimensions and Entropies in Chaotic Systems Quantification of Complex Behavior 2nd Corr. Printing
Editor: G. Mayer-Kress

Selforganization by Nonlinear Irreversible Processes
Editors: W. Ebeling, H. Ulbricht

Instabilities and Chaos in Quantum Optics
Editors: F. T. Arecchi, R. G. Harrison

Nonequilibrium Phase Transitions in Semiconductors Self-Organization Induced by Generation and Recombination Processes By E. Schöll

Temporal Disorder in Human Oscillatory Systems
Editors: L. Rensing, U. an der Heiden, M. C. Mackey

The Physics of Structure Formation Theory and Simulation Editors: W. Guttinger, G. Dangelmayr

Computational Systems – Natural and Artificial Editor: H. Haken

From Chemical to Biological Organization Editors: M. Markus, S. C. Müller, G. Nicolis

Information and Self-Organization A Macroscopic Approach to Complex Systems 2nd Edition By H. Haken

Propagation in Systems Far from Equilibrium Editors: J. E. Wesfreid, H. R. Brand, P. Manneville, G. Albinet, N. Boccara

Neural and Synergetic Computers Editor: H. Haken

Cooperative Dynamics in Complex Physical Systems Editor: H. Takayama

Optimal Structures in Heterogeneous Reaction Systems Editor: P. J. Plath

Synergetics of Cognition Editors: H. Haken, M. Stadler

Theories of Immune Networks Editors: H. Atlan, I. R. Cohen

Relative Information Theories and Applications By G. Jumarie

Dissipative Structures in Transport Processes and Combustion Editor: D. Meinköhn

Neuronal Cooperativity Editor: J. Krüger

Synergetic Computers and Cognition A Top-Down Approach to Neural Nets By H. Haken

Foundations of Synergetics I Distributed Active Systems 2nd Edition By A. S. Mikhailov

Foundations of Synergetics II Complex Patterns 2nd Edition By A. S. Mikhailov, A. Yu. Loskutov

Synergetic Economics By W.-B. Zhang

Quantum Signatures of Chaos 2nd Edition By F. Haake

Rhythms in Physiological Systems Editors: H. Haken, H. P. Koepchen

Quantum Noise 2nd Edition By C. W. Gardiner, P. Zoller

Nonlinear Nonequilibrium Thermodynamics I Linear and Nonlinear Fluctuation-Dissipation Theorems By R. Stratonovich

Self-organization and Clinical Psychology Empirical Approaches to Synergetics in Psychology Editors: W. Tschacher, G. Schiepek, E. J. Brunner

Nonlinear Nonequilibrium Thermodynamics II Advanced Theory By R. Stratonovich

Limits of Predictability Editor: Yu. A. Kravtsov

On Self-Organization An Interdisciplinary Search for a Unifying Principle Editors: R. K. Mishra, D. Maaß, E. Zwierlein

Interdisciplinary Approaches to Nonlinear Complex Systems Editors: H. Haken, A. Mikhailov

Inside Versus Outside Endo- and Exo-Concepts of Observation and Knowledge in Physics, Philosophy and Cognitive Science Editors: H. Atmanspacher, G. J. Dalenoort

Ambiguity in Mind and Nature Multistable Cognitive Phenomena Editors: P. Kruse, M. Stadler

Modelling the Dynamics of Biological Systems Editors: E. Mosekilde, O. G. Mouritsen

Self-Organization in Optical Systems and Applications in Information Technology 2nd Edition Editors: M.A. Vorontsov, W. B. Miller

Principles of Brain Functioning A Synergetic Approach to Brain Activity, Behavior and Cognition By H. Haken

Synergetics of Measurement, Prediction and Control By I. Grabec, W. Sachse

Predictability of Complex Dynamical Systems By Yu. A. Kravtsov, J. B. Kadtke

Interfacial Wave Theory of Pattern Formation Selection of Dentritic Growth and Viscous Fingerings in Hele–Shaw Flow By Jian-Jun Xu

Asymptotic Approaches in Nonlinear Dynamics New Trends and Applications By J. Awrejcewicz, I. V. Andrianov, L. I. Manevitch

Brain Function and Oscillations
Volume I: Brain Oscillations.
Principles and Approaches
Volume II: Integrative Brain Function.
Neurophysiology and Cognitive Processes
By E. Başar

Asymptotic Methods for the Fokker–Planck Equation and the Exit Problem in Applications
By J. Grasman, O. A. van Herwaarden

Analysis of Neurophysiological Brain Functioning Editor: Ch. Uhl

Phase Resetting in Medicine and Biology
Stochastic Modelling and Data Analysis
By P. A. Tass

Self-Organization and the City By J. Portugali

Critical Phenomena in Natural Sciences
Chaos, Fractals, Selforganization and Disorder: Concepts and Tools By D. Sornette

Spatial Hysteresis and Optical Patterns
By N. N. Rosanov

Nonlinear Dynamics of Chaotic and Stochastic Systems
Tutorial and Modern Developments
By V. S. Anishchenko, V. V. Astakhov, A. B. Neiman, T. E. Vadivasova, L. Schimansky-Geier

Synergetic Phenomena in Active Lattices
Patterns, Waves, Solitons, Chaos
By V. I. Nekorkin, M. G. Velarde

Brain Dynamics
Synchronization and Activity Patterns in Pulse-Coupled Neural Nets with Delays and Noise By H. Haken

From Cells to Societies
Models of Complex Coherent Action
By A. S. Mikhailov, V. Calenbuhr

Brownian Agents and Active Particles
Collective Dynamics in the Natural and Social Sciences By F. Schweitzer

Nonlinear Dynamics of the Lithosphere and Earthquake Prediction
By V. I. Keilis-Borok, A. A. Soloviev (Eds.)

Nonlinear Fokker-Planck Equations
Fundamentals and Applications
By T. D. Frank

Patters and Interfaces in Dissipative Dynamics
By L. M. Pismen

MIX
Papier | Fördert gute Waldnutzung
FSC® C083411

Zeitfracht Medien GmbH
Ferdinand-Jühlke-Straße 7
99095 Erfurt, Deutschland
produktsicherheit@kolibri360.de